The Professional and Scientific Literature on Patient Education

HEALTH AFFAIRS INFORMATION GUIDE SERIES

Series Editor: Winifred Sewell, Health Science Information Consultant, currently associated with the University of Maryland and the National Health Planning Information Center

Also in this series:

BIOETHICS—*Edited by Doris Goldstein**

CROSS-NATIONAL STUDY OF HEALTH SYSTEMS: CONCEPTS, METHODS, AND DATA SOURCES—*Edited by Ray H. Elling*

CROSS-NATIONAL STUDY OF HEALTH SYSTEMS: COUNTRIES, WORLD REGIONS AND SPECIAL PROBLEMS—*Edited by Ray H. Elling*

HEALTH CARE ADMINISTRATION—*Edited by Dwight A. Morris and Lynne Darby Morris*

HEALTH CARE COSTS AND FINANCING—*Edited by Rita Keintz**

HEALTH CARE POLITICS, POLICY AND LEGISLATION—*Edited by Joel M. Lee**

HEALTH MAINTENANCE THROUGH FOOD AND NUTRITION—*Edited by Helen Ullrich**

HEALTH PLANNING—*Edited by Lewis Lefko**

HEALTH SCIENCES AUDIOVISUALS—*Edited by Laura P. Barrett**

HEALTH STATISTICS—*Edited by Frieda O. Weise*

HUMAN ECOLOGY—*Edited by Frederick Sargent II**

MEDICAL INFORMATION TRANSFER—*Edited by Winifred Sewell and Marie Dickerman**

QUALITY MAINTENANCE AND EVALUATION OF HEALTH CARE—*Edited by Jay Glasser and Irene Eastling**

SURVEY OF EMERGENCY MEDICAL SERVICES SYSTEM RESOURCES—*Edited by Carlos Fernandez-Caballero and Marianne Fernandez-Caballero**

*in preparation

The above series is part of the
GALE INFORMATION GUIDE LIBRARY
The Library consists of a number of separate series of guides covering major areas in the social sciences, humanities, and current affairs.

General Editor: Paul Wasserman, Professor and former Dean, School of Library and Information Services, University of Maryland

Managing Editor: Denise Allard Adzigian, Gale Research Company

The Professional and Scientific Literature on Patient Education

A GUIDE TO INFORMATION SOURCES

Volume 5 in the Health Affairs Information Guide Series

Lawrence W. Green

*Director, Office of Health Information,
Health Promotion and Physical Fitness, and Sports Medicine
Office of the Assistant Secretary of Health
U.S. Department of Health, Education and Welfare*

Connie Cavanaugh Kansler

Gale Research Company
Book Tower, Detroit, Michigan 48226

Copyright © 1980 by
Lawrence W. Green and Connie Cavanaugh Kansler

ISBN 0-8103-1422-3
Library of Congress Catalog Card Number 80-19649

No part of this book may be reproduced in any form without permission in writing from the publisher, except by a reviewer who wishes to quote brief passages or entries in connection with a review written for inclusion in a magazine or newspaper. Manufactured in the United States of America.

VITAE

Lawrence W. Green is serving as Director of the Office of Health Information, Health Promotion and Physical Fitness and Sports Medicine, Office of the Assistant Secretary for Health, U.S. Department of Health, Education, and Welfare. He is presently on leave as Head, Division of Health Education, School of Hygiene and Public Health, Johns Hopkins University. Dr. Green received his B.S., M.P.H., and Dr.P.H. from the University of California, Berkeley. He has published numerous articles in the health field, has received many distinguished honors during his career, and has served in various positions in many professional associations including the American Public Health Association and the Society for Public Health Education.

Connie Cavanaugh Kansler currently works in the Commissioner's Office for the Nineteenth Judicial District Court in East Baton Rouge Parish. She received her B.A. in English from Louisiana State University. She was the managing editor of HEALTH EDUCATION MONOGRAPHS (vol. 1, no. 33-vol. 6, no. 4) and worked in the Division of Health Education, Department of Health Services Administration, Johns Hopkins University of Hygiene and Public Health (1972-78). Kansler also coedited with L.W. Green and H. Kalmer "Proceedings of the National Conference on Cancer Public Education." HEALTH EDUCATION MONOGRAPHS 1, no. 33 (1973): entire issue.

CONTENTS

Foreword ix
Acknowledgments xiii
Introduction xv

Part I. Background and Organization of Patient Education
    Chapter 1. Background and Organization of Patient
                Education 3
          History and Organization of the Field 3
          Organization of Concepts 4
          Periodicals 5

Part II. The Literature of the Mid-1970s
    Chapter 2. Health Problem Analysis 19
        A. Chronic Disorders 20
        B. Acute Episodes and Dying 27
        C. Mental and Personality Disorders 31
        D. Pregnancy, Childbearing, and the Puerperium 39
    Chapter 3. Behavioral Problem Analysis 45
        E. Preventive Health Practices 46
        F. Illness Behavior and Utilization of Health Services 50
        G. Compliance with Therapeutic Regimens 58
    Chapter 4. Factors Predisposing Patient Behavior 65
        H. Patient Knowledge and Beliefs 70
        I. Patient Attitudes and Values 73
        J. Personality and Other Motivational Factors 77
    Chapter 5. Factors Enabling Patient Behavior 87
        K. Personal Resources Available to Patients 90
        L. Community Resources Available to Patients 91
    Chapter 6. Factors Reinforcing Patient Behavior 97
        M. Family Influences on Patient Behavior 98
        N. Peer Influences on Patient Behavior 102
        O. Other Social Supports and Rewards for Patient
           Behavior 103
        P. Attitudes and Behavior of Health Professionals
           Toward Patients 108

Contents

Chapter 7.	Communications Theory and Practice	119
Q.	Theories and Methods of Interpersonal Communications	124
R.	Small Group Communications in Patient Settings	134
S.	Educational Media and Technology	139
T.	Behavior Modification in Patient Care	146
Chapter 8.	Organizational Theory and Methods for Patient Education	163
U.	Mobilizing Institutional and Volunteer Resources	167
V.	Referral Plans and Community Relations	177
Chapter 9.	Staff Development and Administration	181
W.	In-Service Training and Continuing Education	182
X.	Consultation and Supervision	195
Y.	Professional Preparation and Quality Control	201
Chapter 10.	Research and Evaluation Methods	215
Z.	Methods	217
Appendix 1.	Resources for Patient Education	235
Appendix 2.	Health Newsletters	285
Appendix 3.	Consultation in Patient Education	291

Indexes
Author Index 297
Subject Index 313

FOREWORD

Interest in health and the resources and activities that make it possible is not new. However, the concepts of health as a national resource and as a human right have emerged in the recent past. Legislation during the last two decades has led from these concepts to a complex system with some unsurprising growing pains.

Many people have come into the field bringing a multiplicity of backgrounds to supplement the traditional health sciences. In addition, today's laymen must make decisions on health care at all levels--from voting for or against legislators who will shape health laws, through serving on local health planning boards, to becoming participants in informed decisions on their own health care. The new recruits and the laymen have in common the need for all kinds of information on social, business, legal, ethical, and other aspects of medicine.

Much of this information has previously been unavailable and not readily understandable to the new audiences. Attempts to satisfy the need have resulted in burgeoning publications on a variety of subjects, ranging from the broad to the specific. These new publications are in many forms, from carefully edited, important texts to poorly conceived and executed technical reports, with a vast array in between. There are journals, newsletters, association and university guidebooks and models, statistical reports, and audiovisuals--all with varying quality and format.

Several problems have resulted. In the first place, there has not yet emerged a major bibliographic resource, such as the National Library of Medicine and Excerpta Medica provide for the clinical and research aspects of medicine. Those who have access to MEDLINE services have recently been enabled to find at least a part of the pertinent current periodical references through the new database, Health Planning and Administration, or FILE HEALTH. A second problem is that some of the novices in the field are not accustomed to using published literature in any form, let alone the complex of primary, secondary, and tertiary publications with which their counterparts in clinical medicine and research have become familiar.

Foreword

It is the purpose of the Health Affairs Series in the Gale Information Guide Library to provide a guide for all participants in the complex health care system to information on the system itself--on the process of the delivery of health care. We are concerned with the management of the system and with how researchers, educators, and practitioners assure the best health possible to each individual. We are not concerned with the content of the information with which the researcher, educator, or practitioner deals, but rather with his mode of functioning in a real world of people with different racial, ethnic, sexual, financial, and geographic backgrounds. For example, we are not interested in how a surgical procedure is carried out, but we are concerned with its availability to those who need it. If one understands the social sciences broadly, then the Health Affairs Series covers the social aspects of medicine.

Among those social aspects we include ones which are focused on the individual in relation to his environment, his education for full participation in his own health care, and his social, political, legal, and ethical responsibilities. In addition, we include all components involved with providing facilities and resources necessary for health care delivery. Individual volumes cover: health statistics; human ecology--both man-environment interactions and manipulations and their outcomes and intervention strategies; health maintenance through food and nutrition; the professional and scientific literature of patient education; health care politics, policy, and legislation; cross-national study of health systems; health planning; health care costs and financing; quality maintenance and evaluation of health care; health care administration; emergency medical service systems; preventive, diagnostic, and therapeutic materials in the health care system; education in the health professions; health sciences audiovisuals; medical information transfer; and interface of medicine and the law.

Thus, just as the Gale Information Guide Library encompasses a broad spectrum of all fields of knowledge, the Health Affairs Series covers the social aspects of medicine extensively.

The user may ask, "Why another bibliography when there are already so many?" We agree, but the Health Affairs guides expand the usual bibliography. All contain a careful selection of materials for the intended audience and an evaluation of these materials in annotations and introductory paragraphs. In addition, they usually contain lists of sources of information, such as information centers, schools, publishers, and audiovisuals. The journals which one should read regularly to keep up with the field are mentioned and annotated. In short, the reader will find specific resources for the present, and guidance for the future through the descriptions of periodicals and other sources of information.

Because all of the social sciences dealing with health affairs are interrelated, it is both impossible and undesirable to prescribe strict limits for each individual volume, excluding from one anything included in another. Instead, each volume of the series is complete for the individual who is interested in only one specific subject. At the same time, other volumes serve as excellent supplements when the user wishes to go into related topics in greater depth.

Foreword

A person interested in any of the social aspects of medicine should be able to find a volume that focuses on that interest, though the depth of coverage will vary with the topic. In some instances the amount of literature available determined the division of the series into specific volumes, and coverage of materials is highly selective; in other cases, the choice of volume topic was somewhat arbitrary, determined by the independence or cohesiveness of a subject, and a smaller quantity of literature may have allowed for greater depth of coverage of a topic. The series editor takes responsibility for the general organization and coverage of the series, but has left to the judgment of the volume editors decisions on individual items to be included or excluded. The series should thus be helpful to the user not only for what he is able to find in the individual volumes, but for how it has been sifted to exclude those materials that would send him down blind alleys.

We hope that our audience for the series will consist of newcomers to the fields involved, as well as researchers and practitioners who have worked with health affairs in the past but need to renew their familiarity with resources in some aspect of their current interests. We hope especially that the guides will be useful in the education of students who have chosen one or more of the fields covered for their future careers. And finally, we have tried to make the volumes simple and direct enough that they will provide the informed layperson with access to the information resources needed to make decisions about future procedures and policies in assurance of the best match between national resources and the health care of the nation.

Dr. Green, a recognized authority in the field of health education, and his colleague, Connie Kansler, have brought a historical perspective to this volume. It covers primarily materials in the mid-1970s, when programs and concepts in health and patient education were developing rapidly. Dr. Green's current activity in health information and promotion, plus his continuing interests, have enabled him to include major recent contributions as well.

Those who are familiar with previous bibliographic activity in the Health Education Information Retrieval System at Johns Hopkins will be pleased to find it brought together and summarized in this volume.

Winifred Sewell, Series Editor
Cabin John, Maryland

ACKNOWLEDGMENTS

We are indebted to those responsible for the development of the HEIR System in the 1960s, Frank Holz and Daniel Sullivan, and to Theron Butterworth for his assistance in transferring the system to the Health Education Division at the Johns Hopkins University.

In the years the HEIR System has been at Johns Hopkins, many individuals have assisted in its maintenance and further development. More than anyone, Mrs. Patricia Farr deserves major credit for abstracting and indexing much of the literature published in this volume. Students who have assisted in maintaining the system include Howard Kalmer, Donald Morisky, Judy Chwalow, and Robert Bertera. In meeting the extensive typing requirements of HEIRS, secretaries in the Health Education Division and in the Department of Health Services Administration who generously provided assistance were Susan Bowersox, Theresa Glaser, Louise Lieu, and Loetta Wallace. In the last two years, Jacqui Height has been responsible for all secretarial work for the HEIR System and has provided very helpful assistance to its users.

We are indebted also to Dr. Arthur Bushel, who has permitted and supported this continuing activity of the division during a period of departmental reorganization and shrinking space. All of our colleagues have helped to maintain the system by identifying or contributing documents to be included.

INTRODUCTION

Patient education is a subject of major concern to physicians, nurses, dentists, pharmacists, physical therapists, dieticians, and other direct providers of health care. The literature annotated here is drawn representatively from the journals of these professions. Of equal if not greater importance during this period in the development of patient education has been the research and theoretical contributions of health education specialists and behavioral scientists. The literature of public health education, medical sociology, social psychology, educational psychology, medical anthropology, and behavioral medicine also are sampled representatively.

The other contributors to the recent surge of interest and activity in patient education have been the administrators, consumer groups, health planners, and legislators whose writings have set the agenda for the health professionals and scientists writing on this subject. The sampling of this literature is from the journals and books in hospital administration, public health, and policy analysis.

Within all of these bodies of literature, the selection for inclusion here has been based on an effort to represent the state of the art rather than to epitomize it. The items annotated are not limited to the best, the most cited, the most influential, or the most useful documents. Instead, the lists under each topic include some of the least notable writings on the topic when such writings reflect the prevailing standard or trend in that area of concern.

The selections are from our collection and indexing over the past several years of thousands of published and unpublished documents in the Health Education Information Retrieval System (HEIRS) at the Johns Hopkins University in the Division of Health Education, Department of Health Services Administration, School of Hygiene and Public Health. The HEIR System was developed originally by the Public Health Service, U.S. Department of Health, Education, and Welfare. It was abandoned in the late 1960s as budgetary support for health education within the federal government declined. The accumulated documents and supporting material were transferred to Johns Hopkins in 1970 with the promise that we would revive it and make it more accessible to health professionals, scientists, and students. This book is part of our attempt to fulfill that promise.

Introduction

The index to this book is a highly developed taxonomy of terms related to health education, based on the original thesaurus developed by the Public Health Service, elaborated and extended over time by us. Of the 920 topics in the HEIR System, 576 have been selected as relevant to patient education.

Part I
BACKGROUND AND ORGANIZATION OF PATIENT EDUCATION

Chapter 1
BACKGROUND AND ORGANIZATION OF PATIENT EDUCATION

HISTORY AND ORGANIZATION OF THE FIELD

Patient education has emerged as a subspeciality of health education at the same time that medicine (especially family practice and pediatrics), dentistry, nursing, pharmacy, and dietetics have given increasing attention to professional preparation and practice in patient education. It was not always so. These trends follow a long period of isolation from medical care by health educators, lip service for patient education by nurses, skepticism and disdain by physicians, dentists, and pharmacists. Patient education was considered an incidental by-product of "bedside manner" and good reading material in the waiting room.

Some notable demonstration projects on the organization of hospital based patient education programs from the 1950s and early 1960s are described in:

Richards, R.F., and Kalmer, H., eds. "Patient Education." HEALTH EDUCATION MONOGRAPHS 2 (Spring 1974): whole issue.

This issue of HEALTH EDUCATION MONOGRAPHS also reviews the historical role of the insurance industry and the federal government in supporting patient education. It provides the policy statements on patient education from the Society for Public Health Education and a Task Force commissioned by the President's Committee on Health Education that published its report in 1974. Another article in the same issue contrasts the literature on evaluation of health education in the 1960s with that of the early 1970s. The major difference was the shift from community to clinical sites, which accounted for improved experimental designs for evaluation and the opportunity to submit many assumptions and techniques in health education to a more rigorous test than they had had in preceding decades.

The early 1970s, then, was a turning point for patient education. New sources of support began to develop; research and evaluation began to provide a firmer scientific base for practice; legislators and policymakers became interested; the insurance industry, especially Blue Cross and Blue Shield, pro-

Background and Organization

vided encouragement; the Joint Commission for Accreditation of Hospitals incorporated patient education into its standards for accreditation.

The nursing profession has long recognized the importance of patient education, but the recent emphasis on self-care has given new impetus, as has the "Patient's Bill of Rights" adopted by the American Hospital Association in 1973. These and related developments in nursing are fully developed in five important works:

Nursing Development Conference Group. CONCEPT FORMALIZATION IN NURSING PROCESS AND PRODUCT. Boston: Little, Brown & Co., 1973.

Orem, D.E. NURSING: CONCEPTS OF PRACTICE. New York: McGraw-Hill Book Co., 1971.

Redman, B.K. THE PROCESS OF PATIENT TEACHING IN NURSING. 3d ed. St. Louis: C.V. Mosby Co., 1976.

_____, ed. "Patient Teaching." NURSING DIGEST 6 (Spring 1978): entire issue.

Skydell, B., and Crowder, A.S. DIAGNOSTIC PROCEDURES: A REFERENCE FOR HEALTH PRACTITIONERS AND A GUIDE FOR PATIENT COUNSELING. Boston: Little, Brown & Co., 1975.

The broader field of health education has developed coterminously, partly as a consequence of the developments in patient education. The general field of health education that subsumes patient education is presented comprehensively in the following recent works:

Galli, N. FOUNDATIONS AND PRINCIPLES OF HEALTH EDUCATION. New York: John Wiley, 1978.

Mico, P., and Ross, H.S. HEALTH EDUCATION AND BEHAVIORAL SCIENCE. Oakland, Calif.: Third Party Associates, 1975.

Somers, A.R., ed. PROMOTING HEALTH: CONSUMER EDUCATION AND NATIONAL POLICY. Germantown, Md.: Aspen Systems, 1976.

ORGANIZATION OF CONCEPTS

The scientific and professional literature is organized in the following pages around a framework that places the ultimate health goals or outcomes of patient education at the beginning (part A), the behavioral determinants of those outcomes next (part B), the causes of those behaviors next (part C), and the educational methods of effecting those causes last (part D). Thus, the objects of interest in health education are arranged in causal order from ultimate to immediate. This model of cause-and-effect relationships among variables of concern in health education has been applied to a variety of specific health problems in different populations. These applications and the model are described more extensively in the following reference works:

Green, L.W. "Methods Available to Evaluate the Health Education Components

Background and Organization

of Preventive Health Programs." In PREVENTIVE MEDICINE, U.S.A., pp. 162-71 passim. New York: Prodist, 1976. Also in PROMOTING HEALTH: CONSUMER EDUCATION AND NATIONAL POLICY, edited by A.R. Somers, pp. 167-75 passim. Germantown, Md.: Aspen Systems Corp., 1976.

Green, L.W.; Levine, D.M.; and Deeds, S. "Clinical Trials of Health Education for Hypertensive Outpatients: Design and Baseline Data." PREVENTIVE MEDICINE 4 (1975): 417-25.

Green, L.W., et al. HEALTH EDUCATION PLANNING: A DIAGNOSTIC APPROACH. Palo Alto, Calif.: Mayfield Publishing Co., 1979.

Green, L.W., et al. "Guidelines for Health Education in Maternal and Child Health." INTERNATIONAL JOURNAL OF HEALTH EDUCATION 21, Suppl. 3 (1978): 1-33.

Figure 1 (see page 6) illustrates the model as applied to primary care. The arrows show the direction of cause and effect from left to right, but the proper order of educational diagnosis and planning is to work from right to left. The first step is to specify and clarify the health problem or goal (chapter 2). The second step is to identify the most important behaviors influencing the health problem or goal (chapter 3). For each of the specific behaviors, a further educational diagnosis of the causes of the behavior identifies those causes that have potential for change through health education (chapters 4, 5, 6). Finally, the administrative, organizational, logistical, and technical issues of planning, coordinating, implementing, and evaluating educational resources, methods, and programs are addressed (chapters 7-10).

PERIODICALS

The journals identified most frequently by the articles cited in this bibliography will continue to be the major sources of professional and scientific literature on patient education, but they also reflect the publishing outlets interested in patient education during the mid-1970s. Within the short two-year period between the closing of our initial search and retrieval efforts for this book and its publication, several new journals devoted primarily or substantially to patient education or health education have or will have emerged. These include HEALTH VALUES (Charles B. Slack, Medical Publishers, New Jersey), JOURNAL OF FAMILY AND COMMUNITY HEALTH (Aspen Systems, Germantown, Maryland), PATIENT COUNSELING AND HEALTH EDUCATION (Excerpta Medica, Amsterdam), HEALTH EDUCATION/EDUCATION SANITAIRE (Canadian Ministry of Health and Welfare, Ottawa), and JOURNAL OF BEHAVIORAL MEDICINE (Plenum Press, New York and London).

Other journals have begun to take a more active interest in patient education since the mid-1970s period, as evidenced by increased numbers of articles, editorials, and debate in the letters to the editor. This is particularly notable

Figure 1. Health Education Inputs and Outcomes in Patient Care, as Located in This Guide.

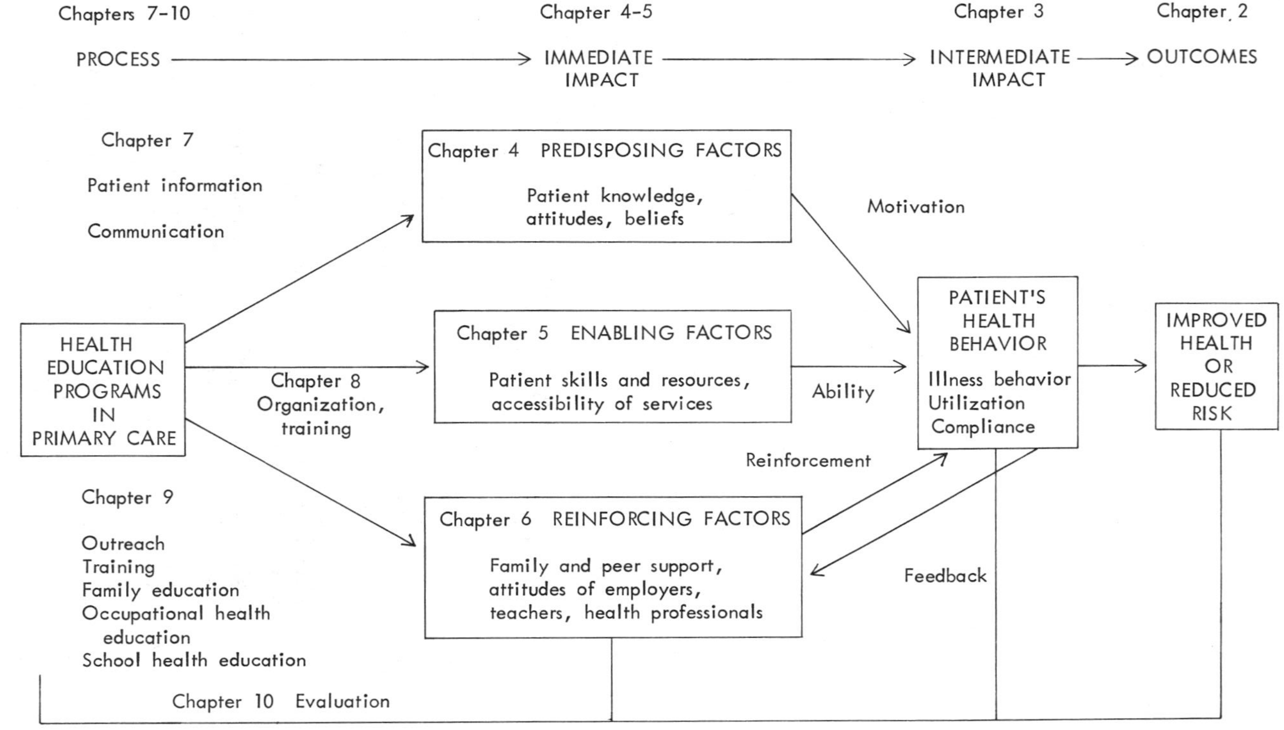

| Chapters 7-10 | Chapter 4-5 | Chapter 3 | Chapter 2 |

PROCESS ⟶ IMMEDIATE IMPACT ⟶ INTERMEDIATE IMPACT ⟶ OUTCOMES

HEALTH EDUCATION PROGRAMS IN PRIMARY CARE

Chapter 7
Patient information
Communication

Chapter 8
Organization, training

Chapter 9
Outreach
Training
Family education
Occupational health education
School health education

Chapter 10 Evaluation

Chapter 4 PREDISPOSING FACTORS
Patient knowledge, attitudes, beliefs

Chapter 5 ENABLING FACTORS
Patient skills and resources, accessibility of services

Chapter 6 REINFORCING FACTORS
Family and peer support, attitudes of employers, teachers, health professionals

PATIENT'S HEALTH BEHAVIOR
Illness behavior
Utilization
Compliance

IMPROVED HEALTH OR REDUCED RISK

Motivation
Ability
Reinforcement
Feedback

6

Background and Organization

in the medical journals, the pharmacy journals, the health policy and health ethics journals, and the publications of other allied health professions.

Some journals always carried articles relevant to patient education but only recently have made a shift to include articles more ostensibly and practically addressed to patient education. This trend is most notable in the public health journals, the psychology journals, and the medical sociology journals.

The current literature, then, will be found concentrated in the following journals, in approximately the order listed:

HEALTH EDUCATION MONOGRAPHS. Thorofare, N.J.: Charles B. Slack, 1957-- . Quarterly.

INTERNATIONAL JOURNAL OF HEALTH EDUCATION. Geneva: International Union for Health Education, 1958-- . Quarterly.

HEALTH VALUES: ACHIEVING HIGH LEVEL WELLNESS. Thorofare, N.J.: Charles B. Slack, 1977-- . Bimonthly.

PATIENT COUNSELING AND HEALTH EDUCATION. Amsterdam: Excerpta Medica, 1978-- . Quarterly.

NURSING OUTLOOK. New York: American Journal of Nursing Co., 1953-- . Monthly.

AMERICAN JOURNAL OF PUBLIC HEALTH. Washington, D.C.: American Journal of Nursing Co., 1911-- . Monthly.

JOURNAL OF COMMUNITY HEALTH. New York: Human Sciences Press, 1975-- . Quarterly.

AMERICAN JOURNAL OF NURSING. New York: American Journal of Nursing Co., 1900-- . Monthly.

MEDICAL CARE. Philadelphia: J.B. Lippincott Co., 1967-- . Monthly.

PUBLIC HEALTH REPORTS. Formerly HEALTH SERVICES REPORTS, HSMHA HEALTH REPORTS. Washington, D.C.: Government Printing Office, 1878-- . Monthly.

NURSING DIGEST. Wakefield, Mass.: 1973-- . Quarterly.

Background and Organization

JOURNAL OF BEHAVIORAL MEDICINE. New York: Plenum Press, 1978-- . Quarterly.

JOURNAL OF THE AMERICAN MEDICAL ASSOCIATION. Chicago: American Medical Association, 1848-- . Weekly.

SOCIAL SCIENCE AND MEDICINE. Elmford, N.Y.: Pergamon Press, 1967-- . 18 issues per year.

PEDIATRICS. Evanston, Ill.: American Academy of Pediatrics, 1948-- . Monthly.

PREVENTIVE MEDICINE. New York: Academic Press, 1972-- . Quarterly.

AMERICAN JOURNAL OF PSYCHIATRY. Washington, D.C.: American Psychiatric Association, 1844-- . Monthly.

NEW ENGLAND JOURNAL OF MEDICINE. Boston: Massachusetts Medical Society, 1812-- . Weekly.

NURSING RESEARCH. New York: American Journal of Nursing Co., 1952-- . Bimonthly.

AMERICAN JOURNAL OF ORTHOPSYCHIATRY. New York: American Orthopsychiatric Association, 1930-- . Quarterly.

JOURNAL OF HEALTH AND SOCIAL BEHAVIOR. Washington, D.C.: American Sociological Association, 1960-- . Quarterly.

JOURNAL OF YOUTH AND ADOLESCENCE. New York: Plenum Press, 1972-- . Quarterly.

CHILD DEVELOPMENT. Chicago: University of Chicago Press, 1930-- . Quarterly.

REHABILITATION COUNSELING BULLETIN. Washington, D.C.: American Personnel and Guidance Association, 1957-- . Quarterly.

BEHAVIOUR RESEARCH AND THERAPY. New York: Pergamon Press; Oxford, Engl.: Headington Hill Hall, 1963-- . Bimonthly.

HOSPITALS: JOURNAL OF THE AMERICAN HOSPITAL ASSOCIATION. Chicago: American Hospital Association, 1936-- . Semimonthly.

Background and Organization

JOURNAL OF MEDICAL EDUCATION. Washington, D.C.: Association of American Medical Colleges, 1926-- . Monthly.

JOURNAL OF THE AMERICAN DIETETIC ASSOCIATION. Chicago: American Dietetic Association, 1925-- . Monthly.

AMERICAN JOURNAL OF HOSPITAL PHARMACY. Washington, D.C.: American Society of Hospital Pharmacists, 1945-- . Monthly.

BULLETIN OF THE MEDICAL LIBRARY ASSOCIATION. Ephrata, Pa.: Science Press, 1911-- . Quarterly.

LANCET. Boston: Little, Brown & Co.; London: Lancet, 1966-- . Weekly.

PEDIATRIC CLINICS OF NORTH AMERICA. Philadelphia: W.B. Saunders Co., 1954-- . Quarterly.

ANNALS OF INTERNAL MEDICINE. Philadelphia: American College of Physicians, 1922-- . Monthly.

BIOSCIENCES COMMUNICATIONS. Basel, Switz.: Arnold Boecklin, 1975-- . Bimonthly.

JOURNAL OF COUNSELING PSYCHOLOGY. Washington, D.C.: American Psychological Association, 1954-- . Monthly.

JOURNAL OF CONSULTING AND CLINICAL PSYCHOLOGY. Washington, D.C.: American Psychological Association, 1937-- . Bimonthly.

JOURNAL OF PERSONALITY AND SOCIAL PSYCHOLOGY. Washington, D.C.: American Psychological Association, 1965-- . Monthly.

JOURNAL OF PSYCHOSOMATIC RESEARCH. Elmsford, N.Y.: Pergamon Press, 1956-- . Bimonthly.

PSYCHOLOGICAL REPORTS. Missouli, Mont.: R.B. and C.H. Ammons, 1955-- . Bimonthly.

PSYCHOSOMATIC MEDICINE. New York: Elsevier North-Holland, 1938-- . Bimonthly.

JOURNAL OF NERVOUS AND MENTAL DISEASE. Baltimore, Md.: Williams & Wilkins Co., 1874-- . Monthly.

Background and Organization

COUNSELOR EDUCATION AND SUPERVISION. Washington, D.C.: American Personnel and Guidance Association, 1961-- . Quarterly.

HEALTH SERVICES REPORTS. Continuing as PUBLIC HEALTH REPORTS (see p. 7).

JOURNAL OF STUDIES ON ALCOHOL. Formerly QUARTERLY JOURNAL OF STUDIES ON ALCOHOL. New Brunswick, N.J.: Rutgers Center of Alcohol Studies, 1940-- . Monthly.

MILBANK MEMORIAL FUND QUARTERLY: HEALTH AND SOCIETY. Formerly MILBANK MEMORIAL FUND QUARTERLY. New York: Milbank Memorial Fund, 1923-- .

ACADEMY OF MANAGEMENT JOURNAL. Mississippi State, Miss.: Academy of Management, 1958-- . Quarterly.

BEHAVIOR THERAPY. New York: Academic Press, 1970-- . Five issues per year.

JOURNAL OF ABNORMAL PSYCHOLOGY. Washington, D.C.: American Psychological Association, 1965-- . Bimonthly.

OMEGA--JOURNAL OF DEATH AND DYING. Farmingdale, N.Y.: Baywood Publishing, 1970-- . Quarterly.

QUARTERLY JOURNAL OF STUDIES ON ALCOHOL. Continuing as JOURNAL OF STUDIES ON ALCOHOL (see above).

SMALL GROUP BEHAVIOR. Beverly Hills, Calif.: Sage Publications, 1970-- . Quarterly.

COMMUNITY MENTAL HEALTH JOURNAL. New York: Human Sciences Press, 1965-- . Quarterly.

EDUCATIONAL TECHNOLOGY. Englewood Cliffs, N.J.: Educational Technology Publications, 1961-- . Monthly.

HOSPITAL AND COMMUNITY PSYCHIATRY. Washington, D.C.: American Psychiatric Association, 1950-- . Monthly.

JOURNAL OF PSYCHOLOGY. Provincetown, Mass.: Journal Press, 1936-- . Bimonthly.

Background and Organization

PROFESSIONAL PSYCHOLOGY. Washington, D.C.: American Psychological Association, 1969-- . Quarterly.

PSYCHOTHERAPY: THEORY, RESEARCH AND PRACTICE. Los Angeles: Division of Psychotherapy, American Psychological Association, 1963-1974. Quarterly.

AMERICAN JOURNAL OF PSYCHOTHERAPY. New York: Association for the Advancement of Psychotherapy, 1946-- . Quarterly.

ARCHIVES OF GENERAL PSYCHIATRY. Chicago: American Medical Association, 1959-- . Monthly.

JOURNAL OF APPLIED PSYCHOLOGY. Washington, D.C.: American Psychological Association, 1917-- . Bimonthly.

JOURNAL OF EXPERIMENTAL AND SOCIAL PSYCHOLOGY. New York: Academic Press, 1965-- . Bimonthly.

JOURNAL OF SCHOOL HEALTH. Kent, Ohio: American School Health Association, 1931-- . Monthly.

JOURNAL OF THE AMERICAN COLLEGE HEALTH ASSOCIATION. Evanston, Ill.: American College Health Association, 1952-- . Bimonthly.

PERSONNEL AND GUIDANCE JOURNAL. Washington, D.C.: American Personnel and Guidance Association, 1922-- . Monthly (September-June).

PSYCHOLOGICAL RECORD. Gambier, Ohio: Kenyon College, 1937-- . Quarterly.

GERONTOLOGIST. Washington, D.C.: Gerontological Society, 1961-- . Bimonthly.

HOSPITAL ADMINISTRATION. Chicago: American College of Hospital Administrators, 1956-- . Quarterly.

HUMAN RELATIONS. New York: Plenum Press, 1947-- . Monthly.

INTERNATIONAL JOURNAL OF PSYCHIATRY IN MEDICINE. Farmingdale, N.Y.: Baywood Publishing, 1970-- . Quarterly.

JOURNAL OF GERONTOLOGY. Washington, D.C.: Gerontological Society, 1946-- . Bimonthly.

Background and Organization

PROGRAMMED LEARNING AND EDUCATIONAL TECHNOLOGY. London: Sweet & Maxwell-Stevens Journals, 1964-- . Quarterly.

PSYCHOLOGICAL BULLETIN. Washington, D.C.: American Psychological Association, 1904-- . Bimonthly.

SOCIAL CASEWORK. New York: Family Service Association of America, 1920-- . Monthly (except August and September).

SOCIAL WORK. Washington, D.C.: National Association of Social Workers, 1956-- . Six issues per year.

AMERICAN JOURNAL OF MENTAL DEFICIENCY. Washington, D.C.: American Association on Mental Deficiency, 1876-- . Bimonthly.

AMERICAN JOURNAL OF SOCIOLOGY. Chicago: University of Chicago Press, 1895-- . Bimonthly.

BRITISH JOURNAL OF ADDICTION. Edinburgh: Longman Group Ltd., Journals Division, 1903-- . Semiannual.

BRITISH JOURNAL OF PREVENTIVE AND SOCIAL MEDICINE. London: British Medical Association, B.M.A. House, 1947-- . Quarterly.

FAMILY COORDINATOR. Minneapolis, Minn.: National Council on Family Relations, 1952-- . Quarterly.

GERONTOLOGICA CLINICA. Continuing as GERONTOLOGIST (see p. 11).

HUMANITAS. Pittsburgh: Duquesne University, 1965-- . Three issues per year.

HUMAN ORGANIZATION. Boulder: University of Colorado, Institute of Behavioral Science, 1941-- . Quarterly.

INTERNATIONAL JOURNAL OF AGING AND HUMAN DEVELOPMENT. Farmingdale, N.Y.: Baywood Publishing, 1973-- . Quarterly.

INTERNATIONAL JOURNAL OF HEALTH SERVICES. Farmingdale, N.Y.: Baywood Publishing, 1971-- . Quarterly.

JOURNAL OF APPLIED BEHAVIORAL SCIENCE. Arlington, Va.: N.T.L. Institute for Applied Behavioral Science, 1965-- . Quarterly.

JOURNAL OF CLINICAL PSYCHOLOGY. Brandon, Vt.: Clinical Psychology Publishing Co., 1945-- . Quarterly.

Background and Organization

JOURNAL OF EDUCATIONAL PSYCHOLOGY. Washington, D.C.: American Psychological Association, 1910-- . Bimonthly.

JOURNAL OF GENERAL PSYCHOLOGY. Provincetown, Mass.: Journal Press, 1927-- . Quarterly.

JOURNAL OF NUTRITION EDUCATION. Berkeley, Calif.: Society for Nutrition Education, 1969-- . Quarterly.

LIFE THREATENING BEHAVIOR. New York: Human Sciences Press, 1971-- . Quarterly.

PERSONNEL JOURNAL. Santa Monica, Calif.: Personnel Journal, Inc., 1922-- . Monthly.

PSYCHOSOMATICS. Irvington, N.J.: Academy of Psychosomatic Medicine, 1960-- . Quarterly.

SOCIAL SERVICE REVIEW. Chicago: University of Chicago Press, 1927-- . Quarterly.

WESTERN JOURNAL OF MEDICINE. San Francisco: California Medical Association, 1878-- . Monthly.

ADULT EDUCATION. Washington, D.C.: Adult Education Association of the United States of America, 1973-- . Quarterly.

AMERICAN JOURNAL OF OBSTETRICS AND GYNECOLOGY. St. Louis, Mo.: C.V. Mosby, 1920-- . Semimonthly.

AMERICAN JOURNAL OF OCCUPATIONAL THERAPY. Rockville, Md.: American Occupational Therapy Association, 1947-- . Ten issues per year.

AMERICAN SOCIOLOGICAL REVIEW. Washington, D.C.: American Sociological Association, 1936-- . Bimonthly.

BRITISH JOURNAL OF EDUCATIONAL TECHNOLOGY. London: Councils and Education Press, 1970-- . Three issues per year.

BULLETIN OF THE PSYCHONOMIC SOCIETY. Austin, Tex.: Psychonomic Society, 1973-- . Monthly.

CANADIAN JOURNAL OF BEHAVIORAL SCIENCE (REVUE CANADIENNE DES SCIENCES DU COMPORTEMENT). Montreal: Canadian Psychological Association, 1969-- . Quarterly.

Background and Organization

CANADIAN JOURNAL OF PUBLIC HEALTH. Ottawa: Canadian Public Health Association, 1909-- . Bimonthly.

CENTRAL STATES SPEECH JOURNAL. West Lafayette, Ind.: Purdue University, 1970-- . Quarterly.

COMMUNICATION RESEARCH. Beverly Hills, Calif.: Sage Publications, 1974-- . Quarterly.

GENETIC PSYCHOLOGY MONOGRAPH. Provincetown, Mass.: Journal Press, 1926-- . Quarterly.

INSTRUCTIONAL SCIENCE. New York: Elsevier North-Holland, 1971-- . Quarterly.

INTERNATIONAL JOURNAL OF EPIDEMIOLOGY. London: Oxford University Press, 1972-- . Quarterly.

INTERNATIONAL JOURNAL OF THE ADDICTIONS. New York: Marcel Dekker Journals, 1966-- . Quarterly.

JOURNAL OF APPLIED BEHAVIOR ANALYSIS. Lawrence, Kans.: Society for the Experimental Analysis of Behavior, 1968-- . Quarterly.

JOURNAL OF APPLIED SOCIAL PSYCHOLOGY. Washington, D.C.: V.H. Winston and Sons, 1971-- . Quarterly.

JOURNAL OF COMPARATIVE FAMILY STUDIES. Calgary, Canada: University of Calgary, Department of Sociology, 1970-- . Three issues per year.

JOURNAL OF PERSONALITY ASSESSMENT. Portland, Oreg.: Society for Personality Assessment, 1970-- . Bimonthly.

JOURNAL OF PUBLIC HEALTH DENTISTRY. Raleigh, N.C.: American Association of Public Health Dentists, 1966-- . Quarterly.

JOURNAL OF RESEARCH AND DEVELOPMENT IN EDUCATION. Atlanta: University of Georgia, 1967-- . Quarterly.

JOURNAL OF RISK AND INSURANCE. Bloomington, Ill.: American Risk and Insurance Association, 1933-- . Quarterly.

JOURNAL OF THE AMERICAN ACADEMY OF CHILD PSYCHIATRY. New Haven, Conn.: Yale University Press, 1962-- . Quarterly.

Background and Organization

MENTAL RETARDATION. Washington, D.C.: American Association on Mental Deficiency, 1963-- . Bimonthly.

ORGANIZATIONAL BEHAVIOR AND HUMAN PERFORMANCE. New York: Academic Press, vol. 5, 1970-- . Bimonthly.

PACIFIC SOCIOLOGICAL REVIEW. Beverly Hills, Calif.: Sage Publications, 1958-- . Quarterly.

PERCEPTUAL AND MOTOR SKILLS. Missoula, Mont.: R.B. and C.H. Ammons, 1949-- . Bimonthly.

PHYSICAL THERAPY. Washington, D.C.: American Physical Therapy Association, 1921-- . Monthly.

PSYCHIATRY IN MEDICINE. Continued as INTERNATIONAL JOURNAL OF PSYCHIATRY IN MEDICINE (see p. 11).

REVIEW OF EDUCATIONAL RESEARCH. Washington, D.C.: American Educational Research Association, 1931-- . Quarterly.

SCANDINAVIAN JOURNAL OF SOCIAL MEDICINE. (Text in English). Stockholm: Almqvist & Wiksell International, 1973-- . Three issues per year.

SOCIOMETRY. Washington, D.C.: American Sociological Association, 1937-- . Quarterly.

Part II

THE LITERATURE OF THE MID-1970S

Chapter 2
HEALTH PROBLEM ANALYSIS

The first step in planning or studying a patient education program or process is to understand the health problem it is intended to reduce. The references annotated in this chapter represent the vast literature in the biomedical and behavioral sciences, where an attempt has been made to understand the human or social aspects of disease, disability, and death. The educational perspective on health and illness is one that views health not as an end in itself but rather as a means to other ends that relate to quality of life, dignity, self-actualization, and other such transcendent ends. Health is an instrumental value rather than a terminal value.

The purposes that health serves in the lives of people give health its various meanings and values to individuals and to society. The health professional must understand these aspects of the health problem or the health goal to develop a meaningful approach to patient education. This perspective requires a conceptual break from the strictly biomedical model in which most clinicians have been trained to think about health. It requires, first, that the health problem or goal of the patient be viewed from the patient's own perspective in addition to a medical perspective. It requires, secondly, that the problem be cast in a broader social context to understand the relative importance, stigma, sympathy, help isolation, ridicule, guilt, and economic threat that will attach to the patient or the problem in the world outside the clinic.

The social and epidemiological diagnosis that is derived from the examination of the health problem, using these perspectives, provides patient education with an ultimate objective or goal that is defined in terms that are concrete and specific, yet sensitive to the needs felt by the patients themselves. Such a diagnosis of the sociomedical problem places the purpose of a patient education program in a broader context than the usual clinical or medicocentric diagnosis.

In the pages that follow, the citations annotated are selected on the basis of their representation of this sociomedical and epidemiological orientation to the assessment of health problems. If the health planner, practitioner, or investigator will begin the process of educational diagnosis by casting the

Health Problem Analysis

health problem into these frameworks rather than a strictly biomedical framework, the later steps in planning, conducting, or evaluating patient education programs will be more properly focused on goals or outcomes that have meaning to the people whose behavior must be influenced.

Current publications of categorical government agencies (e.g., each of the National Institutes of Health), the National Center for Health Statistics, and the voluntary health associations (e.g., American Heart Association) are ready sources of the most recent data on health problems.

A. CHRONIC DISORDERS

A1   American Cancer Society. 1979 CANCER FACTS AND FIGURES. New York: 1979. 31 p.

> The contents of this booklet include data on early detection; the cancer checkup; cancer rates in blacks and whites; cancer and treatment trends; discussion of breast, uterine, colon-rectum, skin, oral, and lung cancers; leukemia; cancer in children; public and professional education; service, rehabilitation, and research; and costs of cancer.

A2   Bartle, S.H., and Bishop, L.F. "Psychological Study of Patients with Coronary Heart Disease with Unexpectedly Long Survival and High Level Function." PSYCHOSOMATICS 15 (1974): 68-69.

> Minnesota Multiphasic Personality Inventories were administered to eighteen patients with coronary heart disease. These patients had survived unexpectedly long with grave prognoses and had maintained a very high level of function that was satisfactory to them. Denial was not seen and the values for the means were in the normal range.

A3   Berkman, B., and Rehr, H. "The Search for Early Indicators of Social Service Need Among Elderly Hospital Patients." JOURNAL OF THE AMERICAN GERIATRICS SOCIETY 22 (1974): 416-21.

> A population of 5,312 elderly patients in the private, semi-private, and teaching wards of a hospital's medical and surgical services during the year 1967 was studied for the possible usefulness of certain sociodemographic characteristics as early indicators of the need for referral of elderly patients to social services.

A4   Borgman, M.F. "Coronary Rehabilitation--A Comprehensive Design." INTERNATIONAL JOURNAL OF NURSING STUDIES 12 (March 1975): 13-21.

> This rehabilitation design includes disease prevention, diagno-

Health Problem Analysis

sis, functional assessment, and the initiation or coordination of nursing and ancillary services. The design is divided into four phases: acute care, remainder of the hospitalization, home convalescence, and recovery maintenance.

A5   Cobb, A.B. MEDICAL AND PSYCHOLOGICAL ASPECTS OF DISABILITY. Springfield, Ill.: Charles C Thomas, 1977. 365 p.

Medical and psychosocial factors related to nine of the major chronic disabilities are reviewed for the student rehabilitation counselor or working practitioner to facilitate interdisciplinary communication in the delivery of rehabilitation services.

A6   Davies, R.K., et al. "Organic Factors and Psychological Adjustment in Advanced Cancer Patients." PSYCHOSOMATIC MEDICINE 35 (1973): 464-71.

Associations between physical factors and psychological adjustment of cancer patients are studied, and implications for improving communications with these patients are reported.

A7   Engel, B.T. "Comment on Self-Control of Cardiac Functioning: A Promise as Yet Unfulfilled." PSYCHOLOGICAL BULLETIN 81 (1974): 43.

The author criticizes Blanchard and Young's definition of a clinically significant heart rate change, indicating that only one tachyarrhythmia, sinus tachycardia, is exclusively a heart rate problem.

A8   Furman, S. "Intestinal Biofeedback in Functional Diarrhea: A Preliminary Report." JOURNAL OF BEHAVIOR THERAPY AND EXPERIMENTAL PSYCHIATRY 4 (1973): 317-21.

The technique of biofeedback was applied to five patients suffering functional disorders of the lower gastrointestinal tract but manifesting no organic findings. It is theorized that the positive responses to treatment were achieved through inhibition of conditioned responses deleteriously controlling autonomic functions.

A9   German, P.S., et al. "Symposium: Health Care of the Aged in Four Ambulatory Settings with a Focus on the Hypertensive Patient." GERONTOLOGIST 15 (1975): 311-32.

This symposium on ambulatory health care delivery to elderly people addresses the characteristics and attitudes toward the elderly by providers of care, providers' attitudes toward drug prescribing, factors associated with the management of hypertension in the elderly, and characteristics and health-seeking behavior of elderly patients.

Health Problem Analysis

A10 Goldberg, R.T. "Adjustment of Children with Invisible and Visible Handicaps: Congenital Heart Disease and Facial Burns." JOURNAL OF COUNSELING PSYCHOLOGY 21 (1974): 428-32.

> Children with congenital heart disease and facial burns were compared on ten measures of adjustment to test the effects of invisible and visible disability upon their social and psychological development. Interview data were coded and analyzed, using sex, grade, and age as covariates.

A11 Gordon, W.A., et al. "Behavioral Correlates of the Coronary Profile." JOURNAL OF CLINICAL PSYCHOLOGY 30 (1974): 343-47.

> A double-blind study using three psychological tasks differentiated between fourteen high and eleven low coronary risk subjects and between those two health groups and twenty-one postmyocardial infarction patients. Gertler et al.'s profile scale was used to define coronary risk.

A12 Grant, I., et al. "Recent Life Events and Diabetes in Adults." PSYCHOSOMATIC MEDICINE 36 (1974): 121-28.

> Fluctuations in diabetic condition and changes in life events were studied in a group of thirty-seven adult diabetic patients over a period of eight to eighteen months.

A13 Hoffman, A.L. "Psychological Factors Associated with Rheumatoid Arthritis: Review of Literature." NURSING RESEARCH 23 (1974): 218-34.

> Investigations of psychological factors relevant to individuals with rheumatoid arthritis are reviewed and classified according to methodological procedures, including case and impressionistic, actuarial, factor analytic, correlational, and natural process studies. Results of investigations are compared, findings are related to a proposed grouping of psychological hypotheses, and directions for future research are recommended.

A14 Horan, P.M., and Gray, B.H. "Status Inconsistency, Mobility and Coronary Heart Disease." JOURNAL OF HEALTH AND SOCIAL BEHAVIOR 15 (1974): 300-310.

> Methodological problems in studies of status inconsistency, mobility, and disease are outlined; and implications for epidemiological research are demonstrated. The literature on status inconsistency, mobility, and coronary heart disease is reviewed. Data from two studies reporting evidence for inconsistency and/or mobility effects on coronary heart disease are reanalyzed using multivariate techniques to deal with the methodological issues.

A15 Idelson, R.K., et al. "Changes in Self-Concept During the Year After a First Heart Attack: A Natural History Approach--Part I." AMERICAN ARCHIVES OF REHABILITATION THERAPY 22 (March 1974): 10-21.

> This paper, based on case analyses, explores major stages in the recovery process of the male heart patient. Using a natural history perspective, it approaches recovery as a form of adult socialization. Changes that take place in the self-concepts of men in the new role of a chronic heart patient and factors that influence these changes are analyzed.

A16 Jenkins, C.D., et al. "Prediction of Clinical Coronary Heart Disease by a Test for the Coronary-Prone Behavior Pattern." NEW ENGLAND JOURNAL OF MEDICINE 290 (1974): 1271-75.

> A prospective study of 2,750 employed men who completed a questionnaire measuring the coronary-prone Type A behavior pattern showed that high scorers had twice the incidence of new coronary heart disease as low scorers over a four-year period.

A17 Keegan, D.L. "Psychosomatics: Toward an Understanding of Cardiovascular Disorders." PSYCHOSOMATICS 14 (1973): 321-25.

> This paper reviews recent research demonstrating the need for continued expansion of cardiology and psychiatric liaison in research and patient care. The areas reviewed include neurophysiology and neurobiochemistry, coronary heart disease, hypertension, and psychological sequelae of cardiovascular disorders. The authors emphasize recent advances made through collaboration of cardiologists, psychiatrists, and basic scientists in understanding the psychosocial links with cardiovascular disorders.

A18 Khan, A.U. "Mechanism of Psychogenic Asthmatic Attack in Children." PSYCHOTHERAPY AND PSYCHOSOMATICS 24 (1974): 137-40.

> The contributions of classical and instrumental conditioning to the occurrence of psychogenic asthmatic attacks in children are discussed.

A19 Khan, A.U., et al. "Hypnotic Suggestibility Compared with Other Methods of Isolating Emotionally-Prone Asthmatic Children." AMERICAN JOURNAL OF CLINICAL HYPNOSIS 17 (July 1974): 50-53.

> This study was designed to determine the hypnotic suggestibility of a group of asthmatic children, to induce asthmatic attacks through hypnotic suggestion, and to compare hypnotic suggestibility with other methods of isolating emotionally-prone asthmatic children. The children and their parents were inter-

Health Problem Analysis

viewed to determine if any emotional precipitants of asthma had existed. Various methods of isolating emotionally-prone asthmatic children are compared.

A20 Koch, M.F., and Molnar, G.D. "Psychiatric Aspects of Patients with Unstable Diabetes Mellitus." PSYCHOSOMATIC MEDICINE 36 (1974): 57-68.

Findings from psychiatric interviews and psychological tests on seven adult patients with unstable diabetes mellitus are discussed, with reference to emotions influencing blood glucose behavior, the impact of diabetes on personality, and the effect of personality on diabetes.

A21 Lambert, G. "Patients with Progressive Neurological Diseases." SOCIAL CASEWORK 55 (1974): 154-59.

Two case summaries are given. One illustrates early intervention of a social worker to help a patient with multiple sclerosis express and clarify feelings about his disability and begin to cope with his situation. The other illustrates the role of the social worker in sustaining a couple during the acute and terminal stages of the wife's progressive illness and prolonged hospitalization.

A22 Levine, E.S. "Psychological Tests and Practices with the Deaf: A Survey of the State of the Art." VOLTA REVIEW 76 (1974): 298-319.

A national survey of psychologists who work with the deaf was undertaken to gather information on their background, orientation, and preparation; the nature of their clientele; the instruments and practices they employ; the difficulties they encounter; and their suggestions and attitudes toward potential improvement in psychological services.

A23 Lucas, R.A. "A Comparative Study of Measures of General Anxiety and Death Anxiety Among Three Medical Groups Including Patient and Wife." OMEGA--JOURNAL OF DEATH AND DYING 5 (1974): 233-43.

A comparative study was made of general anxiety and death anxiety among three different groups of physically ill males (center and home hemodialysis patients and surgery patients) and their wives.

A24 Lynch, J.J., et al. "Effects of Human Contact on Cardiac Arrhythmia in Coronary Care Patients." JOURNAL OF NERVOUS AND MENTAL DISEASE 158 (1974): 88-99.

The effects of human contact on cardiac functions were studied

Health Problem Analysis

by noting on a continuous electrocardiogram each time any person came in contact with coronary care patients.

A25 MacDonald, M.L. "The Forgotten Americans: A Sociopsychological Analysis of Aging and Nursing Homes." AMERICAN JOURNAL OF COMMUNITY PSYCHOLOGY 1 (1973): 272-94.

A sociopsychological analysis of aging as a role label suggests that many of the assumed components of aging are more a function of self-fulfilling social expectations than they are of natural, unavoidable processes. Current environmental conditions are discussed, and alternative strategies grounded in the sociopsychological treatment model are presented.

A26 Mann, G.V. "The Influence of Obesity on Health (First of Two Parts)." NEW ENGLAND JOURNAL OF MEDICINE 291 (1974): 178-85.

Medical knowledge of the influence of obesity on health is reviewed under the headings: metabolic lesions in obesity, brown fat, genetics, feeding controls, metabolic lesions in human subjects, indexes of obesity, experimental human obesity, indexes of obesity in children, and cardiovascular disease--high blood pressure.

A27 _____. "The Influence of Obesity on Health (Second of Two Parts)." NEW ENGLAND JOURNAL OF MEDICINE 291 (1974): 226-32.

Medical knowledge of the influence of obesity on health is reviewed in this second article in a two-part series, under the headings coronary heart disease, the adipocyte-proliferation hypothesis, serum lipids, smoking and obesity, obesity in children, drug treatment, pregnancy, dietary treatment, ileal bypass surgery, group therapy, behavior modification, and the allurement of quacks in obesity.

A28 Oakes, T.W., et al. "Social Factors in Newly Discovered Elevated Blood Pressure." JOURNAL OF HEALTH AND SOCIAL BEHAVIOR 14 (1973): 198-204.

Social characteristics of persons with elevated blood pressure who did not know they had the condition were investigated.

A29 "Sources of Educational Materials on Hypertension." JOURNAL OF THE AMERICAN PHARMACEUTICAL ASSOCIATION NS 14 (1974): 181-85.

This bibliography lists materials on hypertension for use in patient education and materials for use by professionals. Included are printed materials, films and slides, posters, other materials, and ordering information.

Health Problem Analysis

A30 Strupp, H.H., et al. "Effects of Suggestion on Total Respiratory Resistance in Mild Asthmatics." JOURNAL OF PSYCHOSOMATIC RESEARCH 18 (1974): 337-46.

> Thirteen mild asthmatics were given inhalations of saline and Isuprel, with measurements of total respiratory resistance taken before and after each inhalation. These data corroborate the findings in other investigations; that is, one subgroup of the asthmatic population responds mainly to the pharmacological effect of the inhalant, while the other subgroup responds to its suggested effect.

A31 Swartz, H., and Leitch, C.J. "Differences in Mean Adolescent Blood Pressure by Age, Sex, Ethnic Origin, Obesity and Familial Tendency." JOURNAL OF SCHOOL HEALTH 45 (1975): 76-81.

> This study determined the mean values of blood pressures in adolescents by age. Mean differences in systolic and diastolic blood pressures also were tested between age groups, males and females, obese and nonobese subjects, racial groups, and between subjects who had a reported familial history of hypertension and those who did not.

A32 Theorell, T., and Lind, E. "Systolic Blood Pressure, Serum Cholesterol and Smoking in Relation to Sociological Factors and Myocardial Infarction." JOURNAL OF PSYCHOSOMATIC RESEARCH 17 (1973): 327-32.

> In a study of ninety-six randomly selected male subjects between forty and sixty years of age, a set of sociological variables and age were studied in relation to each of the variables systolic blood pressure, cigarette smoking, and serum cholesterol.

A33 Thomas, J.M., Jr., and Weiner, E.A. "Psychological Differences Among Groups of Critically Ill Hospitalized Patients, Noncritically Ill Hospitalized Patients, and Well Controls." JOURNAL OF CONSULTING AND CLINICAL PSYCHOLOGY 42 (1974): 274-79.

> Differences on nine psychological measures were investigated among groups of twenty-five critically ill, hospitalized patients; twenty-five noncritically ill, hospitalized patients; and twenty-five normal, well controls. The instruments used were the Purpose in Life Test, the FIRO-B, and two listening measures on which eye-blink rates were recorded. Four multivariate discriminate function analyses were performed to determine those variables that discriminated groups of subjects. The critically ill group expressed (1) more "purpose in life," (2) an increased need for affection and inclusion, (3) a decreased "wanted control" from others, and (4) an increased rate of eye blinks in responses to disease-related material.

Health Problem Analysis

A34 Train, G.J. "The Aged--with Tender Interest and Concern." PSYCHOMETRICS 16, no. 2 (1975): 79-83.

> A brief survey of the aged is presented, with a discussion of a home for the aged, stress, reactive syndromes, suicide and dying, attitudes of physicians, and treatment.

A35 U.S. Department of Health, Education, and Welfare. National Center for Health Statistics. EDENTULOUS PERSONS, UNITED STATES--1971. Data from the National Health Survey, series 10, no. 89. DHEW Publication no. HRA 74-1516. Rockville, Md.: (1974): 31 p.

> Includes statistics on the prevalence of edentulous persons by age, sex, race, income, education, and place of residence. Data are presented on utilization of dental services and use of dentures by persons who have lost all their natural teeth.

A36 U.S. Department of Health, Education, and Welfare. National Heart and Lung Institute. NEEDS AND OPPORTUNITIES FOR REHABILITATING THE CORONARY HEART DISEASE PATIENT: REPORT OF THE TASK FORCE ON CARDIOVASCULAR REHABILITATION OF THE NATIONAL HEART AND LUNG INSTITUTE. DHEW Publication no. NIH 75-750. Bethesda, Md.: [1974]: 93 p.

> Recommendations of a Task Force on coronary rehabilitation are presented. Subjects include: provision and expansion of rehabilitation services; programs for education and information (public, patient-family, and professional); and patient services.

A37 Vogel, J.M. HOW TO LIVE WITH HEMOPHILIA. New York: Interbook, 1974. 132 p.

> This handbook is designed for use by hemophiliacs and their families and for reference by the family counselor. It contains information to help prevent bleeding episodes and to minimize the effects of episodes that do occur, explains the condition and its genetic background, advises on how to raise a hemophiliac child, discusses psychological aspects of the disease, and gives other general information.

A38 Wiener, C.L. "The Burden of Rheumatoid Arthritis: Tolerating the Uncertainty." SOCIAL SCIENCE AND MEDICINE 9 (1975): 97-104.

> This paper examines disease conditions that produce extreme uncertainty in the lives of patients. It analyzes strategies arthritics develop to tolerate this uncertainty.

B. ACUTE EPISODES AND DYING

B1 Andersen, M.D., and Pleticha, J.M. "Emergency Unit Patients' Perceptions of Stressful Life Events." NURSING RESEARCH 23 (1974): 378-83.

> Perceptions of recent stressful life events of fifty-two emergency unit patients were studied. The readjustment required by the stressful events was measured using the Social Readjustment Rating scale developed by Holmes and Rahe. The extensive degree of recent stress in this sample of patients and its relationship to their perception of severity enhance the theory that health professionals should assess routinely their patients' degree of readjustment to recent life changes.

B2 Asken, M.J. "Psycho-Emotional Aspects of Mastectomy: A Review of Recent Literature." AMERICAN JOURNAL OF PSYCHIATRY 132 (1975): 56-59.

> The recent literature on the psychological aspects of mastectomy, including psychological reactions to the procedure and appropriate intervention strategies, is reviewed. Individual and familial fears and concerns and the importance of preoperative counseling and postoperative rehabilitation are discussed.

B3 Bean, P. "Patterns of Self-Poisoning." BRITISH JOURNAL OF PREVENTIVE AND SOCIAL MEDICINE 28, no. 1 (1974): 24-31.

> A population of 935 hospital admissions for self-poisoning was taken from three general hospitals over a five-year period. All admissions were followed up by use of coroners' records. Demographic and medical characteristics were analyzed to produce a profile of self-poisoning.

B4 Brodland, G.A., and Andreasen, N.J.C. "Adjustment Problems of the Family of the Burn Patient." SOCIAL CASEWORK 55 (1974): 13-18.

> During a one-year study, thirty-two adult burn patients and their families were evaluated psychiatrically on admission and interviewed daily until the time of discharge. Methods of communicating and providing information and understanding for family members and patients are suggested.

B5 Carey, R.G. "Emotional Adjustment in Terminal Patients: A Quantitative Approach." JOURNAL OF COUNSELING PSYCHOLOGY 21 (1974): 433-39.

> Eleven hospital chaplains collected data by interviewing eighty-four dying patients to identify factors relating to emotional adjustment. Emotional adjustment was influenced most by the patient's physical condition (level of discomfort), by previous experiences with dying persons, and by interpersonal relationships.

Health Problem Analysis

B6 Carson, R.A. "Amidst Children and Witnesses: Reflections on Death." HUMANITAS 10 (1974): 9-19

    The Judeo-Christian attitudes toward life, death, and the Resurrection are examined; and theories of death anxiety are compared.

B7 Dohrenwend, B.S., and Dohrenwend, B.P., eds. STRESSFUL LIFE EVENTS: THEIR NATURE AND EFFECTS. New York: Wiley-Interscience, 1974. 340 p.

    Experts from varied disciplines present research findings concerning stressful life events, advances made in clinical and community research, and methodological issues raised by this research. Relationships between stress and physical and psychological illnesses, including heart diseases, depression, suicide, and other problems related to the human response to stress are reviewed.

B8 Kinsman, R.A., et al. "Observations on Patterns of Subjective Symptomatology of Acute Asthma." PSYCHOSOMATIC MEDICINE 36 (1974): 129-43.

    In this study, fifteen patterns of asthma symptomatology based on reported frequency of five symptom categories during asthma attacks were identified within a group of one hundred inpatients with asthma.

B9 Mogielnicki, R.P., et al. "Patient and Bystander Response to Medical Emergencies." MEDICAL CARE 13 (1975): 753-62.

    This study of patient and bystander responses to medical emergencies identifies a significant group of high-risk patients who reacted slowly to emergencies and bypassed the emergency ambulance service entirely.

B10 Neuringer, C., and Harris, R.M. "The Perception of the Passage of Time Among Death-Involved Hospital Patients." LIFE-THREATENING BEHAVIOR 4 (1974): 240-53.

    The extent of the relationship between death involvement and temporal orientation among hospitalized suicidal, terminal, geriatrically ill, and normal patients was evaluated by behavioral estimations of elapsed time and a survey of the patients' opinions concerning the passage of time. The results indicated that the suicidal patients, followed by the geriatrically ill, overestimated time intervals to a much greater extent than did the normal and terminal patients.

B11 Ray, J.J., and Najman, J. "Death Anxiety and Death Acceptance:

Health Problem Analysis

A Preliminary Approach." OMEGA--JOURNAL OF DEATH AND DYING 5 (1974): 311-15.

> A scale to measure death acceptance is described. It was found to be reliable and correlated postively (r = .242 and .263) with two existing death anxiety scales.

B12 Rendtorff, R.C., et al. "Economic Consequences of Gonorrhea in Women: Experience from an Urban Hospital." JOURNAL OF THE AMERICAN VENEREAL DISEASE ASSOCIATION 1 (September 1974): 40-47.

> Extrapolation to the national level of the frequency and costs of gonococcal infection in female patients of an urban hospital's emergency room estimates that the female complications of gonorrhea cost at least $211,893,189 in the United States during 1972. These estimates demonstrated the importance of the gonorrhea problem.

B13 Reynolds, D.K., and Farberow, N.L. "The Suicidal Patient: An Inside View." OMEGA--JOURNAL OF DEATH AND DYING 4 (1973): 229-41.

> This investigation, proposing "experiential research" as a useful supplement in culture and personality studies, focuses on the senior author's experience as a researcher and patient in a psychiatric hospital. His two-week stay as a depressed, suicidal patient provides insight into the effects of sociocultural system pressures on personality functioning and identity reformation. Journal accounts of the first and fourth days of hospitalization describe the experiences of admission and a near-suicide attempt.

B14 Schulz, R., and Aderman, D. "Clinical Research and the Stages of Dying." OMEGA--JOURNAL OF DEATH AND DYING 5 (1974): 137-43.

> Kubler-Ross's claim that terminal patients pass through five psychological stages in a predictable order is examined and found not to be supported by other investigators. Other data show the process of dying to be less rigid and even stageless.

B15 Slote, M.A. "Existentialism and the Fear of Dying." AMERICAN PHILOSOPHICAL QUARTERLY 12 (1975): 17-28.

> An existentialist view of human anxiety about death and responses to that anxiety, based on the work of Pascal, Kierkegaard, Heidegger, and Sartre, is presented.

B16 Sobel, D.E. "Death and Dying." AMERICAN JOURNAL OF NURSING 74 (1974): 98-99.

The author's concepts of the fear of dying and the dying process and the various stages of emotional reaction to dying that a person may go through are described.

B17 Spinetta, J.J. "The Dying Child's Awareness of Death: A Review." PSYCHOLOGICAL BULLETIN 81 (1974): 256-60.

Two controlled studies directly measuring a child's concerns show an awareness by a child as young as six years of the seriousness of his or her illness. Implications for designing future research in this area are discussed.

B18 Thauberger, P.C., and Thauberger, E.M. "A Consideration of Death and a Sociological Perspective in the Quality of the Dying Patient's Care." SOCIAL SCIENCE AND MEDICINE 8 (1974): 437-41.

Emphasis on multivariate analysis as an appropriate research paradigm for the investigation of the dying patient's dynamic social-treatment system is advocated to enhance the quality of the dying patient's care.

B19 Vernon, G.M. "Dying as a Social-Symbolic Process." HUMANITAS 10 (1974): 21-32.

This article analyzes death-related behavior utilizing concepts and orientations that are a part of a perspective called symbolic interactionism, which places strong emphasis upon the social and the symbolic nature of human interaction.

B20 Ward, A.W.M. "Terminal Care in Malignant Disease." SOCIAL SCIENCE AND MEDICINE 8 (1974): 413-20.

This study examines social and medical factors associated with dying at home or in a hospital.

C. MENTAL AND PERSONALITY DISORDERS

C1 Ablon, J. "Al-Anon Family Groups: Impetus for Learning and Change Through the Presentation of Alternatives." AMERICAN JOURNAL OF PSYCHOTHERAPY 28 (1974): 30-45.

This paper describes Al-Anon family groups in a West Coast metropolitan area. Meeting format and group behavior are described and the crucial didactic lesson and operational principles that underlie the Al-Anon program are analyzed.

C2 Browning, P.L. MENTAL RETARDATION: REHABILITATION AND COUNSELING. Springfield, Ill.: Charles C Thomas, 1974. 464 p. Illus. Tables.

Health Problem Analysis

This book is divided into three parts, each preceded by an overview. Part 1 provides a behavioral science orientation to define retardation. Concept, definition, classification, etiology, development, and personality are discussed and serve as a frame of reference for the text. Parts 2 and 3, designed for both the practitioner and researcher, address the areas of rehabilitation and counseling as they pertain to the mental retardation field.

C3 Burt, D.W. "Characteristics of the Relapse Situation of Alcoholics Treated with Aversion Conditioning." BEHAVIOUR RESEARCH AND THERAPY 12 (1974): 121-23.

This study examines the relapse situation of alcoholics in terms of the location of the first-drink episode, the social environment, and type of beverage consumed and compares that situation to the usual drinking situation prior to treatment.

C4 Crisp, A.H., et al. "Sleep Patterns in Obese Patients During Weight Reduction." PSYCHOTHERAPY AND PSYCHOSOMATICS 22 (1973): 159-65.

Findings are presented that support evidence from previous studies of patients with disorders of weight and patients with psychiatric morbidity, that nutritional change is associated with change in sleep patterns independently of mood state.

C5 Evans, M. "Modification of Drinking." JOURNAL OF ALCOHOLISM 8 (1973): 111-13.

The author discusses research on the modification of drinking rather than total abstinence as a treatment approach.

C6 Forrest, G.G. THE DIAGNOSIS AND TREATMENT OF ALCOHOLISM. 2d ed. Springfield, Ill.: Charles C Thomas, 1978. 348 p. Tables.

This approach to alcoholic rehabilitation gives expectancies and procedures for strategies such as residential treatment, Alcoholics Anonymous, and group psychotherapy.

C7 Fort, J. ALCOHOL: OUR BIGGEST DRUG PROBLEM. Health Education Paperback Series. New York: McGraw-Hill, 1973. 180 p.

The effects, history, control, abuse, and treatment of alcohol and alcoholism are described. Education and prevention are presented as solutions to the problem of alcoholism.

C8 Frank, J.D. "Psychotherapy: The Restoration of Morale." AMERICAN JOURNAL OF PSYCHIATRY 131 (1974): 271-73.

Health Problem Analysis

The author suggests that the primary function of all psychotherapies is to combat demoralization through restoring the patient's sense of mastery. Evidence consistent with this hypothesis is reviewed and implications are pointed out for classification of candidates for therapy and for research on sources of the therapist's healing powers.

C9   Geist, R.A. "Some Observations on Adolescent Drug Use: Therapeutic Implications." JOURNAL OF THE AMERICAN ACADEMY OF CHILD PSYCHIATRY 13 (Winter 1974): 54-71.

This article emphasizes the need for long-term, therapeutic programs to effect personality change in adolescent drug users. The author cautions against programs that neglect individual psychological needs.

C10   Gotestam, K.G., and Melin, G.L. "Covert Extinction of Amphetamine Addiction." BEHAVIOR THERAPY 5 (1974): 90-92.

A covert extinction paradigm is developed and applied to four addicts with histories of long-term and heavy intravenous abuse of central stimulants.

C11   Gouge, A.L., and Ekvall, S.W. "Diets of Handicapped Children: Physical, Psychological, and Socioeconomic Correlations." AMERICAN JOURNAL OF MENTAL DEFICIENCY 80 (1975): 149-57.

Diets of patients at a university-affiliated center for developmental disorders were evaluated to identify relationships between nutritional deficiencies and selected physical, psychological, and socioeconomic factors.

C12   Greenblatt, D.J., and Shader, R.I. "Drug Abuse and the Emergency Room Physician." AMERICAN JOURNAL OF PSYCHIATRY 131 (1974): 559-62.

This article discusses some areas of controversy in the uses of barbiturates and major tranquilizers in the emergency room and presents alternative methods of treatment where appropriate. The types of medical conditions discussed in this article are the comotose patient and patients using opiates, alcohol, amphetamines, hallocinogens, and anticholinergics.

C13   Hardy, R.E., and Cull, J.G. DRUG DEPENDENCE AND REHABILITATION APPROACHES. Springfield, Ill.: Charles C Thomas, 1973.

This book includes chapters on the effects of mood altering drugs, causes of drug abuse--a survey of theories, types of narcotic addicts, public vocational rehabilitation programs, clinical and counseling problems, treatment, counseling ap-

Health Problem Analysis

proaches and special programs, a therapeutic approach to rehabilitation of youthful drug abusers, and language of the drug abuser.

C14 Hughson, B., and Lyons, R. "Patient Response to Psychiatric Consultation in a General Hospital." AUSTRALIAN AND NEW ZEALAND JOURNAL OF PSYCHIATRY 7 (1973): 279-82.

Inpatients referred for psychiatric consultation were evaluated during and after consultation to determine their response. The presence of anxiety or depression was associated with a positive response. Patients without a disturbance of mood responded negatively, obtained no direct benefit from consultation, and were likely to have had a diagnosis of pain, alcoholism, or drug dependence.

C15 Jacobson, G.R. "Field Dependence Among Male Alcoholics: Establishing Norms for the Rod-and-Frame Test." PERCEPTUAL AND MOTOR SKILLS 39 (1974): 1015-18.

The performance of 145 hospitalized male alcoholics on the Rod-and-Frame Test for perceptual style was contrasted with that of normal males and male psychiatric inpatients.

C16 Jones, F.H. "A 4-Year Follow-up of Vulnerable Adolescents: The Prediction of Outcomes in Early Adulthood from Measures of Social Competence, Coping Style, and Overall Level of Psychopathology." JOURNAL OF NERVOUS AND MENTAL DISEASE 159 (1974): 20-39.

This study developed predictors of severe psychopathology in early adulthood from behavioral descriptions provided in middle adolescence and early adulthood by parents. The sample was twenty-four males who were first seen by the UCLA Family Project at the UCLA Psychology Department Clinic for behavioral problems in middle adolescence.

C17 Kreisman, J.J. "The Curandero's Apprentice: A Therapeutic Integration of Folk and Medical Healing." AMERICAN JOURNAL OF PSYCHIATRY 132 (1975): 81-83.

Alternative methods for dealing with folk illness in psychotherapy are presented in two case reports describing psychotic Mexican-American patients. The patients were treated successfully with an approach that integrated curanderismo, the Hispanic concept of healing, and traditional therapy. Ethical aspects of this integrated approach are discussed.

C18 Kupfer, D.J., et al. "A Comment on the 'Amotivational Syndrome' in Marijuana Smokers." AMERICAN JOURNAL OF PSYCHIATRY 130 (1973): 1319-22.

Findings of this study, which compared heavy and light smokers of marijuana, suggest that heavy use of marijuana may be related to existing depression, and that impaired motivation may be a manifestation of depression rather than a consequence of frequent marijuana use.

C19 Litman, G.K. "Stress, Affect and Craving in Alcoholics: The Single Case as a Research Strategy." QUARTERLY JOURNAL OF STUDIES ON ALCOHOL 35 (1974): 131-46.

Two alcoholics were studied intensively to examine the relationship between affect and craving for alcohol during treatment. These two preliminary investigations suggest that the strategy of the single-case, design model may prove a useful addition to existing methodologies in alcoholism research.

C20 Luban-Plozza, B., and Comazzi, A. "The Family as a Factor in Psychosomatic Disturbances." PSYCHOTHERAPY AND PSYCHOSOMATICS 22 (1973): 372-77.

A therapeutical method of "psychosomatic confrontation" is presented to assist physicians in recognizing and intervening in disturbed family relationships.

C21 Ludwig, A.M., and Wikler, A. "Craving and Relapse to Drink." QUARTERLY JOURNAL OF STUDIES ON ALCOHOL 35 (1974): 108-30.

A study to assess the importance of internal and external cues in relation to relapse in alcoholics is outlined.

C22 McWilliams, J., et al. "Field Dependence and Self-Actualization in Alcoholics." JOURNAL OF STUDIES ON ALCOHOL 36 (1975): 387-94.

This article reports on the increase of field independence and self-actualization in alcoholic patients following a six-week treatment program. Some causal factors are hypothesized and implications for treatment discussed.

C23 Maddock, J. "Sex Education for the Exceptional Person: A Rationale." EXCEPTIONAL CHILDREN 4 (1974): 273-78.

Guidelines are given for personnel in institutions and residential treatment centers in dealing with the sexuality of the exceptional person.

C24 Mallenby, T.W. "Personal Space: Projective and Direct Measures with Institutionalized Mentally Retarded Children." JOURNAL OF PERSONALITY ASSESSMENT 38 (1974): 28-31.

Two measures of personal space, a projective interaction

measure and a direct physical interaction measure, were employed with institutionalized, mentally retarded children to determine whether they were actually unaware of their abnormality or merely denied its existence.

C25 Maris, R., and Connor, H.E., Jr. "Do Crisis Services Work? A Followup of a Psychiatric Outpatient Sample." JOURNAL OF HEALTH AND SOCIAL BEHAVIOR 14 (1973): 311-22.

> A one-year follow-up of 200 psychiatric outpatients was conducted to determine the relative efficacy of various treatment modalities in a crisis service unit in the emergency room of a city hospital. Some implications and limitations of the data for the utility of crisis services in general are considered.

C26 Marmor, J. PSYCHIATRISTS AND THEIR PATIENTS: A NATIONAL STUDY OF PRIVATE OFFICE PRACTICE. Washington, D.C.: American Psychiatric Association, 1975. 180 p.

> The Joint Information Service surveyed a systematic 10 percent sample of all psychiatrists reporting they spend fifteen or more hours per week seeing private patients. The response rate was 73 percent, a statistically representative sample of the private practice field and its patients. Data on psychiatrists are broken out by analysts, child psychiatrists, and general psychiatrists. Data on patients are categorized by analytic patients, nonanalytic patients of analysts, and those being seen by all other psychiatrists.

C27 Morgenstern, M., and Michal-Smith, H. PSYCHOLOGY IN THE VOCATIONAL REHABILITATION OF THE MENTALLY RETARDED. Springfield, Ill.: Charles C Thomas, 1973. 90 p.

> This book considers the psychosocial aspects of rehabilitation of mentally retarded persons and gives an overview of historical contexts, current insights, and future trends related to vocational rehabilitation and mental retardation.

C28 Moriwaki, S.Y. "Self-Disclosure, Significant Others and Psychological Well-Being in Old Age." JOURNAL OF HEALTH AND SOCIAL BEHAVIOR 14 (1973): 226-32.

> The relationship between the number of significant others reported and the psychological well being of elderly persons is examined.

C29 O'Leary, M.R., et al. "Relationships Between Locus of Control and MMPI Scales Among Alcoholics: A Replication and Extension." JOURNAL OF CLINICAL PSYCHOLOGY 30 (1974): 312-14.

Rotter's Locus of Control scale and the Minnesota Multiphasic Personality Inventory were administered to one hundred male alcoholic veterans. Factor analytically derived subscales of the internal-external scales, which measure personal and sociopolitical control, were found to correlate with a number of MMPI scales.

C30 Ramsay, D.A. SUMMARY OF ARI RESEARCH ON DRUG AND ALCOHOL ABUSE. ARI-RR-1186. Arlington, Va.: Army Research Institute for the Behavioral and Social Sciences, 1975. 24 p.

The major Army Research Institute research on drug and alcohol abuse in the army from 1971 to January 1975 is summarized. An assessment is made of the army's drug prevention programs, and research on the social organization factors that influence drug abuse is reported.

C31 Rosenham, D.L. "On Being Sane in Insane Places." SCIENCE 179 (1973): 250-58.

This article describes an experiment in which eight sane people gained secret admission to twelve different hospitals. Their diagnostic and other experiences in psychiatric institutions are described.

C32 Sargent, J.D., et al. "Psychosomatic Self-Regulation of Migraine Headaches." SEMINARS IN PSYCHIATRY 5 (1973): 415-28.

A historical perspective of research and treatment of the migraine syndrome and studies in animals and humans on control of the autonomic nervous system is given, with implications for psychosomatic medicine.

C33 Skinner, H.A., et al. "Alcoholic Personality Types: Identification and Correlates." JOURNAL OF ABNORMAL PSYCHOLOGY 83 (1974): 658-66.

This study identified the existence of distinct personality types among a large sample of alcoholic psychiatric patients and related its taxonomy to previous research with alcoholic and general psychiatric patients.

C34 Slade, P.D., and Russell, G.F.M. "Experimental Investigations of Bodily Perception in Anorexia Nervosa and Obesity." PSYCHOTHERAPY AND PSYCHOSOMATICS 22 (1973): 359-63.

A technique for obtaining a measurement of body image perception and results of a series of studies on anorexia nervosa and obese patients are described, and implications are discussed.

Health Problem Analysis

C35 Spiegel, J.P. "The Family: The Channel of Primary Care." HOSPITAL AND COMMUNITY PSYCHIATRY 25 (1974): 785-88.

    The author suggests that family therapy is one of the most important channels of primary care in preventive psychiatry. The growth of family medicine and problems existing in the field are described.

C36 Spielberger, C.D., and Sarason, I.G., eds. STRESS AND ANXIETY. Vol. 1. New York: Halstead Press, 1975. 500 p. Append.

    This book is concerned with the effects of psychosocial stress and biographical dispositions on clinical anxiety; the development of psychoneurotic and psychosomatic emotional disorders; results of experimental studies of stress, fear, and anxiety; and recent developments in management of stress and treatment of anxiety.

C37 Strassberg, D.S., and Robinson, J.S. "The Relationship Between Locus of Control and Other Personality Measures in Drug Users." JOURNAL OF CONSULTING AND CLINICAL PSYCHOLOGY 42 (1974): 744-45.

    Sixty individuals who identified themselves as active or recently active heroin addicts were administered a battery of five self-report psychological tests as part of the routine screening process at a drug center. The relationship between locus of control and adjustment for narcotics users was studied.

C38 Tarter, R.E., et al. "Intellectual Competence of Alcoholics." JOURNAL OF STUDIES ON ALCOHOL 36 (1975): 381-86.

    Results of a battery of cognitive tests given to alcoholic and psychiatric inpatients matched in age and education are described.

C39 Weinstein, R.M. "Mental Patients' Perceptions of Illness Etiology." AMERICAN JOURNAL OF PSYCHIATRY 131 (1974): 798-802.

    Hospitalized mental patients' perceptions of factors contributing to their illness in terms of the economic, family, and social problems they experienced are examined.

C40 White, W.C., Jr., et al. "Social Competence and Outcome of Hospitalization: a Preliminary Report." JOURNAL OF HEALTH AND SOCIAL BEHAVIOR 15 (1974): 261-66.

    To evaluate the impact of hospitalization on the posthospital social adjustment of state hospital patients, the Worcester Scale of Social Competence was administered to 159 psychiatric patients subsequent to hospitalization and one year after discharge. Findings are discussed in terms of social stigma.

D. PREGNANCY, CHILDBEARING, AND THE PUERPERIUM

D1   Adler, N.E. "Emotional Responses of Women Following Therapeutic Abortion." AMERICAN JOURNAL OF ORTHOPSYCHIATRY 45 (1975): 446-54.

> Factor analysis of postabortion emotional responses revealed socially and internally based negative emotions as well as positive emotions.

D2   Baizerman, M., et al. "A Critique of the Research Literature Concerning Pregnant Adolescents, 1960-1970." JOURNAL OF YOUTH AND ADOLESCENCE 3 (1974): 61-75.

> A perspective on the research literature from 1960 to 1970 concerning pregnant adolescents is presented. Research methodology and potential content for research are examined and suggestions made for a stronger empirical research base. An analysis of the "risk" concept also is presented.

D3   Barglow, P., and Weinstein, S. "Therapeutic Abortion during Adolescence: Psychiatric Observations." JOURNAL OF YOUTH AND ADOLESCENCE 2 (1973): 331-42.

> This study investigates the psychological reactions of adolescent girls aborted during the first trimester of pregnancy.

D4   Belsky, R. "Vaginal Contraceptives: A Time for Reappraisal?" POPULATION REPORTS. Series H: Barrier Methods, no. 3. Washington, D.C.: George Washington University Medical Center, Department of Medical and Public Affairs, 1975. 20 p. Bibliog.

> This report reviews the literature on vaginal contraceptives. Topics include history, mode of action, effectiveness, use and distribution, and venereal disease prophylaxis.

D5   Bluford, R.J., and Petres, R.E. UNWANTED PREGNANCY: THE MEDICAL AND ETHICAL IMPLICATIONS. New York: Harper & Row, 1973. 116 p.

> A minister and an obstetrician-gynecologist who have team counseled hundreds of women faced with unwanted pregnancies present ten case studies to probe the options, including abortions.

D6   Boston Women's Health Book Collective. OUR BODIES, OURSELVES,

Health Problem Analysis

A BOOK BY AND FOR WOMEN. Rev. 2d ed. New York: Simon & Schuster, 1976. 384 p.

> This collection of articles was prepared to inform women of the physiological, psychological, and social aspects of being female and about the care and control of their bodies. It contains chapters on the anatomy and physiology of reproduction, sexuality, venereal disease, birth control, abortion, childbearing, and health care.

D7 Brashear, D.B. "Abortion Counseling." FAMILY COORDINATOR 22 (1973): 429-35.

> This paper describes factors relevant to the abortion counseling process. Identified populations, psychological cost, the decision-making process, vulnerability of the client, client expectations, information relevant to counseling roles, post-abortion, the guilty client, and implications are major topics.

D8 Davis, N.J. "The Abortion Consumer: Making It Through the Network." URBAN LIFE AND CULTURE 2 (1974): 432-59.

> This study traces the history of the search for an abortion from the perspectives and strategies of women. It describes entrance into the network, the abortion experience, and the aftermath. Features of five stages of client movement are described.

D9 Dickens, H.O., et al. "One Hundred Pregnant Adolescents: Treatment Approaches in a University Hospital." AMERICAN JOURNAL OF PUBLIC HEALTH 63 (1973): 794-800.

> One hundred pregnant teenagers were studied in a hospital's regular Prenatal Clinic and Teen-Obstetrical Clinic to investigate differences between the two groups in prenatal health, clinic attendance, and delivery. Differences in the health of the infants were also examined.

D10 Engelhardt, H.T., Jr. "The Ontology of Abortion." ETHICS 84 (1974): 217-34.

> This paper focuses on the ontological status of the fetus, determining whether or to what extent the fetus is a person.

D11 Figà-Talamanca, I. INDUCED ABORTION IN ITALY. Pacific Health Education Reports, no. 4. Berkeley: University of California School of Public Health, 1974. 139 p.

> An exploratory study was made of the social and psychological factors in the practice of illegally-induced abortion by a group of married women in an urban Italian setting. The au-

thor investigated the preabortion and postabortion contraceptive practices of the subjects, the decision to have an abortion, the attitudes and perceptions of the subjects of the intervention itself, and the consequences of the abortion experience.

D12 Horenstein, D., and Houston, B.K. "The Effects of Vasectomy on Postoperative Psychological Adjustment and Self-Concept." JOURNAL OF PSYCHOLOGY 89 (1975): 167-73.

Twenty men undergoing vasectomies completed the Tennessee Self-Concept Scale shortly before the operation and at six- and eighteen-month postoperative follow-up periods; twenty nonvasectomy comparison subjects completed the test at the same points in time. The effects of vasectomy on psychological adjustment and self-concept are reported.

D13 Huber, S.C., et al. "IUDs Reassessed--A Decade of Experience." POPULATION REPORTS. Series B: Intrauterine Devices, no. 2. Washington, D.C.: George Washington University Medical Center, Department of Medical and Public Affairs, 1975. 28 p. Bibliog. Illus.

The contents of this review on intrauterine devices include a history, mode of action, national family planning programs, insertion procedures, measuring IUD performance, pregnancy, expulsion, removal for bleeding and pain, infection, continuation rates and demographic impact, distribution, and research.

D14 McCarthy, B.W., and Brown, P.A. "Counseling College Women with Unwanted Pregnancies." JOURNAL OF COLLEGE STUDENT PERSONNEL 15 (1974): 442-46.

A four-stage model of counseling for college women with unwanted pregnancies who come to a university counseling center is described. The model is based on a time and goal-limited crisis model.

D15 Miller, W.B. "Relationships Between the Intendedness of Conception and the Wantedness of Pregnancy." JOURNAL OF NERVOUS AND MENTAL DISEASE 159 (1974): 396-406.

The relationship between the intendedness of conception and the wantedness of pregnancy was explored in a population of 221 predominantly white, middle-class women who reported on 379 conceptions and pregnancies.

D16 Opit, L.J., and Brennan, M.E. "Demand for Surgical Sterilization Among Patients in a District Maternity Hospital." BRITISH JOURNAL OF PREVENTIVE AND SOCIAL MEDICINE 28 (1974): 19-23.

Health Problem Analysis

In a large district maternity hospital, 1,079 women patients were interviewed about the perceived need for surgical sterilization. A sample of patients' husbands was contacted by means of a questionnaire about vasectomy.

D17 Penson, A.B., and Mattmiller, E.D. "Birth Control Clinic: Education and Clinical Services." JOURNAL OF THE AMERICAN COLLEGE HEALTH ASSOCIATION 22 (1974): 384-88.

> A birth-control educational program was mandatory before related, available clinical services were offered to women at a university. The medical history form included questions pertaining to the women's sexual activity. The program, in addition to providing contraceptive services, enabled physicians to identify latent medical problems and to take appropriate preventive and/or therapeutic action.

D18 Powell, L.F. "The Effect of Extra Stimulation and Maternal Involvement on the Development of Low-Birth-Weight Infants and on Maternal Behavior." CHILD DEVELOPMENT 45 (1974): 106-13.

> Thirteen low-birth-weight infants were given extra stimulation throughout their hospital stay by hospital staff. Eleven mothers were allowed to handle their low-birth-weight babies in the hospital, and twelve low-birth-weight newborns were placed in a control group, in an attempt to improve the babies' development and improve maternal behavior later.

D19 Rogers, J.M., and Adams, D.W. "Therapeutic Abortion: A Multi-disciplined Approach to Patient Care from a Social Work Perspective." CANADIAN JOURNAL OF PUBLIC HEALTH 64 (1973): 254-59.

> This article describes a beginning approach to meeting the needs of the therapeutic abortion patient. A team process delineating problems related to role relationships and intervention techniques is described.

D20 Sayegh, J., and Green, L.W. "Family Planning Education: Program Design, Training Component and Cost-Effectiveness of a Post-Partum Program in Beirut." INTERNATIONAL JOURNAL OF HEALTH EDUCATION 19, Suppl. (1976): 1-20.

> A prospective control design tested the effect of family planning education during the postpartum lying-in period on recruiting lower socioeconomic contraceptive acceptors. The study determined how women with different life-styles and levels of readiness responded to the education program in the maternity service of the American University Medical Center, Beirut, Lebanon.

Health Problem Analysis

D21 THERAPEUTIC ABORTION IN CALIFORNIA: A BIENNIAL REPORT PREPARED FOR THE 1974 LEGISLATURE. Sacramento: California Department of Health, Health Protection Systems, [1974]. 8 p. Tables.

> This report summarizes therapeutic abortions in California, primarily during 1968 to 1973, under the headings reporting system, patient characteristics, methods of termination, weeks of gestation, legal abortion and fertility, and maternal deaths due to abortion.

D22 Waltz, J.R., and Thigpen, C.R. "Genetic Screening and Counseling: The Legal and Ethical Issues." NORTHWEST UNIVERSITY LAW REVIEW 68 (1973): 696-768.

> Prenatal screening for some genetic diseases is now possible as is screening for nonafflicted individuals to determine whether they are carriers of the traits for certain diseases, but these possibilities have caused considerable controversy. This article discusses the legal and ethical issues in genetic screening and counseling programs.

D23 Williams, T.M. "Childrearing Practices of Young Mothers: What We Know, How It Matters, Why It's So Little." AMERICAN JOURNAL OF ORTHOPSYCHIATRY 44 (1974): 70-75.

> More single women are now keeping and rearing their children, and most of their infants experience multiple care givers. This paper explores the strengths as well as the weaknesses of this type of child rearing. Data comparing infant attachment behaviors toward single mothers and their alternate care givers are discussed.

D24 Wortman, J. "Vasectomy--What Are the Problems?" POPULATION REPORTS. Series D: Sterilization, no. 2. Washington D.C.: George Washington University Medical Center, Department of Medical and Public Affairs, 1975. 16 p. Bibliog.

> Contents of this volume on vasectomies include physical contraindications; discoloration, swelling, and pain; hematoma; infections; sperm granuloma; epididymitis; systemic effects; testicular function and male hormones; sperm antibodies; failure; and psychological effects.

Chapter 3
BEHAVIORAL PROBLEM ANALYSIS

In chapter 2, a rationale was stated for concentrating on the social and psychological aspects of the health problem as a first step in the process of planning or evaluating a patient education program. Once the priority dimensions of the health problem are specified, the second step in planning a patient education program is to identify the specific behaviors that cause the health problem. This step may be called a behavioral diagnosis, a behavioral needs assessment, applied behavioral analysis, or social epidemiology. Different disciplines and professions will prefer somewhat distinct conceptual approaches with their own terminology (called jargon by the other disciplines and professions), but all who concern themselves with health education must address behavior somewhere between educational inputs and health outcomes.

Three broad classes of behavior are variously at issue in patient education: preventive health practices (personal hygiene, nutrition, prenatal care, immunizations, family planning, physical examination, prophylactic use of penicillin or other antibiotic, and physical fitness); illness behavior (utilization of health services, delay or promptness in seeking diagnosis of symptoms, and self-care or home management of symptoms not requiring medical care); and compliance with therapeutic regimens prescribed by a health care provider (taking medications, following diet or exercise recommended, and keeping return appointments).

The references annotated in this chapter represent the professional and scientific literature published in the mid-1970s on these three types of behavior related to patient education. More recent compilations and reviews will be found in the JOURNAL OF BEHAVIORAL MEDICINE, FAMILY AND COMMUNITY HEALTH, HEALTH EDUCATION MONOGRAPHS, JOURNAL OF HEALTH AND SOCIAL BEHAVIOR, SOCIAL SCIENCE AND MEDICINE, and occasionally in the medical, nursing, public health, and allied health journals. The following books and proceedings cover this era most extensively:

Caplan, R., et al. ADHERING TO MEDICAL REGIMENS. Ann Arbor: Research Center for Group Dynamics, Institute for Social Research, University of Michigan, 1976.

Cohen, S., ed. NEW DIRECTIONS IN PATIENT COMPLIANCE. Lexington, Mass.: Lexington Books, D.C. Heath Co., 1979.

Cullen, J.W.; Fox, B.H.; and Isom, R.N., eds. CANCER: THE BEHAVIORAL DIMENSIONS. A National Cancer Institute Monograph. New York: Raven Press, 1976.

Enelow, A.J., and Henderson, J.B., eds. APPLYING BEHAVIORAL SCIENCES TO CARDIOVASCULAR RISK. New York: American Heart Association, 1975.

Gallicchio, J., ed. CONSUMER SELF-CARE IN HEALTH. NCHSR Research Proceedings Series. DHEW Publication no. (HRA) 77-3181. Rockville, Md.: National Center for Health Services Research, August 1977.

Rogers-Warren, A., and Warren, S.F. ECOLOGICAL PERSPECTIVES IN BEHAVIOR ANALYSIS. Baltimore: University Park Press, 1977.

Sackett, D.L., and Haynes, R.B., eds. COMPLIANCE WITH THERAPEUTIC REGIMENS. Baltimore: Johns Hopkins University Press, 1976.

Schottenfeld, D., and Miller, D., eds. PROCEEDINGS OF THE AMERICAN SOCIETY FOR PREVENTIVE ONCOLOGY. New York: Raven Press, 1978.

Schwartz, J., ed. PROCEEDINGS OF AN INTERNATIONAL CONFERENCE ON SMOKING AND HEALTH. New York: American Cancer Society, 1979.

Weiss, S.M., ed. PROCEEDINGS OF THE NATIONAL HEART AND LUNG INSTITUTE WORKING CONFERENCE ON HEALTH BEHAVIOR. DHEW Publication no. (NIH) 77-868. Bethesda, Md.: National Institutes of Health, 1977.

E. PREVENTIVE HEALTH PRACTICES

E1   Anderson, W.F. "Preventive Aspects of Geriatric Medicine." JOURNAL OF THE AMERICAN GERIATRICS SOCIETY 22 (1974): 385-92.

    A program for the practice of preventive medicine in the seventy-and-up age group is outlined. It advocates positive planning of activities, with a purpose clearly understood by the elderly, to improve their physical, mental, and social health.

E2   App, H., et al. "Screening for High Blood Pressure: Programme Results and Implications." INTERNATIONAL JOURNAL OF HEALTH EDUCATION 18 (1975): 39-46.

    A four-month, high blood pressure screening program for state employees is described. The program included counseling and referral to private physicians for employees with abnormal readings. A site-screening method was developed, with implications for other screening programs, and actual screening, counseling, and referral time was computed to measure efficiency and effectiveness.

E3 Coburn, D., and Pope, C.R. "Socioeconomic Status and Preventive Health Behavior." JOURNAL OF HEALTH AND SOCIAL BEHAVIOR 15 (1974): 67-68.

> Hypotheses explaining the link between socioeconomic status and preventive health care were tested, and a preliminary predictive model for preventive health behavior was developed.

E4 Corah, N.L. "The Dental Practitioner and Preventive Health Behavior." HEALTH EDUCATION MONOGRAPHS 2 (1974): 226-35.

> This paper explores the role of the dental practitioner vis-a-vis patient preventive behaviors. Following discussion of the effectiveness of the dentist, the author considers potentially applicable knowledge, in particular as it applies to the areas of behavior versus attitude, behavior modification, the dentist patient relationship, and techniques of behavior change. Future research needs are discussed and an example is given of an evaluation paradigm.

E5 Dales, L.G., et al. "Multiphasic Checkup Evaluation Study: 3. Outpatient Clinic Utilization, Hospitalization, and Mortality Experience after Seven Years." PREVENTIVE MEDICINE 2 (1973): 221-35.

> This paper reports on outpatient clinic utilization, hospitalization, and mortality experience of subjects in a controlled clinical trial aimed at testing the efficacy of periodic Multiphasic Health Checkups in preventing or postponing illness, disability, and death. For a group of causes of death defined as being potentially postponable or preventable, the study group mortality rate has been significantly lower (p 0.05).

E6 Dalzell-Ward, A.J. "The Contribution of Physical Education to Health Education." ASPECTS OF EDUCATION, no. 16, 1974, pp. 66-69.

> The author describes the task of the physical educator in integrating the development of mind and body to give adolescents a basic knowledge of bodily functions and proper care. Prevention of deleterious conditions associated with lack of physical activity could result from appropriate physical habits founded in early life.

E7 Elwood, T.W., and Oakes, T.W. "Failure by a Group of Elderly Men to Use a Preventive Health Service." JOURNAL OF AMERICAN GERIATRICS SOCIETY 23 (1975): 74-76.

> The perceptions of twenty-five males (age range 53-62) were studied in relation to nonuse of multiphasic screening examinations. Reasons for these men not taking the screening examination were classified as social influences, health care setting,

beliefs about health, use of alternative forms of care, the time involved for testing, and disinterest.

E8 Haefner, D.P. "The Health Belief Model and Preventive Dental Behavior." HEALTH EDUCATION MONOGRAPHS 2 (1974): 420-32.

This paper focuses on studies concerning the hypothesized relationship between health beliefs and preventive dental behavior. The studies are grouped under the headings susceptibility; seriousness; perceived benefits; perceived salience; combinations of health benefits; fear appeals; beliefs, behavior, and socioeconomic level; and behavior, beliefs, and habits.

E9 LaDou, J., et al. "Health Hazard Appraisal in Patient Counseling." WESTERN JOURNAL OF MEDICINE 122 (1975): 177-80.

A program of annual health examinations was expanded to include counseling based on a computerized appraisal of individual patient's specific health hazard factors. Data obtained from a specially designed questionnaire, laboratory tests, and a physical examination were reviewed with each patient, and methods of correcting health hazards were stressed.

E10 Miller, W.B. "Psychological Antecedents to Conception Among Abortion Seekers." WESTERN JOURNAL OF MEDICINE 122 (1975): 12-19.

At a university hospital, 642 women seeking induced abortion were surveyed before the procedure regarding their perception of what psychological and behavioral factors influenced their becoming pregnant. Implications for educational and counseling programs are discussed.

E11 Neeman, R.L., and Neeman, M. "Cancer Prevention Education for Youth--A Key for Control of Uterine and Breast Cancer?" JOURNAL OF SCHOOL HEALTH 44 (1974): 543-47.

Cancer prevention education for adolescent girls, both in and out of high school, aimed at developing lifetime habits reflecting the woman's responsibility for her own cancer prevention actions. The educational program is presented as a potential for controlling malignancies of the breast and uterus. Barriers to accepting responsibility for preventive health practices and behaviors resulting in delay in seeking medical care are discussed.

E12 Oja, P., et al. "Feasibility of an 18 Months' Physical Training Program for Middle-aged Men and Its Effect on Physical Fitness." AMERICAN JOURNAL OF PUBLIC HEALTH 64 (1974): 459-65.

This study evaluated an exercise program designed for use in the primary prevention of coronary heart disease (CHD). Ninety middle-aged male executives with relatively high values for certain CHD risk factors were subjected to a vigorous physical training program of three weekly exercise sessions for eighteen months. The organization and content of the program and the methods of the exercise prescription are described.

E13 Ramcharan, S., et al. "Multiphasic Checkup Evaluation Study." PREVENTIVE MEDICINE 2 (1973): 207-20.

A controlled trial to evaluate the efficacy of the periodic health examination utilizing automated multiphasic testing was conducted. A study group of 5,156 members of the Kaiser Foundation Health Plan was urged to have annual Multiphasic Health Checkups. Evaluation was done by comparing outcomes between the study group and a similarly selected control group of 5,557 subjects who were not urged to have periodic health checkups. After five to seven years, a favorable impact on the health of the older males compared to the older control males was evidenced by a reduction in self-rated disability and reported time lost from work, a greater proportion working, and a lower self-reported utilization of medical services.

E14 Rose, K. "To Keep the People in Health." JOURNAL OF THE AMERICAN COLLEGE HEALTH ASSOCIATION 22 (1973): 80-83.

The preventive philosophy in health care is discussed, with a practical application illustrated in relation to cardiovascular diseases.

E15 Rosenstock, I.M. "The Health Belief Model and Preventive Health Behavior." HEALTH EDUCATION MONOGRAPHS 2 (1974): 354-86.

This paper reviews studies on preventive health behavior. Studies are reviewed under headings that include how people use health services, demographic variations, attributing cause to demographic correlates, perception of symptoms as an intermediate variable in utilization, evidence for and against the Health Belief Model, retrospective and prospective studies, and health education as an independent variable.

E16 Schwarz, K. "Health Education and the Provision of Medical Care." HEALTH EDUCATION JOURNAL 33 (1974): 3-7.

This paper discusses health education as it relates to primary and secondary prevention, presymtomatic and early diagnosis, and management and rehabilitation.

Behavioral Problem Analysis

E17 Smith, J.M. "An Evaluation of the Applicability of the Rosenstock-Hochbaum Health Behavior Model to the Prevention of Periodontal Disease in English Schoolgirls." JOURNAL OF CLINICAL PERIODONTOLOGY 1 (1974): 222-31.

> Middle-class schoolgirls who attended their dentists regularly were questioned about their knowledge and behavior relating to periodontal disease. The author contends that findings from this study cast some doubt on the applicability of the Rosenstock-Hochbaum health belief model to preventive dental behavior related to periodontal disease.

E18 Spillane, W.H., and Ryser, P.E. FERTILITY KNOWLEDGE, ATTITUDES AND PRACTICES OF MARRIED MEN. Cambridge, Mass.: Ballinger Publishing Co., 1975. 196 p.

> This book reports family planning research on the role of married men in decisions involving contraceptive use and family size.

F. ILLNESS BEHAVIOR AND UTILIZATION OF HEALTH SERVICES

F1 Aday, L.A., and Andersen, R. DEVELOPMENT OF INDICES OF ACCESS TO MEDICAL CARE. Ann Arbor, Mich.: Health Administration Press, 1975. xvii, 306 p. Bibliog. Tables.

> This book is an overview of access to medical care, based on a 1970 nationwide survey of health services utilization and expenditures. Analyses of selected process and outcome indices are made of the survey data, and a review of the literature is presented. Data sources judged to be useful to construct and update national measures of access are evaluated.

F2 Anderson, J.G., and Bartkus, D.E. "Choice of Medical Care: A Behavioral Model of Health and Illness Behavior." JOURNAL OF HEALTH AND SOCIAL BEHAVIOR 14 (1973): 348-62.

> A behavioral model of sociodemographic, economic, ecological, need, and social psychological variables is developed and used to account for differential patterns of health and illness behavior among members of a prepaid medical group, a sample of students enrolled in a university health plan.

F3 Antonovsky, A., and Hartman, H. "Delay in the Detection of Cancer: A Review of the Literature." HEALTH EDUCATION MONOGRAPHS 2 (1974): 98-128.

> This review covers literature on the consequences, the definition, and the extent of delay, and includes a table summarizing studies reporting delay. The literature on reasons for de-

lay is also reviewed and includes types of studies done, cancer-related variables, psychological variables, the relationship of the individual to the health care system, and the sociodemographic profile of the delayer.

F4 Berkanovic, E., and Reeder, L.G. "Can Money Buy the Appropriate Use of Services? Some Notes on the Meaning of Utilization Data." JOURNAL OF HEALTH AND SOCIAL BEHAVIOR 15 (1974): 93-99.

This paper notes the convergence of research and opinion supporting the assertion that perceived symptoms and ability to pay are the major determinants of health services utilization. Four reservations pertaining to this body of evidence are offered, and research supporting the role of cultural factors in determining the utilization of services is cited. Several cultural barriers to the full utilization of services are discussed.

F5 Bridges-Webb, C. "The Traralgon Health and Illness Survey: Part 3. Illnesses and Their Medical and Hospital Care." INTERNATIONAL JOURNAL OF EPIDEMIOLOGY 3 (1974): 233-46.

An interview survey with 371 randomly selected families provided information about 2,603 illnesses. The survey was supplemented with information about diseases from a random sample of general practitioner consultations and with details of all conditions responsible for hospital admissions in the study population during a one-year period. The nature of the illnesses reported, the proportion of different classes receiving medical treatment, and the effects of age and sex are discussed.

F6 Brown, J.S., and Rawlinson, M. "Relinquishing the Sick Role Following Open-Heart Surgery." JOURNAL OF HEALTH AND SOCIAL BEHAVIOR 16 (1975): 12-27.

A semantic differential instrument was used with 150 patients one or more years after their open-heart surgery. The semantic difference between the concepts "most persons who are sick" and "myself after heart surgery" constituted the measure of the patient's tendency to relinquish the sick role. Five variables: depression, preoperative tendency to reject the sick role, duration of illness prior to surgery, age, and sex were significant predictors of the tendency to relinquish the sick role following surgery.

F7 Burgess, A.W., and Holmstrom, L.L. "Crisis and Counseling Requests of Rape Victims." NURSING RESEARCH 23 (1974): 196-202.

A study of 146 adult and pediatric victims of sexual assault in a metropolitan area revealed a positive response by the victims to psychological intervention during the crisis period. Victims were interviewed at a hospital immediately following the assault and were followed up with telephone calls from the investigators and counselors.

F8 Burke, W.M. "Attitudes and the Utilization of Health Services." JOURNAL OF THE AMERICAN COLLEGE HEALTH ASSOCIATION 22 (1974): 320-24.

The results of a mailed questionnaire survey in a student population are reported. Perceived morbidity was the best predictor of health service utilization. There was a positive relationship between attitudes and both the decision to utilize professional medical services and the type of services utilized.

F9 Carlton, B. "Adults' Knowledge of, Needs for, Attitudes Towards and Utilization of Health and Medical Resources in Two Southeastern Kentucky Counties." Ed.D. dissertation, University of Tennessee, 1973.

A survey of the population of Bell and Harlan counties, Kentucky, collected information on attitudes and opinions toward health and medical care, health needs, health knowledge, and the utilization of medical services and facilities. Those surveyed perceived sickness to be an inherent part of their life-style; were pessimistic about the state of their health; had low knowledge of diseases and disease symptoms; did not obtain from physicians, clinics, and hospitals the health information needed to improve their knowledge; did not hold their physicians in high esteem; and were dissatisfied with many aspects of the system that provided sick care to them. Insufficient money influenced the decisions of many not to seek care.

F10 Creer, T.L., et al. "Managing a Hospital Behavior Problem: Malingering." JOURNAL OF BEHAVIOR THERAPY AND EXPERIMENTAL PSYCHIATRY 5 (1974): 259-62.

A technique employed at a children's asthma research institute and hospital to modify the behavioral pattern of malingering is described.

F11 CURRENT ESTIMATES FROM THE HEALTH INTERVIEW SURVEY, UNITED STATES--1973. Vital and Health Statistics, Data from the National Health Survey, series 10, no. 95. DHEW Publication no. HRA 75-1522. Rockville, Md.: National Center for Health Statistics, 1974. 77 p.

This report, based on data collected in the Health Interview Survey during 1973, gives estimates of incidence of acute conditions, number of persons reporting limitation of activity, number of persons injured, number of hospital discharges, number of persons with hospital episodes, number of disability days, and frequency of dental and physician visits.

F12 Galvin, M.E., and Fan, M. "The Utilization of Physicians' Services in Los Angeles County, 1973." JOURNAL OF HEALTH AND SOCIAL BEHAVIOR 16 (1975): 74-94.

This paper reports preliminary findings from research on utilization of physicians' services. Data were obtained by interviews from a sample of Los Angeles County households in 1973. Results reported here are for adult respondents.

F13 Green, L.W., and Roberts, B.J. "The Research Literature on Why Women Delay in Seeking Medical Care for Breast Symptoms." HEALTH EDUCATION MONOGRAPHS 2 (1974): 129-77.

This paper reviews published research and evaluative studies on motivational factors and social influences as they relate to the educational approach in achieving early diagnosis and treatment among women with signs or symptoms of breast cancer.

F14 Herzlich, C. HEALTH AND ILLNESS: A SOCIAL PSYCHOLOGICAL ANALYSIS. Translated by D. Graham. European Monographs in Social Psychology 5. New York: Academic Press, 1973. xvi, 159 p. Append. Bibliog. Indexes.

Contents include the individual, the way of life, and the genesis of illness; nature, constraint and society; health and illnesses; illnesses--dimensions and limits; the sick and the healthy; the social representation of health and illness; conceptions of illness and illness behavior; and the invalid and self-identity.

F15 Hirschman, R. "Utilization of Mental Health Consultation and Self-Perceptions of Intraorganizational Importance and Influence." JOURNAL OF CONSULTING AND CLINICAL PSYCHOLOGY 42 (1974): 916.

This study tested the notion that self-perceived intraorganizational power of individuals is inversely related to their willingness to utilize the services of a mental health consultant.

F16 Huffine, C.L., and Craig, T.J. "Social Factors in the Utilization of an Urban Psychiatric Emergency Service." ARCHIVES OF GENERAL PSYCHIATRY 30 (1974): 249-55.

Admissions to the psychiatric emergency service of an inner-city university hospital are analyzed, and suggestions are

Behavioral Problem Analysis

offered for ways in which services may be altered to meet treatment needs of inner-city populations.

F17 Kegeles, S.S. "Behavioral Science Data and Approaches Relevant to the Development of Education Programs in Cancer." HEALTH EDUCATION MONOGRAPHS 1, no. 36 (1973): 18-33.

Concepts, approaches, and findings of behavioral scientists that have potential relevance for certain problems of cancer are presented. Program efforts for screening for presymptomatic cancer detection, particularly activities directed toward increasing utilization of cervical cytology and breast cancer examination, are reviewed. Briefly described are self-care behavior, preparation of patients for cancer surgery, and rehabilitation of patients after cancer surgery.

F18 Kirscht, J.P. "The Health Belief Model and Illness Behavior." HEALTH EDUCATION MONOGRAPHS 2 (1974): 387-408.

This review article examines psychosocial components of individual initiative in coping with distress episodes, especially the health beliefs that appear useful in understanding how health decisions are made. The article is organized into sections on health beliefs and illness behavior, models of illness behavior and their relationships to the Health Belief Model, illness behavior and health beliefs, response to symptoms, decisions to seek care, social-structural factors and beliefs, and problems and issues.

F19 Leyhe, D.L., et al. "Medi-Cal Patient Satisfaction in Watts." HEALTH SERVICES REPORTS 88 (1973): 351-59.

This paper concerns a household interview survey of a sample of Medi-Cal beneficiaries living in the target area of an OEO neighborhood health center in south central Los Angeles. The objective was to determine what proportion of the interviewees and their families used the center, the interviewees' other sources of care, and the nature of their satisfactions and dissatisfactions with the health care resources available to them.

F20 Lurie, O.R. "Parents' Attitudes Toward Children's Problems and Toward Use of Mental Health Services: Socioeconomic Differences." AMERICAN JOURNAL OF ORTHOPSYCHIATRY 44 (1974): 109-20.

This paper explores attitudes of parents of socially and economically diverse groups toward their children's problems and toward seeking help through the utilization of mental health resources.

F21 Mayo, J.A. "Utilization of a Community Mental Health Center by

Blacks: Admission to Inpatient Status." JOURNAL OF NERVOUS AND MENTAL DISEASE 158 (1974): 202-7.

>This study tests the hypothesis that utilization of the community mental health center is influenced more by demographic and psychosocial characteristics and by perceptions and values of purveyors of mental health resources than by availability and organization of services.

F22 THE MEDICAL AIDE PROJECT--CASE MANAGEMENT IN A MEDICAL ASSISTANCE PROGRAM THROUGH THE USE OF THE SUBPROFESSIONAL. SRS-11-57106-001. Portsmouth, Va.: Portsmouth Department of Public Health, 1974. 40 p.

>This project was to determine if the physical and mental health of low income persons would be improved by no-cost physical examinations and by medical and general counseling by indigenous medical case aides. Data on health status, utilization of medical services, and environmental status were compared for two groups of experimental subjects (1,497) and 831 control subjects. Intermediate data are given.

F23 Miller, P.M., et al. A Retrospective Analysis of Alcohol Consumption on Laboratory Tasks as Related to Therapeutic Outcome." BEHAVIOR RESEARCH AND THERAPY 12 (1974): 73-76.

>This article examines the relationship between drinking on laboratory-analogue measures at the onset of inpatient alcoholism treatment and abstinence after treatment.

F24 Oakes, T.W., et al. "Health Service Utilization by Smokers and Nonsmokers." MEDICAL CARE 12 (1974): 958-66.

>The authors studied differences in the utilization of several types of health services by cigarette smokers, nonsmokers, and exsmokers. The study sample consisted of white men and women, aged twenty years or older, who were members of a prepaid health plan.

F25 Pritchard, M. "Dimensions of Illness Behavior in Long-Term Haemodialysis." JOURNAL OF PSYCHOSOMATIC RESEARCH 18 (1974): 351-56.

>A principle components (factor) analysis was performed on the Pearson correlation matrix of twenty-two items concerned with illness behavior as measured by a response to illness questionnaire in fourteen patients undergoing long-term haemodialysis. Seven factors were labeled: hopeless defeat, anxious preoccupation, outward hostility, helpless loss, challenging appraisal, illness as enemy, and paranoid withdrawal.

Behavioral Problem Analysis

F26 Pritchard, M. "Meaning of Illness and Patients' Response to Long Term Haemodialysis." JOURNAL OF PSYCHOSOMATIC RESEARCH 18 (1974): 457-64.

> This study extends previous research on analyses of data from a response-to-illness questionnaire completed by patients undergoing long-term haemodialysis. Analysis produced seven factors that represent dimensions of illness behavior.

F27 Salloway, J.C. SOCIAL NETWORKS AND HEALTH CARE CONSUMERSHIP: APPLICATIONS OF MODELS OF HEALTH SERVICE UTILIZATION. Urbana: University of Illinois, Department of Sociology, 1974. 98 p. Append. Bibliog.

> Topics covered in this report include the process of seeking medical care, sociocultural variation and health behavior, a review and critique of pertinent literature, a social networks model of health care utilization, research methods, hypotheses, and a discussion of extended friendship and family systems and isolates.

F28 Scott, C.S. "Competing Health Care Systems in an Inner City Area." HUMAN ORGANIZATION 34 (1975): 108-10.

> This paper reviews hypotheses that attempt to explain the variable use of folk and orthodox health care systems by ethnic groups. The need for research into underlying factors involved in making choices between or among therapies and healers in competitive situations is pointed out.

F29 Selman, P., and Stephens, C. "Missed Appointments at a Family Planning Clinic." JOURNAL OF BIOSOCIAL SCIENCES 5 (1973): 421-42.

> Home visits were made by a social worker to 152 women who failed to keep appointments to discuss or be fitted with an IUD at a hospital clinic. The visits were to encourage the women to accept a new appointment or make alternative plans for contraception. Reasons for missing appointments were classified as immediate, practical problems; marital circumstances and husband's attitude; questions relating to the IUD; and choice of an alternative method.

F30 Spiro, H.R., et al. "Cost Financed Mental Health Facility: II. Utilization Profile of a Labor Union Program." JOURNAL OF NERVOUS AND MENTAL DISEASE 160 (1975): 241-48.

> Data are presented on the utilization of a cost-financed mental health facility. The utilization is enhanced by a mental health education campaign.

Behavioral Problem Analysis

F31 Twaddle, A.C. "The Concept of Health Status." SOCIAL SCIENCE AND MEDICINE 8 (1974): 29-38.

> This paper reviews the literature on the concept of health status and studies linking social stratification, ethnicity, and situational factors with differences in the designation of individuals as "sick." A model focusing on the process of status designation, with special attention to the circumstances under which individuals are defined as well or ill, is presented.

F32 U.S. Department of Health, Education, and Welfare. Alcohol, Drug Abuse, and Mental Health Administration. National Institute of Mental Health. UTILIZATION OF MENTAL HEALTH FACILITIES 1971. DHEW Publication no. NIH 74-657. Rockville, Md.: 1974.

> This report shows changes in the types of mental health facilities used since 1955 and in rates of admission. Detailed tables on the data analyzed in the study are included.

F33 Vacalis, T.D. "Determination of Vital Areas of Knowledge Needed for Wise Consumer Use of Health Care Services." JOURNAL OF SCHOOL HEALTH 44 (1974): 390-94.

> To determine the knowledge necessary for the public to effectively utilize a health care delivery system, a questionnaire was sent to thirty-nine health professionals for review and comment. A revised questionnaire of twenty-seven items was sent to a random sample of 150 physicians and 150 teachers.

F34 Veeder, N.W. "Health Services Utilization Models for Human Services Planning." JOURNAL OF THE AMERICAN INSTITUTE OF PLANNERS 41 (1975): 101-9.

> Models explaining health service utilization patterns developed by Rosenstock, Suchman, Anderson, and Gross are reviewed for the purpose of stimulating further research about service utilization in relation to health and other services.

F35 Weiner, H. "Birth Order and Illness Behavior." JOURNAL OF INDIVIDUAL PSYCHOLOGY 29 (1973): 173-75.

> This study investigates the relationship of birth order to illness behavior by comparing oldest born and youngest born. A sample of freshmen were followed through four years in college to compare utilization of the college health service.

F36 Weiner, H. "A Comparison of Frequent and Non-Frequent Health Service Users." JOURNAL OF THE AMERICAN COLLEGE HEALTH ASSOCIATION 22 (1974): 315-19.

> Entering freshmen's use of a university health service over

their four years at the university was tabulated and compared with sixteen characteristics for male students and with seventeen characteristics for female students. Overall, male and female students visited the health service in proportion to their number, but this appeared to mask different responses to different conditions. There was a stable pattern of use over the four years of college attendance.

F37   Woodruff, R.A., Jr., et al. "Alcoholics Who See a Psychiatrist Compared to Those Who Do Not." QUARTERLY JOURNAL OF STUDIES ON ALCOHOL 34 (1973): 1162-71.

Interviews were conducted with twenty-eight alcoholic patients at a psychiatric clinic and twenty-five alcoholics who had never sought psychiatric treatment for their alcoholism to determine what factors bring patients to the clinic.

F38   Zarit, S.H., and Kahn, R.L. "Aging and Adaptation to Illness." JOURNAL OF GERONTOLOGY 30 (1975): 67-72.

The relation of age, adaptation to illness, and the severity of cerebral dysfunction was evaluated for persons who had suffered cerebrovascular accidents. All eighty-nine subjects were interviewed at least one month after the onset of illness.

G. COMPLIANCE WITH THERAPEUTIC REGIMENS

G1   Becker, M.H. "The Health Belief Model and Sick Role Behavior." HEALTH EDUCATION MONOGRAPHS 2 (1974): 409-19.

This paper reviews studies that have attempted to evaluate the ability of one or more of the Health Belief Model dimensions to explain and predict compliance. A somewhat wider literature is also examined to expand the traditional model by incorporating findings related to motivations and doctor-patient interactions.

G2   Becker, M.H., et al. "A New Approach to Explaining Sick-Role Behavior in Low-Income Populations." AMERICAN JOURNAL OF PUBLIC HEALTH 64 (1974): 205-16.

This study examines the value of a behavioral model, derived from social psychological theory, that employs health motivations, perceptions, and attitudes of mothers as predictors of compliance with regimens prescribed for their children.

G3   Bellack, A.S., et al. "The Contribution of External Control to Self-Control in a Weight Reduction Program." JOURNAL OF BEHAVIOR

THERAPY AND EXPERIMENTAL PSYCHIATRY 5 (1974): 245-49.

> A weight-reduction program emphasizing self-control was presented to twenty volunteers under three different conditions. Results are discussed in terms of information value or signaling function of external contact in treatment programs emphasizing self-control.

G4 Blackwell, B. "Drug Therapy: Patient Compliance." NEW ENGLAND JOURNAL OF MEDICINE 289 (1973): 249-52.

> The problem of noncompliance is discussed under the headings: frequency and types, factors associated with noncompliance, the illness, the patient, the physician, the medication regimen, the treatment milieu, prevention of noncompliance, recognition of the "at-risk" patient, treatment planning, and treatment explanation.

G5 Brand, F.N., and Smith, R.T. "Medical Care and Compliance Among the Elderly After Hospitalization." INTERNATIONAL JOURNAL OF AGING AND HUMAN DEVELOPMENT 5 (1974): 331-46.

> Chronically ill patients (N= 114) were studied six months after discharge from a general hospital for readmission rates; utilization of hospital services, outpatient clinics, and family practitioners; and compliance with physician's recommendations.

G6 Deykin, E., et al. "Participation in Therapy: A Study of Attendance Patterns in Depressed Outpatients." JOURNAL OF NERVOUS AND MENTAL DISEASE 160 (1975): 42-47.

> The attendance patterns of thirty-six predominantly lower class, depressed women receiving outpatient psychotherapy were studied to determine the actual amount and frequency of therapy received in an eight-month period.

G7 Drouin, B., et al. "Entrainement Modere et Readaptation des Coronariens: Resultats Obtenus chez 27 Patients Convalescents d'Infarctus de Myocarde" [Moderate training and rehabilitation after coronary occlusion: results obtained in 27 patients recovering from myocardial infarction]. COEUR 5 (1974): 663-64, 667-76.

> Moderate physical training under medical supervision was given for a period of three months to twenty-seven patients who had sustained coronary occlusions. Resulting positive changes were studied on the basis of clinical examinations and exercise tests at the beginning and end of the training period. Possible causes of the changes are discussed.

G8 Emrick, C.D. "A Review of Psychologically Oriented Treatment of

Alcoholism: II. The Relative Effectiveness of Different Treatment Approaches and the Effectiveness of Treatment Versus No Treatment." JOURNAL OF STUDIES ON ALCOHOL 36 (1975): 88-108.

> This review of 384 studies of psychologically oriented alcoholism treatment focuses on the effects of different treatment methods on long-term outcome.

G9   Etzwiler, D.D. "Why Not Put Your Patients Under Contract?" PRISM 2 (1974): 26-28.

> This paper describes a contract approach to patient care, in which the patient agrees to make an effort to improve his or her health by following a specific therapeutic regimen.

G10 Fiester, A.R., et al. "Shaping a Clinic Population: The Dropout Problem Reconsidered." COMMUNITY MENTAL HEALTH JOURNAL 10 (1974): 173-79.

> This study involved a comparison of dropout and nondropout community outpatients on selected demographic variables. Findings suggest that outpatient clinics shape patients into two groups: those who make repeated use of a variety of mental health services and those who quickly turn away from such services and are unlikely to reapply at a later date.

G11 Gillum, R.F., and Barsky, A.J. "Diagnosis and Management of Patient Noncompliance." JOURNAL OF THE AMERICAN MEDICAL ASSOCIATION 228 (17 June 1974): 1563-67.

> A review of the literature shows factors most consistently related to noncompliance: psychological, environmental, and social factors; characteristics of the therapeutic regimen; and properties of the physician-patient interaction.

G12 Haynes, R.B.; Taylor, W.; and Sackett, D.L., eds. COMPLIANCE IN HEALTH CARE. Baltimore: Johns Hopkins University Press, 1979.

> Contents of the bibliography include original articles on measurement, reviews, commentaries, factors studied in relation to compliance, objective methods of measuring compliance, clinical perspectives of compliance studies, tested and suggested methods of improving compliance, and an appendix of methodologic standards for compliance studies.

G13 Huber, E.G., et al. "Ein neuer Weg zur Behandlung der Fettsucht im Kindesalter: Erfahrungen mit zwei Therapielagern" [A new method of treating obesity in children: experience from two therapy camps]. PAEDIATRIE AND PAEDOLOGIE 10 (1975): 88-96.

> A report is given on "therapy camps," a new method of treating obesity in children.

Behavioral Problem Analysis

G14 Kasl, S.V. "The Health Belief Model and Behavior Related to Chronic Illness." HEALTH EDUCATION MONOGRAPHS 2 (1974): 433-54.

   This review article considers the following areas: seeking medical attention in the presence of symptoms, especially referral behavior and delay; compliance with medical regimen; staying in treatment; and modification of life-style habits to reduce risk. This review emphasizes outpatient behavior and specifically omits sick role behavior of hospitalized patients.

G15 Kline, J., and King, M. "Treatment Dropouts from a Community Mental Health Center." COMMUNITY MENTAL HEALTH JOURNAL 9 (1973): 354-60.

   Treatment dropouts from a mental health center who had left treatment without staff approval were compared on thirty-nine demographic, mental status, and social history variables.

G16 Malmquist, A., and Hagberg, B. "A Prospective Study of Patients in Chronic Hemodialysis--V: A Follow-up Study of Thirteen Patients in Home-Dialysis." JOURNAL OF PSYCHOSOMATIC RESEARCH 18 (1974): 321-26.

   Thirteen patients in home dialysis were psychiatrically evaluated before the start of home dialysis, and both psychiatrically and psychologically evaluated at follow-up. The patients, selected for home dialysis on the basis of positive predictive personality characteristics, were rehabilitated and well-adjusted at follow-up.

G17 Mitchell, J.H. "Compliance with Medical Regimens: An Annotated Bibliography." HEALTH EDUCATION MONOGRAPHS 2 (1974): 75-87.

   This analysis of papers describing patient compliance to prescribed drug therapy or medical regimens lists articles that discuss the social, economic, or demographic factors associated with compliance or noncompliance.

G18 Rodin, J., and Slocohower, J. "Fat Chance for a Favor: Obese-Normal Differences in Compliance and Incidental Learning." JOURNAL OF PERSONALITY AND SOCIAL PSYCHOLOGY 29 (1974): 557-65.

   The effects of manipulations of salient external cues on incidental learning and compliance were tested for overweight and normal subjects.

G19 Sackett, D.L., and Haynes, R.B., eds. COMPLIANCE WITH THERAPEUTIC REGIMENS. Baltimore: Johns Hopkins University Press, 1976. 293 p.

Behavioral Problem Analysis

This first edition (see G12 for second) of proceedings from the McMasters University Symposia on Patient Compliance includes reviews of the extent and dimensions of the compliance problem, the state of the art in patient education, and other strategies to increase compliance.

G20 Seeman, M.V. "Patients Who Abandon Psychotherapy: Why and When." ARCHIVES OF GENERAL PSYCHIATRY 30 (1974): 486-91.

The author examines the problem of patients leaving psychotherapy prematurely. Recommendations are given to therapists for avoiding this problem.

G21 Sikes, S., and Singh, D. "Obesity and Compliance." BULLETIN OF THE PSYCHONOMIC SOCIETY 4 (September 1974): 176.

Obese and normal-weight subjects were tested to determine differences in motivational strength, self-esteem, and compliance.

G22 Skoloda, T.E., et al. "Treatment Outcome in a Drinking-Decisions Program." JOURNAL OF STUDIES ON ALCOHOL 36 (1975): 365-80.

A six-week, inpatient treatment program is described, in which alcoholics were free to decide whether or not to drink during the program.

G23 Stimson, G.V. "Obeying Doctor's Orders: A View from the Other Side." SOCIAL SCIENCE AND MEDICINE 8 (1974): 97-104.

An alternative to viewing compliance, approaching the problem from the perspective of the patient, is recommended. The focus is then on the social context in which illnesses are lived and treatments used. A more active view of patients is entailed in which patients have expectations of the doctor, evaluate the doctor's actions, and are able to make their own treatment decisions.

G24 Weintraub, M., et al. "Compliance as a Determinant of Serum Digoxin Concentration." JOURNAL OF THE AMERICAN MEDICAL ASSOCIATION 224 (1973): 481-85.

Serum digoxin concentrations were measured in 101 outpatients by radio-immunoassay with radioactive iodine 125 to evaluate factors affecting these concentrations. Patient compliance, ascertained by asking how often patients missed taking digoxin, was the most important determinant of serum digoxin concentrations.

G25 Whelan, E.M. "International Committee on Applied Research in Population: Compliance with Contraceptive Regimens." STUDIES IN FAMILY PLANNING 5 (1974): 349-55.

> The author proposes two indirect measures of compliance in contraceptive use: for orals, extrapolation from data on breakthrough bleeding during use of the pill; and for vaginal methods, extrapolation from use-effectiveness data and estimates of the probability of conception from a single unprotected coitus. Method- and client-related factors that affect compliance are discussed.

Chapter 4
FACTORS PREDISPOSING PATIENT BEHAVIOR

In this and the following two chapters, the antecedents or causes of patient behavior are surveyed. For any specific behavior identified in the previous chapter, there will be at least three types of influence that should be considered in planning a patient education program. These three classes of influence or causation are grouped according to the major types of strategies or methods employed in patient education programs, as was shown in the overall framework in figure 1.

The specific components and relationships of the three classes of behavioral determinants are shown in figure 2. The recent literature on predisposing factors is represented in this chapter.

Predisposing factors are those qualities or characteristics of a patient related to his or her motivation to act. As figure 2 illustrates, these might include knowledge, attitudes, beliefs, and values. In a general sense, predisposing factors are the "personal preferences" that an individual or group brings to an educational experience. These preferences may either support or resist health behavior; in any case they are influential. Although demographic factors such as socioeconomic status, age, gender, and family size also are important, they are beyond the direct influence of a patient education program.

Knowledge. Although knowledge gain does not always cause behavior to change, positive associations between the two variables have been demonstrated in numerous studies. Health knowledge of some nature is necessary before most personal health actions will occur. The desired health action will not likely occur unless a person receives a cue strong enough to motivate him or her to act on the knowledge he or she has. The implication is that knowledge is a necessary but not sufficient factor in changing health behavior.

An introductory comment about beliefs, values, and attitudes may be helpful here. Beliefs, values, and attitudes are independent constructs, yet the differences between them are often fine and complex. As a result, the literature on these several factors, which is vast, is quite technical and some of it is understood by only a few scholars in psychology.

Factors Predisposing Patient Behavior

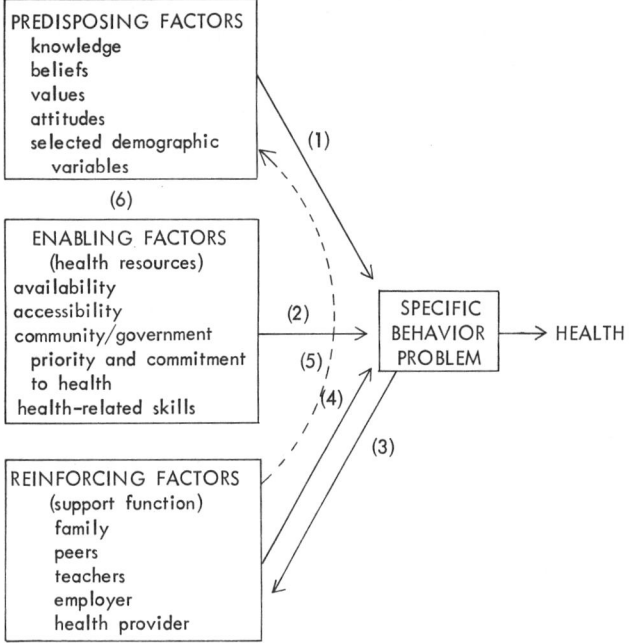

Figure 2

Three Categories of Factors Contributing to Health Behavior

Note: Solid lines imply contributing influence and the dotted line implies secondary effects. The numbers indicate the approximate order in which they usually occur. L.W. Green, et al. HEALTH EDUCATION PLANNING: A DIAGNOSTIC APPROACH. Palo Alto, Calif.: Mayfield Publishing Co., 1979.

Factors Predisposing Patient Behavior

For a more thorough analysis, the reader is referred to the theoretical and research literature in cognitive psychology, especially as reflected in the ANNUAL REVIEW OF PSYCHOLOGY chapters on attitude theory and measurement, the HANDBOOK OF SOCIAL PSYCHOLOGY edited by Elliot Aaronson, and journals such as COGNITIVE PSYCHOLOGY, COUNSELING PSYCHOLOGIST, DEVELOPMENTAL PSYCHOLOGY, JOURNAL OF APPLIED PSYCHOLOGY, JOURNAL OF COUNSELING PSYCHOLOGY, JOURNAL OF EDUCATIONAL PSYCHOLOGY, JOURNAL OF GERONTOLOGY, JOURNAL OF PERSONALITY AND SOCIAL PSYCHOLOGY, JOURNAL OF YOUTH AND ADOLESCENCE, PERCEPTION, AND PSYCHOLOGICAL REPORTS.

Although the relationship between behavior and constructs such as attitudes, beliefs, and values is not clearly understood, there is ample evidence for the practitioner to be confident that there is indeed an association between them. There is a circular relationship between behavior and attitudinally related constructs in that attitudes are to some degree the determinants, components, and consequences of behavior.*

Beliefs. A belief is a conviction that some phenomenon or object is true or real. Faith, trust, and truth are words commonly used to express or imply belief. Belief statements in the context of health might include: "I don't believe that medication can work"; "If this diet won't work for him, it sure isn't going to work for me"; "Exercise won't make any difference"; "When your time is up, your time is up and there's nothing you can do about it." If beliefs such as these are strongly held, to what extent will they limit positive behavior? Can they be changed?

The most extensively documented model relating beliefs to health and illness behavior is the Health Belief Model, notably employed by Hochbaum, Rosenstock, Leventhal, Kegeles, Kirscht, and Becker. This model attempts to explain and predict health-related behavior in terms of certain belief patterns. The principle of the model is based on the following propositions: For behavior change to occur (1) a person must believe that his or her health is in jeopardy; (2) the person must also perceive how potentially serious the condition might be (that belief is usually based on the anticipation of personal pain or discomfort, time lost from work, potential economic difficulties, and so forth); (3) after assessing the circumstances, the person must then believe that benefits are indeed possible and within his or her grasp; and (4) finally, fundamental to the entire model, there must be a "cue to action" or precipitating behavior.

Attitudes. Attitude is one of the most ambiguous yet most widely used words in the behavioral sciences. Attitude is a rather constant feeling that is directed toward an object. Evaluation, or a good-bad dimension, is inherent in the structure of an attitude. One of the techniques frequently used to

---

*See L.W. Green. "Should Health Education Abandon Attitude-Change Strategies: Perspectives from Recent Research." HEALTH EDUCATION MONOGRAPHS 1, no. 30 (1970): 25-47.

Factors Predisposing Patient Behavior

measure attitudes is the Semantic Differential. In this approach, bipolar adjectives are selected so that a response to a given concept can be placed along a continuum between the two adjectives. For example, a person's attitude toward cigarette smoking would be inferred from a series of responses in which cigarette smoking would be placed on a continuum between extreme opposites for each of several evaluative dimensions.

```
CONCEPT:  Cigarette Smoking
good   :_____:__X__:_____:_____:_____:_____:_____: bad
pretty :_____:_____:__X__:_____:_____:_____:_____: ugly
happy  :__X__:_____:_____:_____:_____:_____:_____: sad
```

In patient education, the tasks are to determine (1) which attitudes and beliefs are most important in affecting the behavior in question, (2) how prevalent such attitudes and beliefs are in the patient population to be educated, and (3) how changeable such attitudes and beliefs may be, based on previous experience and research. The first two tasks are diagnostic. The third is a strategic question based on the context of the patient population and the patient education program. If the attitudes and beliefs are deeply rooted in cultural tradition, they are going to be more difficult to change and will require communications directed not only at the patients themselves but also at their families, friends, employers, and others. These considerations become the concern of chapter 6 on reinforcing factors.

Knowledge, beliefs, and attitudes are analyzed and reviewed usually together as a set of cognitive or "internal" factors influencing health behavior. They are typically viewed as subject to change through social and psychological processes. Textbooks on psychology and cognitive learning theory provide theoretical foundations for patient education in relation to these predisposing factors. Classic textbooks used in the professional preparation of health education specialists include:

Becker, M.H., ed. THE HEALTH BELIEF MODEL AND PERSONAL BEHAVIOR. Thorofare, N.J.: Charles B. Slack, 1976.

Fishbein, M., ed. READINGS IN ATTITUDE THEORY AND MEASUREMENT. New York: John Wiley and Sons, 1967.

Fishbein, M., and Ajzen, I. BELIEF, ATTITUDE, INTENTION, AND BEHAVIOR: AN INTRODUCTION TO THEORY AND RESEARCH. Reading, Mass.: Addison-Wesley, 1975.

Freeman, H.E.; Levine, S.; and Reeder, L.G., eds. HANDBOOK OF MEDICAL SOCIOLOGY. 3d ed. Englewood Cliffs, N.J.: Prentice Hall, 1978.

Jaco, E.G., ed. PATIENTS, PHYSICIANS AND ILLNESS. 2d ed. New York: The Free Press, 1972.

Klausmeier, H.J., and Ripple, R. LEARNING AND HUMAN ABILITIES: EDUCATIONAL PSYCHOLOGY. 3d ed. New York: Harper and Row, Publishers, 1971.

Knowles, M. THE MODERN PRACTICE OF ADULT EDUCATION. New York: Association Press, 1970.

Knutson, A.L. THE INDIVIDUAL, SOCIETY AND HEALTH BEHAVIOR. New York: Russell Sage Foundation, 1965.

Kosa, J., et al., eds. POVERTY AND HEALTH: A SOCIOLOGICAL ANALYSIS. Cambridge, Mass.: Harvard University Press, 1969.

Lewin, K. FIELD THEORY IN SOCIAL SCIENCE: SELECTED THEORETICAL PAPERS. New York: Harper and Bros., 1951.

Markle, S. GOOD FRAMES AND BAD. 2d ed. New York: John Wiley and Sons, 1969.

Maslow, A.H. TOWARD A PSYCHOLOGY AND BEING. Princeton, N.J.: D. Van Nostrand Co., 1962.

Mechanic, D. PUBLIC EXPECTATIONS AND HEALTH CARE. New York: John Wiley and Sons, 1972.

Parsons, T. THEORIES OF SOCIETY. New York: The Free Press, 1965.

Rogers, C.R. FREEDOM TO LEARN. Columbus, Ohio: Charles E. Merrill Publishing Co., 1969.

Skipper, J.K., and Leonard, R.C. SOCIAL INTERACTION AND PATIENT CARE. Philadelphia: Lippincott Co., 1965.

<u>Values</u>. As another class of predisposing factors, values influence health behavior by providing a moral, ethical, or philosophical justification or rationale for the behavior. The standard works relating values to education processes are:

Rath, L., et al. VALUES AND TEACHING. Columbus, Ohio: Charles E. Merrill, 1966.

Simon, S.B.; Howe, L.W.; and Kirshenbaum, H. VALUES CLARIFICATION. New York: Hart Publishing Co., 1972.

More recent applications to health and patient education are:

Berger, B., et al. "Values Clarification and the Cardiac Patient." HEALTH EDUCATION MONOGRAPHS 3 (1975): 191-99.

Dalis, G.T., and Strasser, B.B. TEACHING STRATEGIES FOR VALUES AWARENESS AND DECISION-MAKING IN HEALTH EDUCATION. Thorofare, N.J.: Charles B. Slack, 1977.

Faden, R.R., and Faden, A.I., eds. "Ethical Issues in Public Health Policy: Health Education and Lifestyle Interventions." HEALTH EDUCATION 6 (Summer 1978): entire issue.

Rockeach, M. THE NATURE OF HUMAN VALUES. New York: The Free Press, 1973.

Weber, C. "A Comparison of Values Clarification and Lecture Methods in Health Education." JOURNAL OF SCHOOL HEALTH 48 (May 1978): 269-73.

The foregoing represent values as analyzed and applied at a level of personal

Factors Predisposing Patient Behavior

decision-making behavior. Values are understood and utilized also at a cultural level. Anthropologists and others have contributed to this perspective on factors predisposing patient behavior. The classic textbooks and monographs on this aspect of predisposition are:

Clark, M. HEALTH IN THE MEXICAN AMERICAN CULTURE. Berkeley: University of California Press, 1959.

Foster, G.M. TRADITIONAL CULTURES AND THE IMPACT OF TECHNOLOGICAL CHANGE. New York: Harper and Bros., 1962.

Freire, P. PEDAGOGY OF THE OPPRESSED. New York: Herder and Herder, 1971.

Hagen, E.E. ON THE THEORY OF SOCIAL CHANGE. Homewood, Ill.: The Dorsey Press, 1962.

Kluckhohn, F.R., and Strodtbeck, F.L. VARIATIONS IN VALUE ORIENTATIONS. Evanston, Ill.: Row, Peterson and Co., 1961.

Mead, M. CULTURE AND COMMITMENT. New York: Natural History Press, 1970.

Means, R.L. THE ETHICAL IMPERATIVE: THE CRISIS IN AMERICAN VALUES. Garden City, N.Y.: Doubleday, 1969.

Paul, B.D., ed. HEALTH, CULTURE AND COMMUNITY. CASE STUDIES OF PUBLIC REACTIONS TO HEALTH PROGRAMS. New York: Russell Sage Foundation, 1955.

Rogers, E.M., and Shoemaker, F.F. COMMUNICATION OF INNOVATION: A CROSS-CULTURAL APPROACH. New York: Free Press, 1971.

Spicer, E.H., ed. HUMAN PROBLEMS IN TECHNOLOGICAL CHANGE: A CASEBOOK. New York: Russell Sage Foundation, 1952.

Zobrowski, M. PEOPLE IN PAIN. San Francisco: Jossey-Bass, Publishers, 1969.

H. PATIENT KNOWLEDGE AND BELIEFS

H1   Campbell, J.D. "Illness Is a Point of View: The Development of Children's Concepts of Illness." CHILD DEVELOPMENT 46 (1975): 92-100.

>   To study development of concepts of illness, this researcher interviewed 264 children and their mothers. Group profiles were compared for typical definitions of illness given.

H2   Cobliner, W.G. "Pregnancy in the Single Adolescent Girl: The Role of Cognitive Functions." JOURNAL OF YOUTH AND ADOLESCENCE 3 (1974): 17-29.

>   Interviews were conducted with 211 single, adolescent girls, free of known psychiatric disturbance, who had undergone

an elective abortion at a metropolitan municipal hospital. While providing help for the possible emotional stress of the girls' situation, the researchers examined the girls' knowledge of birth control methods and their attempts to avoid pregnancy.

H3   Elder, R.G. "Social Class and Lay Explanations of the Etiology of Arthritis." JOURNAL OF HEALTH AND SOCIAL BEHAVIOR 14 (1973): 28-38.

> Interviews of 160 middle-aged Americans with specific symptoms of arthritis, representing five different social strata, reflected the continued existence of folk theories about the cause and prevention of arthritis, particularly in the lower class.

H4   Kirscht, J.P. "Research Related to the Modification of Health Beliefs." HEALTH EDUCATION MONOGRAPHS 2 (1974): 455-69.

> This paper reviews research on the modification of health beliefs. Topics discussed include the change context of the Health Belief Model, terminology, attitude-belief change and behavior, health beliefs and their interrelationship, belief change and behavioral outcomes, and interaction of changes in beliefs with social and situational forces.

H5   Knopf, A., and Wakefield, J. "Effect of Medical Education on Smoking Behaviour." BRITISH JOURNAL OF PREVENTIVE AND SOCIAL MEDICINE 28 (1974): 246-51.

> At Manchester University, medical students (N = 658) and law students (N = 245) were compared on their knowledge of the hazards of smoking and on differences in their smoking behavior.

H6   Logan, M.H. "Humoral Medicine in Guatemala and Peasant Acceptance of Modern Medicine." HUMAN ORGANIZATION 32 (1973): 385-95.

> When physicians prescribe medicines or dietary regimens that conflict with a patient's belief in the humoral concept, the successful treatment of that patient can be adversely affected. From the patient's point of view, treatment can only be effective if the prescribed medicines or foods are of an opposite temperature quality of his or her disorder. By analyzing the cognitive system underlying humoral classifications and demonstrating how commitment to humoral medicine can impede effective patient care, suggestions for improving health programs in Latin America are presented for medical personnel, illustrating the applicability of anthropological research in the medical sciences.

Factors Predisposing Patient Behavior

H7   Luscutoff, S.A., and Elms, A.C. "Advice in the Abortion Decision." JOURNAL OF COUNSELING PSYCHOLOGY 22 (1975): 140-46.

   Subjects in this study were 224 therapeutic abortion patients, 71 obstetrics patients, and 201 nonhospitalized control subjects. They reported the number of contacts-for-advice they had made when forming decisions to have a therapeutic abortion, to carry a pregnancy to term, or to engage in a significant behavior that did not involve a pregnancy.

H8   Mackie, M. "Lay Perception of Heart Disease in an Alberta Community." CANADIAN JOURNAL OF PUBLIC HEALTH 64 (1973): 445-54.

   A probability sample of 982 Calgary, Alberta, residents was interviewed to determine knowledge about and attitudes toward cardiovascular diseases and relevant preventive health practices.

H9   Peters, E.N., and Hoekelman, R.A. "A Measure of Maternal Competence." HEALTH SERVICES REPORTS 88 (1973): 523-26.

   This paper reports on the construction of an initial measure of knowledge of infant care to be applied before pediatric care is given. The relationship of this measure to other maternal characteristics is explored.

H10   Powers, W.T. BEHAVIOR: THE CONTROL OF PERCEPTION. Chicago: Aldine, 1973. 276 p.

   The theory set forth and developed in this book proposes a testable model of behavior based on feedback relationships between organisms and their environment. It attempts to reconcile the conflict between behaviorists and humanists and to aid in an understanding of ourselves that is both scientific and humane.

H11   Pritchard, M. "Reaction to Illness in Long Term Haemodialysis." JOURNAL OF PSYCHOSOMATIC RESEARCH 18 (1974): 55-67.

   The attitudes and reactions of fourteen dialysis patients to their illness were studied, using a questionnaire designed to measure variables in perception, explanation, and results of illness; relationship with others; and affective, cognitive, and behavioral response.

H12   Rosenstock, I.M. "Historical Origins of the Health Belief Model." HEALTH EDUCATION MONOGRAPHS 2 (1974): 328-35.

   This paper describes the circumstances surrounding the development and emerging theory of the Health Belief Model.

H13 Smith, M.B. "Competence and Adaptation: A Perspective on Therapeutic Ends and Means." AMERICAN JOURNAL OF OCCUPATIONAL THERAPY 28 (1974): 11-15.

> The author discusses the concept of competence as a replacement for the catchword, adjustment. Competence, it is felt, connotes the dispositions and capabilities of a successfully coping person, a capable agent rather than a patient.

H14 Vernon, D.T., and Bigelow, D.A. "Effect of Information About a Potentially Stressful Situation on Responses to Stress Impact." JOURNAL OF PERSONALITY AND SOCIAL PSYCHOLOGY 29 (1974): 50-59.

> Irving Janis's explanation for the ameliorative effect of information on stress impact was tested in a study of eighty hernia-repair patients. One experiment assessed the effects of detailed, accurate information on preoperative attitudes and affective states, while another assessed the effects on postoperative affect and behavior.

H15 Yangdon, T. "Patterns of Knowledge About Treatment and Outcomes of Hypertension Among Hypertensive Outpatients of Two Johns Hopkins Hospital Clinics." Master of Science thesis, Johns Hopkins University School of Hygiene and Public Health, 1975. 61 p.

> Trained interviewers questioned 150 hypertensive patients at two outpatient clinics of an inner-city hospital. Questions were designed to test patients' knowledge about hypertension in general, treatment for hypertension, and outcomes of hypertension. The major hypotheses tested were that the patient's level of knowledge about hypertension is positively related to number of years of formal education and to the length of time the patient has known he or she has hypertension.

I. PATIENT ATTITUDES AND VALUES

I1 Ajzen, I., and Fishbein, M. "Factors Influencing Intentions and the Intention-Behavior Relation." HUMAN RELATIONS 27 (1974): 1-15.

> Communicative and compliance behaviors of subjects working in three-person groups were predicted. Measuring intentions after intervening events had occurred or taking the intervening events into account improved behavioral prediction.

I2 Anderson, H.E., Jr., and Long, R.M. "Clients in a Comprehensive Rehabilitation Center: An Attitudinal Study." REHABILITATION COUNSELING BULLETIN 17 (1974): 232-38.

> An opinion survey was conducted among ninety-two student-

patients at a rehabilitation center to examine their attitudes toward various aspects of the institution and other concepts. A table of results and a discussion of methods, outcome, and implications are included.

13 Becker, M.H., and Maiman, L.A. "Sociobehavioral Determinants of Compliance with Health and Medical Care Recommendations." MEDICAL CARE 12 (1975): 10-24.

This paper reviews the literature on patient acceptance of recommended health behaviors to find social-psychological and related variables that have proved to be consistent predictors of compliance. An hypothesized model is presented for explaining and predicting compliance behavior.

14 de Araujo, G., et al. "Life Change, Coping Ability and Chronic Intrinsic Asthma." JOURNAL OF PSYCHOSOMATIC RESEARCH 17 (1973): 359-63.

This study examines the association between psychosocial assets, environmental change, and dosage of adrenocorticosteroids required to control chronic intrinsic asthma.

15 Einhorn, R.F. "An Investigation of Attitudes Toward Follow-up Care and Birth Control in an Abortion Population." Master's thesis, Johns Hopkins University School of Hygiene and Public Health, 1973. viii, 114 p.

This investigation determined factors responsible for low return rates to a postabortion clinic. A questionnaire was administered to 133 patients immediately after the termination of pregnancy to uncover social and psychological variables influencing a patient's decision of whether or not to return for a postabortion checkup. The questionnaire was based on the Health Belief Model.

16 Himmelfarb, S., and Eagly, A., eds. READINGS IN ATTITUDE CHANGE. New York: John Wiley and Sons, 1974. 655 p.

This book is a collection of forty-nine empirical articles on attitude change. Seven articles deal with attitude change by means of exposure, association, and reinforcement; thirty-one articles consider attitude change via persuasive communication; and eleven articles describe attitude change through self-discrepant behavior. The conclusion discusses current trends in attitude theory and research.

17 Jaccard, J. "A Theoretical Analysis of Selected Factors Important to Health Education Strategies." HEALTH EDUCATION MONOGRAPHS 3 (1975): 152-67.

Based on the theoretical work of Dulaney and Fishbein, two major problems for understanding health behaviors are delineated. An analysis of factors affecting the relation between health intentions and behavior is presented, and implications for health education are discussed. A social psychological model specifying two major determinants of behavioral intentions is contrasted with the Health Belief Model.

18 Jenny, J., et al. "Parents' Satisfaction and Dissatisfaction with Their Children's Dentist." JOURNAL OF PUBLIC HEALTH DENTISTRY 33 (1973): 211-21.

Parents of 838 Caucasian school children were surveyed to assess whether they thought their child's dentist was a good dentist; whether they had considered changing dentists; and whether they agreed with some aspects of delivering dental services. The answers were cross-tabulated with the socio-economic status of each respondent.

19 Liska, A.E. "Attitude-Behavior Consistency as a Function of Generality Equivalence Between Attitude and Behavior Objects." JOURNAL OF PSYCHOLOGY 86 (1974): 217-28.

This study tested the hypothesis that attitude-behavior consistency is affected by the extent to which attitude and behavior are measured at an approximately equivalent level of generality.

I10 _____. "Emergent Issues in the Attitude-Behavior Consistency Controversy." AMERICAN SOCIOLOGICAL REVIEW 39 (1974): 261-72.

This paper reviews recent efforts to examine attitude-behavior inconsistency as a multivariate research problem.

I11 _____. "Impact of Attitude on Behavior: Attitude-Social Support Interaction." PACIFIC SOCIOLOGICAL REVIEW 17 (1974): 83-97.

The relationship between attitude, social support, and behavior, as reported in three studies, is reviewed.

I12 McCleaf, J.E., and Colby, M.A. "The Effects of Students' Perceptions of a Speaker's Role on Their Recall of Drug Facts and Their Opinions and Attitudes About Drugs." JOURNAL OF EDUCATIONAL RESEARCH 68 (1975): 382-86.

This article describes an experiment to determine what effect students' perceptions of the social role of a video-taped speaker and the speaker's personal experiences with drugs had on their recall of drug information presented and their opinions and attitudes about drugs.

113 McKinney, J.P. "The Development of Values: A Perceptual Interpretation." JOURNAL OF PERSONALITY AND SOCIAL PSYCHOLOGY 31 (1975): 801-7.

> A perceptual model is proposed as a theory of the development of values. The hypothesis tested is that values will develop more strongly in individuals whose locus of control is internal. Subjects responded to a locus of control measure, rating six attitude concepts on the semantic differential.

114 Reich, J. "Factors Influencing Patient Satisfaction with the Results of Esthetic Plastic Surgery." PLASTIC AND RECONSTRUCTIVE SURGERY 55 (1975): 5-13.

> Various factors that influence a patient's attitudes toward esthetic plastic surgery are discussed, and methods to increase his or her satisfaction with a good result are presented.

115 Roberto, E.L. "Marital and Family Planning Expectancies of Men Regarding Vasectomy." JOURNAL OF MARRIAGE AND THE FAMILY 36 (1974): 698-706.

> Prospective vasectomees were studied to determine the organization of their expectancy beliefs about vasectomy and the structural relationships between their attitude and expectancy beliefs. Implications for the screening and counseling of prospective vasectomees are discussed.

116 Shaffer, D.R. "Attitude Extremity as a Determinant of Attitude Change in the Forced-Compliance Experiment." BULLETIN OF THE PSYCHONOMIC SOCIETY 3, no. 1B (1974): 51-53.

> Subjects having either extreme or moderate initial attitudes wrote counterattitudinal essays in a test of contradictory hypotheses derived from Festinger's cognitive dissonance theory and Bem's self-perception theory.

117 Uhlenhuth, E.H., et al. "Symptom Intensity and Life Stress in the City." ARCHIVES OF GENERAL PSYCHIATRY 31 (1974): 759-64.

> In a probability sample of urban adults, relationships among self-rated symptom intensity, life stress of recent undesirable events, and demographic characteristics are reported.

118 Weisenberg, M., et al. "Pain: Anxiety and Attitudes in Black, White, and Puerto Rican Patients." PSYCHOSOMATIC MEDICINE 37 (1975): 123-35.

> Reactions of black, white, and Puerto Rican patients were studied in an outpatient dental emergency clinic, using the State-Trait Anxiety Inventory, palmar sweat prints, an inter-

view to obtain patient characteristics and attitudes toward pain, the Dental Anxiety Scale, and posttreatment dentist rating.

J. PERSONALITY AND OTHER MOTIVATIONAL FACTORS

J1 Adamson, J.D., et al. "Measures Associated with Outcome on One Year Follow-up of Male Alcoholics." BRITISH JOURNAL OF ADDICTION 60 (1974): 325-37.

Fifty-two alcoholics were studied while receiving treatment as hospital inpatients or as residents in a rehabilitation facility. A particular focus was to investigate whether anomy would be associated with poor prognosis.

J2 Adler, R., and Lomazzi, F. "Psychological Factors and the Relationship Between Perceptual Style and Pain Tolerance." PSYCHOTHERAPY AND PSYCHOSOMATICS 22 (1973): 347-50.

Thirty healthy, paid male volunteers, physicians, and medical students were tested to confirm the positive correlation between perceptual style and pain tolerance. Perception was assessed by measuring a subject's field-orientation with the Embedded Figures Test and pain was inducted by using the Submaximum Effort Tourniquet Technique, a modified ischaemic muscle pain test.

J3 Balch, P., and Ross, A.W. "Predicting Success in Weight Reduction as a Function of Locus of Control: A Unidimensional and Multidimensional Approach." JOURNAL OF CONSULTING AND CLINICAL PSYCHOLOGY 43 (1975): 119.

The relationship between locus of control and completion and success in a self-control weight reduction program is examined using a unidimensional and multidimensional approach to locus of control. The implications of these findings for selection criteria for applicants to weight reduction programs and possibilities for modification of the external orientation of such applicants are discussed.

J4 Bond, M.R. "Personality Studies in Patients with Pain Secondary to Organic Disease." JOURNAL OF PSYCHOSOMATIC RESEARCH 17 (1973): 257-63.

This paper reports the results of two studies in which the relation between personality structure and pain experienced in organic disease was examined. Results of the first study, on women with carcinoma of the cervix, reveal a close relation between the neuroticism of personality and the presence or absence of pain, and also its intensity. The second study concerns the changes in personality associated with the complete relief of severe intractable pain by surgical means.

J5 Butts, S.V., and Chotlos, J. "A Comparison of Alcoholics and Nonalcoholics on Perceived Locus of Control." QUARTERLY JOURNAL OF STUDIES ON ALCOHOL 34 (1973): 1327-32.

Rotter's Internal-External Control scale and the Marlowe-Crowne Social Desirability scale were administered to seventy-four male alcoholics and to sixty-eight male nonalcoholics of similar social class. Scores indicated that alcoholics had significantly more external control.

J6 Costello, R.M., et al. "Measurement of Subjective Responses to Alcohol and Nonalcohol Slides by Alcoholic Respondents." BEHAVIOUR RESEARCH AND THERAPY 12 (1974): 35-40.

Although anxiety traditionally has been hypothesized as the cognitive process that results from aversive conditioning, alternate hypotheses have not been studied adequately. This investigation of adjective ratings on alcohol-related and non-alcohol related stimuli reveals four alcohol-related constructs: (1) danger versus safety, (2) approach or appetitiveness, (3) avoidance or aversiveness, and (4) general evaluation, good versus bad.

J7 Craig, K.D., and Neidermayer, H. "Autonomic Correlates of Pain Thresholds Influenced by Social Modeling." JOURNAL OF PERSONALITY AND SOCIAL PSYCHOLOGY 29 (1974): 246-52.

Autonomic measures were examined in forty subjects whose pain thresholds, willingness to accept high levels of shock and expressions of distress, were influenced by exposing them to models who simulated different levels of discomfort and pain susceptibility, ostensibly in response to the same shocks the subjects were accepting.

J8 D'Onofrio, C.A. "Motivational and Promotional Factors Associated with Acceptance of a Contraceptive Method in the Postpartum Period." Dr.P.H. dissertation, University of California at Berkeley, 1973.

This exploratory study asks if the acceptance of a birth-control method in the postpartum period is associated with underlying motivational factors predisposing women to take this action, to formalized educational efforts by hospital staff, or to some combination of these variables. Within a theoretical framework based on the work of Lewin and a motivational model of health behavior developed by Hochbaum, Rosenstock, and others, three specific hypotheses relating to these variables are proposed.

J9   Dudley, D.L., et al. "Quantification of Psychosocial Variables in Intrinsic Asthma; Relationship to Physiologic Variability." PSYCHOTHERAPY AND PSYCHOSOMATICS 24 (1974): 129-31.

   A pilot study of thirty-six patients with intrinsic asthma indicated that routine testing of psychosocial assets and life change is of value in clinical management and that asthma patients should receive additional medical attention and be taught to reduce life change to a level they can tolerate.

J10 Edwards, G., et al. "A Comparison of Female and Male Motivation for Drinking." INTERNATIONAL JOURNAL OF ADDICTIONS 8 (1973): 577-87.

   Stated motivations for drinking of 281 women and 306 men who drank more than "once or twice a month" are compared. The basic motivational factor structure for men and women appears rather similar, but the findings suggest the need for close analysis of social controls that differently model male and female drinking in different societies.

J11 English, G.E., and Curtin, M.E. "Personality Differences in Patients at Three Alcoholism Treatment Agencies." JOURNAL OF STUDIES ON ALCOHOL 36 (1975): 52-61.

   Personality inventories were completed by patients in three alcoholism treatment programs and two control groups. Differences in the personality profiles of these groups were compared.

J12 Felton, B., and Kahana, E. "Adjustment and Situationally-Bound Locus of Control Among Institutionalized Aged." JOURNAL OF GERONTOLOGY 29 (1974): 295-301.

   The relationship between perceived locus of control and adjustment among institutionalized aged was examined, using residents' solutions to hypothetical problems as the measure of perceived control.

J13 Glass, D.C., et al. "Time Urgency and the Type A Coronary-Prone Behavior Pattern." JOURNAL OF APPLIED SOCIAL PSYCHOLOGY 4 (1974): 125-40.

   Two experiments examined behavioral consequences of a sense of time urgency, which presumably characterizes individuals classified as having Type A coronary-prone behavior patterns. Consideration was also given to possible physiological mechanisms mediating the relationship between psychological variables such as the Type A pattern and actual occurrence of coronary heart disease.

Factors Predisposing Patient Behavior

J14 Gove, W.R., and Lester, B.J. "Social Position and Self-Evaluation: A Reanalysis of the Yancey, Rigsby, and McCarthy Data." AMERICAN JOURNAL OF SOCIOLOGY 79 (1974): 1308-14.

> The effect of sex and marital status on psychiatric symptoms and self-esteem is analyzed, as well as the way race interacts with sex and marital status and the way city interacts with sex, marital status, and race.

J15 Haan, N. "The Adolescent Antecedents of an Ego Model of Coping and Defense and Comparisons with Q-Sorted Ideal Personalities." GENETIC PSYCHOLOGY MONOGRAPHS 89 (1974): 273-306.

> Two conceptualizations of the "ideal" personality--one concerned with coping, arising from ego-cognitive theory and the other from a methodology wherein psychologists Q-sorted their ideal personality--were studied with a longitudinal sample of ninety-nine subjects. The adolescent antecedents of the ego model--personality, interpersonal behavior, socioeconomic status, IQ, and family milieu--were investigated and the results were compared with those previously reported by others for the Q-sorted ideals.

J16 Herman, C.P. "External and Internal Cues as Determinants of the Smoking Behavior of Light and Heavy Smokers." JOURNAL OF PERSONALITY AND SOCIAL PSYCHOLOGY 30 (1974): 664-72.

> The role of external and internal smoking cues as determinants of smoking behavior was investigated as an analogy to Schachter's model of eating behavior.

J17 Hewitt, J., and Goldman, M. "Self-Esteem, Need for Approval, and Reactions to Personal Evaluations." JOURNAL OF EXPERIMENTAL AND SOCIAL PSYCHOLOGY 10 (1974): 201-10.

> The reactions of high and low self-esteem individuals to negative and positive evaluators of self were measured.

J18 Johnson, J.E., and Rice, V.H. "Sensory and Distress Components of Pain: Implications for the Study of Clinical Pain." NURSING RESEARCH 23 (1974): 203-9.

> In the laboratory, fifty-two male subjects experienced ischemic pain in their arms to test the hypothesis that the intensity of the reactive component of pain experienced is a function of the congruency between expected and experienced physical sensations. Findings suggest that in clinical settings patients who receive a partial description of sensations they may experience will have as much reduction in distress as those who receive a complete description. The potential usefulness

of the measurement of each component of pain versus pain threshold measures is discussed in terms of clinical management of pain.

J19 Kane, F.J., et al. "Motivational Factors in Pregnant Adolescents." DISEASES OF THE NERVOUS SYSTEM 35 (1974): 131-34.

Etiologic factors in fifty-two pregnant adolescent residents in New Orleans maternity homes were studied through the use of the Minnesota Multiphasic Personality Inventory.

J20 Kenigsberg, D., et al. "The Coronary-Prone Behavior Pattern in Hospitalized Patients with and without Coronary Heart Disease." PSYCHOSOMATIC MEDICINE 36 (1974): 344-51.

Hospitalized male and female patients aged twenty-two to sixty-four with coronary heart disease and noncardiovascular diseases were compared in terms of selected behavioral variables. The major behavioral assessment was the Jenkins Activity Survey designed to measure the coronary-prone behavior pattern (Type A).

J21 Kiell, N., ed. THE PSYCHOLOGY OF OBESITY: DYNAMICS AND TREATMENT. Springfield, Ill.: Charles C Thomas, 1973. xxi, 458 p. Bibliog.

This book is a collection of previously published articles by experts in the field of obesity. Theoretical papers, reviews of the literature, reports of experimental studies, and case histories are included, as well as an extensive bibliography.

J22 Kilmann, R.H., and Taylor, V. "A Contingency Approach to Laboratory Learning: Psychological Types Versus Experiential Norms." HUMAN RELATIONS 27 (1974): 891-909.

This study investigates the psychological dynamics and situational factors that determine whether an individual will experience support and/or confrontation in a laboratory setting. Implications are given for (1) identification of the rejectors of particular laboratory experiences, (2) design of different laboratories via alternative experiential norms, and (3) intervention strategies for staff and trainers according to the laboratory setting and goals.

J23 Kimball, C.P., et al. "The Experience of Cardiac Surgery: V. Psychological Patterns and Prediction of Outcome." PSYCHOTHERAPY AND PSYCHOSOMATICS 22 (1973): 310-19.

One hundred and eighty patients undergoing cardiac surgery were evaluated psychiatrically prior to surgery and followed afterwards for up to thirty months. Analyses of interviews led to the identification of variables identifying four groups having prognostic value for long-term survival.

J24 Krantz, D.S., et al. "Helplessness, Stress Level, and the Coronary-Prone Behavior Pattern." JOURNAL OF EXPERIMENTAL AND SOCIAL PSYCHOLOGY 10 (1974): 284-300.

Two experiments were conducted to examine the relationship between stress level and learned helplessness in human subjects. The results are interpreted in terms of differing perceptions of threat imposed by lack of environmental control.

J25 McFarland, R.A., and Coombs, R. "Anxiety and Feedback as Factors in Operant Heart Rate Control." PSYCHOPHYSIOLOGY 11 (1974): 53-57.

This study was conducted to determine the relationships among manifest anxiety, amount of feedback, and heart rate control. Subjects were chosen on the basis of scores on the Taylor Manifest Anxiety scale.

J26 Maiman, L.A., and Becker, M.H. "The Health Belief Model: Origins and Correlates in Psychological Theory." HEALTH EDUCATION MONOGRAPHS 2 (1974): 336-53.

To trace the development and identify correlates of the Health Belief Model, six psychological models of decision making in a choice situation are examined. Decision-making theory is extended to the area of motivation by including a general motivational concept wherein the individual desires to achieve success or to avoid failure. Analogies are presented between the basic concepts of the Health Belief Model and components of the general psychological model.

J27 Matsuno, A.S., et al. "Four Factors Affect Weight Control for Obese Children." JOURNAL OF NUTRITION EDUCATION 6 (1974): 104-7.

A weight control program for obese children was developed by using a team approach. Individualized counseling, nutrition education, and physical activity were program components.

J28 Mellett, P. "Psychological States of Asthmatics After Treatment." JOURNAL OF PSYCHOSOMATIC RESEARCH 17 (1973): 299-302.

A measurement to demonstrate the general psychological improvement of asthmatics who had been relieved of symptoms is described.

J29 Miller, P.M., et al. "Relationship of Alcohol Cues to the Drinking

Behavior of Alcoholics and Social Drinkers: An Analogue Study." PSYCHOLOGICAL RECORD 24 (Winter 1974): 61-66.

> The effects of visual alcohol cues on the operant drinking behavior of twenty alcoholics and twenty social drinkers were evaluated.

J30 Neufeld, R.W.J., and Davidson, P.O. "Sex Differences in Stress Response: A Multivariate Analysis." JOURNAL OF ABNORMAL PSYCHOLOGY 83 (1974): 178-85.

> This study examined the configuration of ten behavioral, physiological, and subjective measures of stress among subjects classified on the basis of sex and exposure to a stressor (mutilated bodies) and a benign stimulus. Implications for assessing stress-reducing treatments are briefly discussed.

J31 Nowicki, S., and Hopper, A.E. "Locus of Control Correlates in an Alcoholic Population." JOURNAL OF CONSULTING AND CLINICAL PSYCHOLOGY 42 (1974): 735.

> This study relates locus of control of reinforcement orientation to behavior for both male and female alcoholics. Based on an evaluation of previous studies in this area and using a measure of locus of control orientation other than the Rotter scale, it was predicted and shown that externality was related to dysfunction behavior. Implications for treatment are discussed.

J32 Orford, J., and Hawker, A. "An Investigation of an Alcoholism Rehabilitation Halfway House: II. The Complex Question of Client Motivation." BRITISH JOURNAL OF ADDICTION 69 (1974): 315-23.

> The hypothesis that relatively low levels of motivation for change are responsible for the link between youth and premature departure from an alcoholism halfway house was tested using two measures of motivation. An alternative set of concepts, based on Janis and Mann's decision-making model, is proposed.

J33 Parbrook, G.D., et al. "Personality Assessment and Post-Operative Pain and Complications." JOURNAL OF PSYCHOSOMATIC RESEARCH 17 (1973): 277-85.

> A group of fifty male patients having elective peptic ulcer surgery and a group of fifty female patients having elective cholecystectomies were selected for this trial. The purpose was to see if any correlation existed between psychological and other factors, assessed preoperatively, and the patients' pain and progress after operation.

Factors Predisposing Patient Behavior

J34   Plutchik, R., et al. "Studies of Body Image III: Body Feelings as Measured by the Semantic Differential." INTERNATIONAL JOURNAL OF AGING AND HUMAN DEVELOPMENT 4 (1973): 375-80.

> A semantic differential questionnaire was designed to assess the connotative meanings of the word "head" as an important aspect of body image. The questionnaire was given to 203 individuals, including geriatric patients in a home and hospital for the aged, geriatric psychiatric patients, middle-aged schizophrenics, and university students. Aging was found to be less disruptive to body image than was mental illness.

J35   Roback, H.B., et al. "Personality Differences Between Fee Paying and Non-Fee Paying Patients Seen for Psychological Testing." JOURNAL OF CONSULTING AND CLINICAL PSYCHOLOGY 42 (1974): 734.

> This study investigated personality differences between fee-paying and nonfee-paying male patients who were seen for psychological testing.

J36   Rodin, J. "Causes and Consequences of Time Perception Differences in Overweight and Normal Weight People." JOURNAL OF PERSONALITY AND SOCIAL PSYCHOLOGY 31 (1975): 898-904.

> This article reports on three experiments investigating differences between overweight and normal weight subjects in time perception. The differences would represent a lack of internal responsiveness in the obese as well as external reactivity in a noneating setting. Implications for a theory about the development and consequences of obesity are discussed.

J37   Sarason, I.G., and Spielberger, C.D., eds. STRESS AND ANXIETY. Vol. 2. Washington, D.C.: Hemisphere Publishing Corp., 1975. 350 p.

> This volume deals with issues and problems currently attracting growing inquiry among researchers. The topics discussed include whether coping skills needed for successful adaptation can be taught to persons susceptible to the unwanted effects of stress.

J38   Stern, G.S., et al. "Perceived Aversiveness of Recommended Solution and Locus of Control as Determinants of Response to Danger." JOURNAL OF RESEARCH IN PERSONALITY 9 (March 1975): 37-47.

> As determined by scores on Rotter's Locus of Control scale, internal and external subjects were exposed either to a high-, low-, or no-fear message on the dangers of exposure to a sunlamp. Results are interpreted in terms of the perceptual processes that may mediate the relationship between fear and action.

J39 Thomae, H. "Cognition and Motivation: Modern Aspects of an Ancient Problem." PSYCHOLOGIA 16 (1973): 179-90.

> After an analysis of dominant theories regarding motivation-cognition relationships, the author introduces several principles that regulate cognitive-motivation sequences in the behavioral continuum.

J40 Thornhill, M.A., et al. "A Computerized and Categorized Bibliography on Locus of Control." PSYCHOLOGICAL REPORTS 36 (1975): 505-6.

> Over 1,200 published and unpublished references on locus of control are categorized in this computerized bibliography.

J41 Tobacyk, J.J., et al. "Effects of Congruence-Incongruence Between Locus of Control and Field Dependence on Personality Functioning." JOURNAL OF CONSULTING AND CLINICAL PSYCHOLOGY 43 (1975): 81-85.

> Lefcourt and Telegdi's concepts of congruence and incongruence between perceptual skills and expectancies were studied with four groups of subjects, using Rotter's Locus of Control Scale and the Rod-and-Frame Test.

J42 Wunderlich, R.A. "Personality Characteristics of Super-Obese Persons as Measured by the California Psychological Inventory." PSYCHOLOGICAL REPORTS 35 (1974): 1029-30.

> In a six-month weight reduction program, twenty-three super-obese males (M = 341 lb.) and females (M = 263 lb.) were compared with normative groups on the California Psychological Inventory. Females scored significantly different from the norm group on six of the fifteen predicted scales, males on nine. A characteristic personality of the obese could not be described from the data.

J43 Zuckerman, M. "Attribution Processes, Placebo Effect, and Anxiety over Dental Treatment." REPRESENTATIVE RESEARCH IN SOCIAL PSYCHOLOGY 5 (1974): 35-46.

> The two hypotheses of this study are that attribution of arousal to the anesthetic injection will reduce anxiety and that unexpected arousal, which is attributed to the dental treatment, will increase anxiety.

Chapter 5
FACTORS ENABLING PATIENT BEHAVIOR

It is almost tautological to say that a patient cannot be expected to do something unless he or she is able to do it. Yet, patient education sometimes suffers for lack of attention to enabling factors. A patient education program that focuses exclusively on predisposing factors often will result in heightened motivation for a behavior that cannot occur because it is blocked by physical circumstances at home or at work, or the resources necessary for the behavior (health insurance, community services or facilities) are unavailable or inaccessible. Heightened motivation for behavior that cannot occur will lead to frustration.

The literature on enabling factors related to patient behavior is concentrated in medical sociology, rehabilitation, and medical care organization. This literature overlaps substantially with the public health literature on community development, community organization, and health planning. The journals most likely to carry current articles on enabling factors related to patient behavior are AMERICAN JOURNAL OF PUBLIC HEALTH, HEALTH EDUCATION MONOGRAPHS, HEALTH SERVICES RESEARCH, HOSPITALS: JOURNAL OF THE AMERICAN HOSPITAL ASSOCIATION, INQUIRY, INTERNATIONAL JOURNAL OF HEALTH EDUCATION, INTERNATIONAL JOURNAL OF REHABILITATION RESEARCH, INTERNATIONAL JOURNAL OF HEALTH SERVICES, JOURNAL OF COMMUNITY HEALTH, JOURNAL OF FAMILY AND COMMUNITY HEALTH, JOURNAL OF HEALTH AND SOCIAL BEHAVIOR, MEDICAL CARE, MILBANK MEMORIAL FUND QUARTERLY, NEW ENGLAND JOURNAL OF MEDICINE, PUBLIC HEALTH REPORTS, REHABILITATION COUNSELING BULLETIN, REHABILITATION LITERATURE, and SOCIAL WORK IN HEALTH CARE.

Supplementing the books and articles listed and annotated in the next two sections, several earlier and more recent books that represent the subject of resources enabling patient compliance and self-care most comprehensively are the following:

Anderson, C.L., et al. COMMUNITY HEALTH. 3d ed. St. Louis: C.V. Mosby Co., 1978.

Anderson, R. A BEHAVIORAL MODEL OF FAMILIES' USE OF HEALTH SERVICES. Research Series No. 25. Chicago: University of Chicago, Center for Health Administration Studies, 1968.

Factors Enabling Patient Behavior

Bennis, W.G., et al., eds. THE PLANNING OF CHANGE. 3d ed. New York: Holt, Rinehart and Winston, 1976.

Borman, L.D. EXPLORATIONS IN SELF-HELP AND MUTUAL AID. Evanston, Ill.: Center for Urban Affairs, Northwestern University, 1975.

Caplan, G., and Killilea, M. SUPPORT SYSTEMS AND MUTUAL HELP: A MULTIDISCIPLINARY EXPLORATION. New York: Grune and Stratton, 1976.

Fonoroff, A., and Levin, L.S., eds. "Self Care." HEALTH EDUCATION MONOGRAPHS 5 (Summer 1977): entire issue.

Gartner, A., and Riessman, F. SELF HELP IN THE HUMAN SERVICES. San Francisco: Jossey-Bass Publishers, 1977.

Grosser, C.G. NEW DIRECTIONS IN COMMUNITY ORGANIZATION: FROM ENABLING TO ADVOCACY. New York: Praeger Publishers, 1973.

Health Resources Associates. MARKETING HEALTH MAINTENANCE ORGANIZATIONS TO LOW INCOME PERSONS. Contract no. HSM 110-72-316. Rockville, Md.: Health Maintenance Organization Service, U.S. Department of Health, Education and Welfare, July 1973.

Howard, J., and Strauss, A., eds. HUMANIZING HEALTH CARE. New York: Wiley-Interscience, 1975.

Illich, I. MEDICAL NEMESIS: THE EXPROPRIATION OF HEALTH. New York: Pantheon, 1976.

Katz, A., and Bender, E.I. THE STRENGTH IN US. New York: Franklin Watts, 1976.

Kosa, J.; Antonovsky, A.; and Zola, I.K. POVERTY AND HEALTH: A SOCIOLOGICAL ANALYSIS. Cambridge, Mass.: Harvard University Press, 1969.

Levin, L.S.; Katz, A.; and Holtz, E. SELF-CARE: LAY INITIATIVES IN HEALTH. New York: Prodist, 1976.

Little, Arthur D., Co. FINAL REPORT OF A SURVEY OF CONSUMER HEALTH EDUCATION PROGRAMS. NTIS Order No. P.B. 251775. Washington, D.C.: U.S. Department of Commerce, 1976.

Mechanic, D., ed. THE GROWTH OF BUREAUCRATIC MEDICINE. New York: Wiley-Interscience, 1976.

Moynihan, D.P., ed. ON UNDERSTANDING POVERTY: PERSPECTIVES FROM THE SOCIAL SCIENCES. New York: Basic Books, 1969.

Navarro, V. MEDICINE UNDER CAPITALISM. New York: Prodist, 1976.

President's Committee on Health Education. REPORT OF THE PRESIDENT'S COMMITTEE ON HEALTH EDUCATION. New York: Public Affairs Institute, 1973.

Salber, E.J. CARING AND CURING: COMMUNITY PARTICIPATION IN HEALTH SERVICES. New York: Prodist, 1975.

Simmons, J.J., ed. "Making Health Education Work." AMERICAN JOURNAL OF PUBLIC HEALTH 65, Suppl. (October 1975): vi, 1-49.

SOURCE BOOK OF HEALTH INSURANCE DATA, 1977-1978. Washington, D.C.: Health Insurance Institute, 1978.

Warren, R.B., and Warren, D.I. THE NEIGHBORHOOD ORGANIZER'S HANDBOOK. Notre Dame, Ind.: University of Notre Dame Press, 1977.

In addition to resources for self-care, the other aspect of enabling factors to which patient education must be addressed is the development of self-care skills. This has been a concept in nursing theory that has evolved steadily since the earliest Nightingale model of helping the helpless. The current emphasis in nursing theory on self-care skills is reflected in several recent works, including:

Nursing Development Conference Group. CONCEPT FORMALIZATION IN NURSING PROCESS AND PRODUCT. Boston: Little, Brown and Co., 1973.

Orem, D.E. NURSING: CONCEPTS OF PRACTICE. New York: McGraw-Hill Book Co., 1971.

Redman, B.K. THE PROCESS OF PATIENT TEACHING IN NURSING. 3d ed. St. Louis: C.V. Mosby Co., 1976.

Travelbee, J. INTERPERSONAL ASPECTS OF NURSING. 2d ed. Philadelphia: F.A. Davis Co., 1971.

Self-care skills have had their most significant scientific development from an extension of psychological research on behavior modification into "self-control," "self-monitoring," and "self-directed behavior." These applications of behavior-modification principles have added a dimension to enabling factors in patient education that will be covered more extensively in chapter 6, section T, but their development as enabling factors should be noted here. This development is traced in the following recent reviews:

Craighead, W.E., et al., eds. BEHAVIOR MODIFICATION: ISSUES, PRINCIPLES, AND APPLICATIONS. New York: Houghton Mifflin, 1976.

Foreyt, J.P., ed. BEHAVIOR MODIFICATION APPROACHES TO OBESITY. New York: Pergamon, 1976.

Green, L.W., et al. "Research and Demonstration Issues in Self-Care: Measuring the Decline of Medicocentrism." In CONSUMER SELF-CARE IN HEALTH, edited by J. Gallicchio, pp. 20-26. Research Proceedings Series. DHEW Publication no. (HRA) 77-3181. Washington, D.C.: National Center for Health Sciences Research, August 1977.

Jeffrey, D.B., and Katz, R.C. HOW TO HELP YOURSELF LOSE WEIGHT: A SELF-CONTROL APPROACH TO DIETARY, EXERCISE, AND PSYCHOLOGICAL MANAGEMENT. New York: Prentice-Hall, 1977.

Katz, R.C., and Zlutnick, S., eds. BEHAVIOR THERAPY AND HEALTH CARE: PRINCIPLES AND APPLICATIONS. New York: Pergamon, 1975.

Factors Enabling Patient Behavior

Mahoney, M.J., and Thoresen, C.E., eds. SELF-CONTROL: POWER TO THE PERSON. Belmont, Calif.: Wadsworth Publishing Co., 1974.

Watson, D.L., and Thorp, R.G. SELF-DIRECTED BEHAVIOR: SELF-MODIFICATION FOR PERSONAL ADJUSTMENT. 2d ed. Monterey, Calif.: Brooks/Cole, 1977.

Williams, B.J., et al., eds. OBESITY: BEHAVIORAL APPROACHES TO DIETARY MANAGEMENT. New York: Brunner/Mazel, 1977.

The most comprehensive directory of resources for patients beyond the clinical setting is prepared and revised periodically by the American Public Health Association for the Health Resources Administration: American Public Health Association. CONSUMER HEALTH EDUCATION: A DIRECTORY. Rev. ed. DHEW Pub. No. (HRA) 77-607. Rockville, Md.: National Center for Health Services Research, Health Resources Administration, October 1976.

K. PERSONAL RESOURCES AVAILABLE TO PATIENTS

K1   Aday, L.A. "Economic and Noneconomic Barriers to the Use of Needed Medical Services." MEDICAL CARE 13 (1975): 447-56.

> In this paper, an index of access to medical care describes use of services in relation to actual need for care. Implications of the findings for evaluating existing and proposed national health policies are discussed.

K2   Frazier, P.J., et al. "Parents' Descriptions of Barriers Faced and Strategies Used to Obtain Dental Care." JOURNAL OF PUBLIC HEALTH DENTISTRY 34 (1974): 22-38.

> This paper reports a study of problems of parents in gaining dental treatment for their children. Strategies for obtaining dental treatment were identified from interviews with families who were successful in obtaining treatment during the twelve-month study, and barriers were identified from interviews with unsuccessful families.

K3   Freidson, E. "Prepaid Group Practice and the New 'Demanding Patient.'" MILBANK MEMORIAL FUND QUARTERLY 51 (1973): 473-88.

> Based on an extensive field study of the practitioners in a large, prepaid, service-contract group practice, this paper discusses how a prepaid service contract and closed-panel practice brings a new dimension into doctor-patient relations and how physicians respond. Physicians were particularly upset by a new type of "demanding patient" who claimed services on the basis of contractual rights and threatened appeal to higher bureaucratic authority. Modes of dealing with such patients are discussed briefly.

K4   Miller, M.H. "Who Receives Optimal Medical Care?" JOURNAL OF HEALTH AND SOCIAL BEHAVIOR 14 (1973): 176-82.

> A comparison is made of medical care received by lower-class and upper-class cancer patients. The author questions the belief that upper-class patients receive "optimal" care by distinguishing specialized from personalized care.

K5   Monteiro, L.A. "Expense Is No Object: Income and Physician Visits Reconsidered." JOURNAL OF HEALTH AND SOCIAL BEHAVIOR 14 (1973): 99-115.

> The relationship between income and physician visits and the assumption that the poor underutilize physician services are reexamined and tested with data from a survey of Rhode Island residents.

K6   Perkoff, G.T., et al. "Medical Care Utilization in an Experimental Prepaid Group Practice Model in a University Medical Center." MEDICAL CARE 12 (1974): 471-85.

> Medical care utilization is reported for the first two years of an experimental prepaid group practice. Methodological considerations are emphasized, and comparisons are made between prospectively selected study and control families.

K7   Phelps, C.E. DEMAND FOR HEALTH INSURANCE: A THEORETICAL AND EMPIRICAL INVESTIGATION. R-1054-OEO. Santa Monica, Calif.: The Rand Corporation, 1974. 210 p.

> This study, taken from the author's doctoral dissertation, contributes to the theory of demand for insurance and provides empirical estimates of demand for insurance.

K8   Rimm, I.J., and Rimm, A.A. "Association Between Socioeconomic Status and Obesity in 59,556 Women." PREVENTIVE MEDICINE 3 (1974): 543-72.

> A study of 59,556 weight-conscious women in the United States (members of TOPS Clubs) is reported. Relationships between obesity and respondents' education, weight, age, family income, and husband's education were analyzed.

L. COMMUNITY RESOURCES AVAILABLE TO PATIENTS

L1   Angers, W.P., and Haffly, J.E. "Vocational Rehabilitation Counseling of the Epileptic." PSYCHOLOGIA 16 (1973): 201-8.

> The occupations of 1,374 epileptics are presented for the pro-

fessional rehabilitation counselor who assists epileptics in developing their potential for reaching realistic vocational goals.

L2 Berkanovic, E., et al. "The Effects of Prepayment on Access to Medical Care: The PACC Experience." MILBANK MEMORIAL FUND QUARTERLY 53 (1975): 241-54.

   Data reported in this article are from a larger study in which a prepaid medical foundation was compared with a nonprepaid fee-for-service system. Comparisons were made on factors pertaining to perceptions of health care of Medicaid recipients and of physicians. The impact of prepayment on Medicaid recipients' perceptions of their <u>access</u> to health care was analyzed.

L3 Bodenheimer, T., et al. "Capitalizing on Illness: The Health Insurance Industry." INTERNATIONAL JOURNAL OF HEALTH SERVICES 4 (1974): 583-98.

   The history of the private health insurance industry in the United States and its importance as a profit-making sector of the economy are described. The effects of the domination of Blue Cross by hospital representatives, of Blue Shield by physicians, and of commercial insurance companies by banks and industrial corporations are discussed in relation to discrimination against the elderly, the sick, and the poor, and to rapidly rising medical costs.

L4 THE FEDERAL HEALTH DOLLAR, 1969-1976: A CHARTBOOK ANALYSIS OF ACTIVITIES SUPPORTED AND STRATEGIES PURSUED IN FEDERAL EXPENDITURES FOR HEALTH. Washington, D.C.: National Planning Association, Center for Health Policy Studies, 1977. 72 p. Charts.

   This chartbook provides an overview of the scope and magnitude of the federal government's involvement in the American health care system.

L5 Fuchs, V.E. WHO SHALL LIVE? HEALTH, ECONOMICS, AND SOCIAL CHOICE. New York: Basic Books, 1974. 168 p.

   Major problems of health and medical care (cost, accessibility, inequities, and disparities in health levels within the United States and between the United States and other countries) are approached from an economic viewpoint. The central theme of this text is the necessity for choices at both the individual and social level concerning the amount, quality, and kind of health care the United States will have. Relationships among health, economics, and social choice are examined.

L6 Gentry, J.T., et al. "Promoting the Adoption of Social Work Services by Hospitals and Health Departments." AMERICAN JOURNAL OF PUBLIC HEALTH 63 (1973): 117-25.

> This paper identifies the level at which social work services have been implemented by acute general hospitals and health departments in the United States and some of the community, organizational, and personal variables associated with implementation.

L7 Glaser, F.B., and Greenberg, S.W. "Relationship Between Treatment Facilities and Prevalence of Alcoholism and Drug Abuse." JOURNAL OF STUDIES ON ALCOHOL 36 (1975): 348-58.

> Data are presented suggesting that in Pennsylvania the availability of treatment for drug abuse and alcoholism is inversely related to the prevalence of the problems. It is hypothesized that, as the treatment response to one increases, the treatment response to the other declines. Possible causes and consequences are outlined.

L8 Graning, H.M. "It's Long Overdue." HEALTH EDUCATION MONOGRAPHS 2, Suppl. 1 (1974): 65-72.

> Concepts underlying reimbursement for patient education and eight criteria under which Blue Cross, Medicare, and Medicaid should pay for patient education in a hospital setting are outlined.

L9 Hardy, W.E., Jr., et al. "Health Education Spans Outreach Clinics: A Concept to Consider." HEALTH EDUCATION MONOGRAPHS 3 (1975): 89-99.

> This article discusses accessibility of health services in rural and low-income metropolitan areas and changes needed in the health care system. It also presents the concept of a Health Outreach Clinic to make service reasonably accessible within any given geographic area.

L10 HEALTH EDUCATION MATERIALS AND THE ORGANIZATIONS WHICH OFFER THEM. New York: Health Insurance Institute, 1977. 25 p.

> This compilation of national organizations that distribute materials on health education lists organizations which offer materials on general aspects of health and those which offer materials on specific health problems.

L11 Hiatt, H.H. "Protecting the Medical Commons: Who Is Responsible?" NEW ENGLAND JOURNAL OF MEDICINE 293 (1975): 235-41.

> Finite resources for medical care and growing demands on

medical resources are discussed. The author proposes that medical practices must be evaluated in terms of social and medical priorities, particularly in light of the establishment of national health insurance.

L12   Jones, E.W., et al. "HIP Incentive Reimbursement Experiments: Utilization and Costs of Medical Care, 1969 and 1970." SOCIAL SECURITY BULLETIN 37 (December 1974): 3-34.

The Health Insurance Plan of Greater New York carried out a three-year experiment with financial incentives to reduce the total cost of care for its Medicare enrollment. Data on characteristics of the study populations and utilization and charges for 1969 and 1970 are presented.

L13   Kakalik, J.S., et al. SERVICES FOR HANDICAPPED YOUTH: A PROGRAM OVERVIEW. R-1220-HEW. Santa Monica, Calif.: The Rand Corp., 1974. 354 p.

This first of two reports describes a twenty-two-month, cross-agency evaluation of federal and state programs for assistance to handicapped youth. Purposes of the study were to describe current federal and state programs for service to mentally and physically handicapped youth in the United States, to estimate the resources devoted to various classes of handicapped youth, and to identify major problems of the present service system. The study was used to assist DHEW officials by evaluating current policies and providing information on alternative future policies to improve the delivery of services to youth with hearing or vision handicaps.

L14   Kane, R.L., et al., eds. THE HEALTH GAP: MEDICAL SERVICES AND THE POOR. New York: Springer Publishing Co., 1976. 321 p. Bibliog.

Health professionals discuss factors contributing to the health system's failure to reach the needy and offer programs designed to bring about their participation.

L15   Kelly, A., and Munan, L. "Epidemiological Patterns of Childhood Mortality and Their Relation to Distance from Medical Care." SOCIAL SCIENCE AND MEDICINE 8 (1974): 363-67.

In a field survey defining causes of death among preschoolers in a semirural region of Canada, families of deceased children were compared with a probability sample of families with living children with respect to distance to the nearest source of medical care. The study underscores the need for considering distance to medical care as an important entry in decision-making processes for the allocation of health resources.

Factors Enabling Patient Behavior

L16 Magnuson, W.G., and Segal, E.A. HOW MUCH FOR HEALTH? Washington, D.C.: Robert B. Luce, 1974. xxi, 210 p. Bibliog.

   Topics in this book include today's budget, priorities and directions, the conquest of cancer, the environment, consumer safety and protection, research, standards and quality, manpower and facilities, health care delivery, health financing, and environmental health.

L17 Newhouse, J.P., et al. POLICY OPTIONS AND THE IMPACT OF NATIONAL HEALTH INSURANCE. Santa Monica, Calif.: The Rand Corp., 1974. 68 p.

   This report examines prototypical plans for national health insurance and the impact of these plans on the health care system. The authors list estimates of increased costs for each type of plan and discuss costs in relation to health care goals.

L18 Pearson, C.E. "An Historical Case for the Role of the Insurance Industry in Patient Education." HEALTH EDUCATION MONOGRAPHS 2 (1974): 39-43.

   A national health care program to guarantee every American access to quality health care regardless of income is outlined. The objectives of the program, which proposes government and private industry cooperation, are detailed. The history of the Metropolitan Life Insurance Company's program in health education is given as an example.

L19 Rabin, D.L., and Schach, E. "Medicaid, Morbidity, and Physician Use." MEDICAL CARE 13 (1975): 68-78.

   A household interview on use of health services compared use of physicians and of preventive services by Medicaid recipients and two other income groups.

L20 Roemer, M.I., and Shonick, W. "HMO Performance: The Recent Evidence." MILBANK MEMORIAL FUND QUARTERLY 51 (1973): 271-317.

   Health maintenance organizations are assessed as alternatives in modifying the U.S. health care delivery system toward more economical patterns and in encouraging preventive and ambulatory care rather than costly hospital services.

L21 Russell, L.B., et al. FEDERAL HEALTH SPENDING, 1969-74. Washington, D.C.: National Planning Association, Center for Health Policy Studies, 1974. 138 p. Append. Tables.

   This publication describes the health expenditures and programs of the federal government for the period 1969 to 1974 and the changing nature and dimensions of the problems addressed.

L22 Smart, R.G. "Employed Alcoholics Treated Voluntarily and Under Constructive Coercion: A Follow-up Study." QUARTERLY JOURNAL OF STUDIES ON ALCOHOL 35 (1974): 196-209.

Behavior changes after treatment were compared in alcoholics who were coerced by their employers to seek treatment (mandatory referral) and in employed alcoholics who sought treatment on their own (voluntary referral). All patients participated in similar three-week group and occupational inpatient therapy programs. An overall improvement rating scale (a composite of drinking, work, family, social behavior, financial, residence, and incarceration scales) was used to assess posttreatment changes.

L23 Tessler, R., and Mechanic, D. "Consumer Satisfaction with Prepaid Group Practice: A Comparative Study." JOURNAL OF HEALTH AND SOCIAL BEHAVIOR 16 (1975): 95-113.

This study compares satisfaction of consumers participating in a prepaid group practice with that of consumers of alternative health insurance plans in a large metropolitan area.

L24 _____. "Factors Affecting the Choice Between Prepaid Group Practice and Alternative Insurance Programs." MILBANK MEMORIAL FUND QUARTERLY 53 (1975): 149-72.

This paper examines the basis for the selection of prepaid group practice in a dual-choice situation and the social, attitudinal, and health characteristics of populations choosing prepaid programs in contrast to other plans.

Chapter 6
FACTORS REINFORCING PATIENT BEHAVIOR

A behavior that is motivated and enabled still will not persist if the patient receives or even anticipates no social support or reward for the behavior. It will fail to develop as a habit or pattern of patient compliance and self-care if family, friends, colleagues, employers, teachers, and health professionals fail to encourage or at least condone the behavior. The next four sections provide annotations of works that have addressed these reinforcing factors from the standpoint of assessing their influence on patient behavior and patient adaptation to illness.

Patient education that fails to take these social forces beyond the patient into account can expect only short-term changes in behavior. In addition to the annotated references in this chapter, the following recent reviews of literature and conceptual papers provide a background for the study of this aspect of patient education:

Becker, M.H., and Maiman, L.A. "Sociobehavioral Determinants of Compliance with Health and Medical Care Recommendations." MEDICAL CARE 13 (1975): 10-23.

Cobb, S. "Social Support as a Moderator of Life Stress." PSYCHOSOMATIC MEDICINE 38 (1976): 300-314.

Green, L.W. "Educational Strategies to Improve Compliance with Therapeutic and Preventive Regimens: The Recent Evidence." In COMPLIANCE IN HEALTH CARE, edited by R.B. Haynes et al., pp. 157-73. Baltimore: Johns Hopkins University Press, 1979.

Green, L.W., and Green, P.F. "Intervening in Social Systems to Make Smoking Education More Effective." In SMOKING AND HEALTH: HEALTH CONSEQUENCES, EDUCATION, CESSATION ACTIVITIES, AND GOVERNMENT ACTION, edited by J. Steinfeld et al., pp. 393-401. Proceedings of the Third World Conference on Smoking and Health. USDHEW Publication no. (NIH) 77-1413. Washington, D.C.: Government Printing Office, 1977.

Langlie, J.K. "Social Networks, Health Beliefs, and Preventive Health Behavior." JOURNAL OF HEALTH AND SOCIAL BEHAVIOR 18 (1977): 244-60.

Mechanic, D. "Illness Behavior, Social Adaptation, and the Management of Illness: A Comparison of Educational and Medical Models." JOURNAL OF NERVOUS AND MENTAL DISEASE 165 (1977): 79-87.

Minkler, M. "The Use of Incentives in Family Planning Programmes: A Study of Competing Theories Regarding Their Influence on Attitude Change." INTERNATIONAL JOURNAL OF HEALTH EDUCATION 19, Suppl. (1976): 1-12.

Mullen, P.D. "Cutting Back After a Heart Attack: An Overview." HEALTH EDUCATION MONOGRAPHS 6 (1978): 295-311.

Mushkin, S.J., ed. CONSUMER INCENTIVES FOR HEALTH CARE. New York: Prodist, 1974.

Norr, K.L., et al. "Explaining Pain and Enjoyment in Childbirth." JOURNAL OF HEALTH AND SOCIAL BEHAVIOR 18 (1977): 260-75.

Pratt, L. "Changes in Health Care Ideology in Relation to Self-Care by Families." HEALTH EDUCATION MONOGRAPHS 5 (1977): 121-35.

Ross, J. "Influence of Experts and Peers upon Negro Mothers of Low Socioeconomic Status." JOURNAL OF SOCIAL PSYCHOLOGY 89 (1973): 79-84.

Salloway, J.C., and Dillon, P.B. "A Comparison of Family Networks and Friend Networks in Health Care Utilization." JOURNAL OF COMPARATIVE FAMILY STUDIES 4 (1973): 131-42.

Stein, L.I., et al. "Alternative to the Hospital: A Controlled Study." AMERICAN JOURNAL OF PSYCHIATRY 132 (1975): 517-22.

M. FAMILY INFLUENCES ON PATIENT BEHAVIOR

M1 Becker, M.H., and Green, L.W. "A Family Approach to Compliance with Medical Treatment: A Selective Review of the Literature." INTERNATIONAL JOURNAL OF HEALTH EDUCATION 18 (1975): 173-82.

> This paper demonstrates the usefulness of a family approach to understanding and enhancing compliance with medical treatment. Relationships are documented between extent of patient cooperation and family members' assumption of responsibility for the sick member's care; evaluation of the illness and the recommended treatment; existing patterns of illness behavior; health beliefs, sympathy, support, and encouragement; willingness to engage in "environmental control"; compatibility of normal roles and patterns with the patient's sick role or regimen; and interspousal communication and attitudinal concordance.

M2 Bracken, M.B., et al. "The Decision to Abort and Psychological Sequelae." JOURNAL OF NERVOUS AND MENTAL DISEASE 158 (1974): 154-62.

> The importance of the level of support of significant others in the decision to abort is examined as a predictor of the reac-

tion to the abortion among a sample of 489 women aborting at a New York clinic. Reaction to the abortion was measured within an hour of the procedure using an instrument consisting of nine psychological, social, and intrapsychic items.

M3 Cline, F.W., and Rothenberg, M.B. "Preparation of a Child for Major Surgery: A Case Report." JOURNAL OF THE AMERICAN ACADEMY OF CHILD PSYCHIATRY 13 (Winter 1974): 78-94.

The experience of a seven-year-old, middle-class Caucasian boy with open-heart surgery is described. Parental conflicts, destructive parent-child interactions, and the chronicity of the illness caused a request for special preoperative psychiatric evaluation.

M4 Eyberg, S.M., and Johnson, S.M. "Multiple Assessment of Behavior Modification with Families: Effects of Contingency Contracting and Order of Treated Problems." JOURNAL OF CONSULTING AND CLINICAL PSYCHOLOGY 42 (1974): 594-606.

An evaluation of behavior modification training for parents of children with behavior problems is presented. Treatment outcome was measured by criteria designed to reflect the degree of parental cooperation and of actual changes in both attitudes and behaviors.

M5 Forehand, R., et al. "Parent Behavior Training: Effects on the Non-Compliance of a Deaf Child." JOURNAL OF BEHAVIOR THERAPY AND EXPERIMENTAL PSYCHIATRY 5 (1974): 281-83.

In this study, effects of a parent-centered behavioral training program on the noncompliance of a deaf child are examined. The program, designed to alter general parent-child interactions, initially taught the mother reinforcement skills for desirable behavior and, subsequently, a time-out procedure for deviant behavior.

M6 Gliva, G.E., and Lesser, A.L. "Involving the Family in the Case Conference." CANADA'S MENTAL HEALTH 22, no. 2 (1974): 5-8.

This paper addresses the involvement of the family in the conference process as a major intervention. The traditional case conference is reviewed, and an innovation and a primer for the implementation of this intervention are given.

M7 Huberty, D.J. "Adapting to Illness Through Family Groups." INTERNATIONAL JOURNAL OF PSYCHIATRY IN MEDICINE 5 (1974): 231-42.

This paper describes practical measures for assuring family in-

volvement in physical rehabilitation and discusses possible
methods of establishing such groups in a hospital setting.
Case examples are given in the areas of stroke, cancer, diabetes, and coronary care.

M8 Mannino, F.V., and Shore, M.F. "Family Structure, Aftercare, and Post-Hospital Adjustment." AMERICAN JOURNAL OF ORTHOPSYCHIATRY 44 (1974): 76-85.

This follow-up study of an aftercare program compares a group of former patients with a control group. It explores the effects of family structure and sex on posthospital adjustment and evaluates the relationship of family structure to treatment outcome.

M9 Mullen, P.D. "Health Education for Heart Patients in Crisis." HEALTH SERVICES REPORTS 88 (1973): 669-75.

Data from observations and interviews with myocardial infarction patients and their spouses are presented in terms of educational program applications involving the MI patients, their families, and the hospital staff.

M10 Pinkston, E.M., and Herbert-Jackson, E.W. "Modification of Irrelevant and Bizarre Verbal Behavior Using Parents as Therapists." SOCIAL SERVICE REVIEW 49 (1975): 46-63.

In two cases of bizarre and irrelevant verbalization, parents were trained to use a fifteen-second time-out procedure contingent on bizarre verbalization and to reinforce desirable behaviors. These methods were employed in a tutorial setting with two problem children who had been rejected from public school.

M11 Pratt, L. "The Significance of the Family in Medication." JOURNAL OF COMPARATIVE FAMILY STUDIES 4 (1973): 13-31.

This paper analyzes the extent of the family's involvement in medication activity and the social forces that appear to foster the family's involvement in and control over medication.

M12 Robinson, L.H. "Group Work with Parents of Retarded Adolescents." AMERICAN JOURNAL OF PSYCHOTHERAPY 28 (1974): 397-408.

A combination of education, counseling, and therapy is suggested for parents who are rejecting or fearful of the future of their retarded children and who thus overprotect and infantalize them. Advantages are cited for providing these services for groups of parents.

M13 Roghmann, K.J., et al. "Family Coping with Everyday Illness: Self

Reports from a Household Survey." JOURNAL OF COMPARATIVE FAMILY STUDIES 4 (1973): 49-62.

    This study, using data from a cross-sectional survey, focuses on the extent to which family structure and functioning affect everyday coping with illness or other disruptions.

M14   Rose, S.D. "Training Parents in Groups as Behavior Modifiers of Their Mentally Retarded Children." JOURNAL OF BEHAVIOR THERAPY AND EXPERIMENTAL PSYCHIATRY 5 (1974): 135-40.

    Natural and foster parents' groups were taught a number of principles of behavior modification and how to apply them to their retarded children. Group leaders used programmed booklets, modeling and behavior rehearsal, lecturing and discussion, weekly assignments, and positive reinforcement.

M15   Schaefer, J.W., et al. "Group Counseling for Parents of Hyperactive Children." CHILD PSYCHIATRY AND HUMAN DEVELOPMENT 5 (Winter 1974): 89-94.

    This article describes a procedure in which parents of hyperactive children learned how to make and enforce rules and influence their children's behavior, using the principles of learning theory.

M16   Sternlicht, M., and Sullivan, I. "Group Counseling with Parents of the MR: Leadership Selection and Functioning." MENTAL RETARDATION 12 (October 1974): 11-13.

    This paper describes the roles and functions of the leader of a group of parents of the mentally retarded. The concept of several leaders is explored, and diverse types of group objectives are considered.

M17   Tarver, J., and Turner, A.J. "Teaching Behavior Modification to Patient's Families." AMERICAN JOURNAL OF NURSING 74 (1974): 282.

    Behavior therapy was selected as a treatment for psychiatric patients and subsequently taught to their family members in a six-session training period.

M18   Tavormina, J.B. "Basic Models of Parent Counseling: A Critical Review." PSYCHOLOGICAL BULLETIN 81 (1974): 827-35.

    This article defines the structure of parent counseling procedures as therapeutic strategies for behavior problems in children. It evaluates research evidence on the effectiveness of two basic counseling models: The behavioral and the reflective. Analyses of design, methodology, and outcome from each method are reviewed.

M19 Wiig, E.H. "Counseling the Adult Aphasic for Sexual Readjustment."
REHABILITATION COUNSELING BULLETIN 16 (1973): 110-19.

> This article presents observations of sexual readjustment problems and counseling experiences in one hundred ambulatory adults with chronic aphasia of various degrees and types and with various educational backgrounds, life-styles, and ages.

N. PEER INFLUENCES ON PATIENT BEHAVIOR

N1 Alterman, A.I., et al. "Social Modification of Drinking by Alcoholics." QUARTERLY JOURNAL OF STUDIES ON ALCOHOL 35 (1974): 917-24.

> During a six-week experimental treatment program in which patients could choose to drink, forty-four male alcoholics attended group discussion prior to the drinking decisions phase of the program. It is concluded that social reinforcement of abstinence within a group setting can effectively reduce the number of patients drinking.

N2 Erickson, R.C., and Hyerstay, B.J. "The Dying Patient and the Double-bind Hypothesis." OMEGA--JOURNAL OF DEATH AND DYING 5 (1974): 287-98.

> The double-bind hypothesis is applied to the communication patterns of significant others surrounding the dying patient. The potential for psychologically destructive social interactions is documented by drawing parallels with the schizophrenogenic double-bind situation.

N3 Hagberg, B., and Malmquist, A. "A Prospective Study of Patients in Chronic Hemodialysis IV: Pretreatment Psychiatric and Psychological Variables Predicting Outcome." JOURNAL OF PSYCHOSOMATIC RESEARCH 18 (1974): 315-19.

> Twenty-three patients with chronic renal failure, who were psychiatrically and psychologically evaluated before beginning hemodialysis, were studied. Psychiatric and psychological data were combined to predict rehabilitation ability. Prognostically favorable variables were regular social contacts, adequate reaction toward kidney disease, expectation of fast rehabilitation, and selected defense mechanisms.

N4 Lindberg, F.H., et al. "Group Therapy with Hospitalized Patients: Increasing Therapeutic Interaction Using a Feedback-Escape Technique." SMALL GROUP BEHAVIOR 5 (1974): 486-94.

> A group of seven chronic, hospitalized mental patients was measured for the total amount of time spent in a leaderless

group until they reached a criterion of forty-five minutes of work, as measured by the Hill Interaction Matrix. They were compared under two conditions: a baseline condition where no feedback was given and an experimental condition where feedback was accompanied by an escape condition.

N5 Wagner, P. "Children Tutoring Children." MENTAL RETARDATION 12 (October 1974): 52-55.

> This paper reviews literature of tutorial models with non-mentally retarded youngsters. Relevant retarded-tutoring-retarded literature is reviewed in the second part of the article.

N6 Webster, M., Jr., and Sobieszek, B.I. SOURCES OF SELF-EVALUATION: A FORMAL THEORY OF SIGNIFICANT OTHERS AND SOCIAL INFLUENCE. New York: Wiley-Interscience, 1974. 189 p.

> This book reviews a large body of classical literature on the self: how individuals come to hold evaluations of self and how they act on those evaluations. It presents recent experimental data, tests several versions of the original concepts, and demonstrates the building of formal theory using experiments in a step-by-step process.

O. OTHER SOCIAL SUPPORTS AND REWARDS FOR PATIENT BEHAVIOR

O1 Cassell, J.L. "The Function of Humor in the Counseling Process." REHABILITATION COUNSELING BULLETIN 17 (1974): 240-45.

> This article develops an awareness of the concept of humor as a significant behavioral response for understanding the client and focusing on his or her problem. The usefulness of humor in monitoring, in therapeutic processes, and in the coping process in general is discussed.

O2 Chiles, J.A. "A Practical Therapeutic Use of the Telephone." AMERICAN JOURNAL OF PSYCHIATRY 131 (1974): 1030-31.

> The use of the telephone is described as a planned, often daily, part of psychotherapy in which positive behavior is reinstated. Two cases and a discussion of the technique involved are presented.

O3 Deliege, D. "The Sociological Framework Surrounding Inpatients." INTERNATIONAL NURSING REVIEW 21 (1974): 16-20.

> Conflicts as an aspect of the sociological setting around inpatients are discussed in relation to their impact on the emotional life and health of patients. Sources of conflict,

general as well as specific, in hospital life are analyzed, and psychosociological techniques for resolving conflicts are discussed.

O4 French, A.P., and Tupin, J.P. "Therapeutic Application of a Simple Relaxation Method." AMERICAN JOURNAL OF PSYCHOTHERAPY 28 (1974): 282-87.

This paper describes the use of a relaxation method by patients with serious medical problems. The method consists of muscular relaxation followed by use of a pleasant, relaxing memory as a center of attention for meditationlike effect.

O5 Hawley, B.P. "The First Dental Visit for Children from Low Socioeconomic Families." JOURNAL OF DENTISTRY FOR CHILDREN 41, no. 5 (1974): 46-51.

The first dental appointment of black and white, low socioeconomic children in a child and youth center of a large university hospital was used: to establish an incidence of behavior problems for children and youth center children; reduce the incidence of behavior problems by writing a preappointment letter to the child emphasizing that the visit would be an enjoyable experience; reduce the frequency of broken appointments by such a letter; and correlate behavior problems with the mother's dental knowledge, anxiety level, and expectations for the child's first dental visit.

O6 Hornstra, R.K., and Lubin, B. "Relationship of Outcome of Treatment to Agreement About Treatment Assignment by Patients and Professionals." JOURNAL OF NERVOUS AND MENTAL DISEASE 158 (1974): 420-23.

Applicants to an urban community mental health center were asked what was the best possible treatment they could have right now. Intake clinicians subsequently made assignments without knowledge of the applicants' responses to the question. Three different categorizations of the match or nonmatch between treatment preferred and treatment assigned were not significantly related at three-month patient interviews.

O7 Hulka, B.S., et al. "Correlates of Satisfaction and Dissatisfaction with Medical Care: A Community Perspective." MEDICAL CARE 13 (1975): 648-58.

Attitudes toward physicians and medical services were surveyed among residents of a probability sample of households in a city of 200,000 people. The attitude questionnaire was completed by 1,713 adults in 1,112 households. Professional competency, personal qualities, and accessibility (including cost and convenience) were rated.

O8 Jenny, J., et al. "Explaining Variability in Caries Experience Using an Ecological Model." JOURNAL OF DENTAL RESEARCH 53 (1974): 554-64.

> A model including diet, oral hygiene, and dental treatment and three ecological levels (community, family, individual) was tested to study variability in caries experience.

O9 Kaplan, De-Nour, A., and Czaczkes, J.W. "Team-Patient Interaction in Chronic Hemodialysis Units." PSYCHOTHERAPY AND PSYCHOSOMATICS 24 (1974): 132-36.

> Findings are presented on team-patient interaction and its influence on patient behavior.

O10 Levitz, L.S., and Stunkard, A.J. "A Therapeutic Coalition for Obesity: Behavior Modification and Patient Self-Help." AMERICAN JOURNAL OF PSYCHIATRY 131 (1974): 423-27.

> Sixteen chapters of TOPS (Take Off Pounds Sensibly), with a total of 234 members, received one of four treatments: behavior modification conducted by a professional therapist, behavior modification conducted by the TOPS leader, nutrition education conducted by the TOPS leader, and continuation of the usual TOPS program. Positive results of lower attrition and greater weight loss were attributed to the behavior modification treatment.

O11 Mahoney, M.J., and Thoreson, C.E. SELF-CONTROL: POWER TO THE PERSON. Belmont, Calif.: Wadsworth Publishing Co., 1974. 368 p.

> This text, written for professionals and students, presents an overview of self-control theory and research. Part 1 discusses basic principles and processes in self-regulation. Social learning theory is reviewed, and a functional behavioral model for learning the skill of self-regulation is proposed. The authors describe and evaluate several self-control strategies and discuss the relevance of behavioral self-control for humanistic goals. Part 2 contains fourteen selected resource articles, representing both theoretical and applied approaches to self-control.

O12 Peterson, J.M., and Farley, F.H. "Concept Learning, Informative Feedback, and Individual Differences in Motivation." JOURNAL OF GENERAL PSYCHOLOGY 90 (1974): 179-86.

> Resultant Achievement Motivation scores for sixty female college students were computed from achievement motivation and test anxiety measures. Scores obtained were used as

predictors of performance under conditions of differing rates of informational feedback in a concept-learning task.

O13 Rutzen, S.R. "The Social Importance of Orthodontic Rehabilitation: Report of a Five Year Follow-Up Study." JOURNAL OF HEALTH AND SOCIAL BEHAVIOR 14 (1973): 233-39.

> Five years after finishing orthodonic treatment for malocclusions, 250 persons were interviewed and compared to 67 persons with untreated malocclusions. The research was to determine differences that might be due to orthodontic treatment in social rank variables, courtship, self-concept, and personality measures.

O14 Shafii, M., et al. "Meditation and Marijuana." AMERICAN JOURNAL OF PSYCHIATRY 131 (1974): 60-63.

> The authors, using a questionnaire survey, found that the longer a person had practiced meditation, the more likely it was that he or she had decreased or stopped the use of marijuana.

O15 Suchotliff, L., and Meyer, S. "The Educational Process as a Treatment Modality in a Drug Rehabilitation Program." AMERICAN JOURNAL OF PSYCHIATRY 132 (1975): 195-97.

> An education program that was an integral part of a residential drug rehabilitation program is described and advocated as a treatment modality to supplement and complement other aspects of the therapeutic community by providing vocational skills acquisition.

O16 Suedfeld, P., and Ikard, F.F. "Use of Sensory Deprivation in Facilitating the Reduction of Cigarette Smoking." JOURNAL OF CONSULTING AND CLINICAL PSYCHOLOGY 42 (1974): 889-95.

> Twelve months after a twenty-four-hour period in a socially isolated, monotonous environment, subjects had reduced their rate of cigarette smoking by an average of 48 percent compared with 16 percent for control subjects. Sensory deprivation is viewed as a facilitator of long-term behavioral change in human beings.

O17 Thies, A.P., and Chance, J. "Potential Losses Versus Potential Gains as Determinants of Behavior." JOURNAL OF PSYCHOLOGY 89 (1975): 81-88.

> Rotter's hypothesis of the power of the negative sign of anticipated reinforcement to change psychological situations to elicit generalized expectancies leading to avoidant behavior, was tested on ten-year-old school boys.

O18 U.S. Department of Health, Education, and Welfare. National Institute of Mental Health, Alcohol, Drug Abuse, and Mental Health Administration. A SOCIAL WORK GUIDE FOR LONG-TERM CARE FACILITIES. DHEW Publication no. HSM 73-9156. Rockville, Md.: 1974. vi, 216 p. Bibliog.

> The major focus of this book is social work services to ease problems faced by the aging and their families when long-term care seems indicated. Emphasis is placed on interview and assessment procedures that can alleviate the stress of entry into a facility and help staff provide the care most suited to each person's needs. The volume also deals with services involved in aftercare, special needs of the dying, planning, research and training programs, and administration.

O19 Visintainer, M.A., and Wolfer, J.A. "Psychological Preparation for Surgical Pediatric Patients: The Effect of Children's and Parent's Stress Responses and Adjustment." PEDIATRICS 56 (1975): 187-202.

> A clinical experiment tested variations of psychological preparation and supportive care designed to increase the adjustment of children (and their parents) hospitalized for elective surgery. Eighty-four children, aged three to twelve, admitted for tonsillectomies, were assigned randomly to one of three treatment conditions or to a control group. Posthospital adjustment was assessed, using Vernon et al.'s Post-Hospital Behavior Inventory. Parent outcome measures included self-ratings for anxiety and satisfaction with information and care.

O20 Volicer, B.J. "Patients' Perceptions of Stressful Events Associated with Hospitalization." NURSING RESEARCH 23 (1974): 235-38.

> Hospital patients on cancer, surgical, and medical wards were asked to rate forty-five stress-producing events related to the experience of their hospitalizations in terms of the relative amount of adaptation required to cope with each event, using an arbitrary standard. A high consensus about the order of events was found among patients from the three wards.

O21 Wiesenthal, D.L. "Some Effects of the Confirmation and Disconfirmation of an Expected Monetary Reward on Compliance." JOURNAL OF SOCIAL PSYCHOLOGY 92 (1974): 39-52.

> This study tested the dissonance theory prediction of an energization of either compliance or noncompliance and the social-exchange-theory predictions of conformity as a direct function of pay level. Subjects had their expectation of pay for serving in a conformity experiment disconfirmed by receiving amounts either greater or lesser than were expected, while control subjects received their expected pay.

Factors Reinforcing Patient Behavior

O22 Wright, B.A. "Changes in Attitudes Toward People with Handicaps." REHABILITATION LITERATURE 34 (1973): 354-58.

> This review cites examples of increased emphasis on human and civil rights and other social development for people with physical and mental handicaps.

O23 Yen, S. "Availability of Activity Reinforcers in a Drug Abuse Clinic: A Preliminary Report." PSYCHOLOGICAL REPORTS 34, no. 3, pt. 1 (1974): 1021-22.

> To determine the availability of effective reinforcers of activity that could be delivered in an outpatient drug abuse clinic, a survey was administered to twenty-five methadone maintenance patients. The privilege of taking methadone (not yet part of the program) had a low reinforcing property.

P. ATTITUDES AND BEHAVIOR OF HEALTH PROFESSIONALS TOWARD PATIENTS

P1 Abramson, R., and Block, B. "Ego-Supportive Care in Open-Heart Surgery." PSYCHIATRY IN MEDICINE 4 (1973): 427-37.

> Treatment for remission of psychopathological phenomena in patients undergoing open-heart surgery is described. Ego-supportive measures include (1) establishment of a relationship, (2) reassurance, (3) environmental support, (4) consultative relationships with staff, (5) ventilation of feelings, (6) medical measures in support of somatic function, and (7) antianxiety drugs.

P2 Aiba, F.H., et al. "The Alcoholic Patient-Nurse Relationship Viewed in a Transcultural Context." JAPANESE PSYCHOLOGICAL RESEARCH 15 (July 1973): 82-91.

> The self-concept, self-ideal, perceived self, and role expectation of seventy-three nurses and seventy-three alcoholic patients chosen from three populations (rural Japan, urban Japan, and California) were measured by questionnaire. The purpose was to clarify and explore the alcoholic patient-nurse interrelationship on a transcultural, sociological, and psychological level.

P3 Annas, G.J., and Healey, J.M., Jr. "The Patient Rights Advocate: Redefining the Doctor-Patient Relationship in the Hospital Context." VANDERBILT LAW REVIEW 27 (1974): 243-69.

> Rights of patients and problems with the traditional model of the doctor-patient relationship in the hospital context are discussed. The authors advocate a complete statement defining

patient rights, both those legally recognized and those granted as a matter of hospital policy, and adoption of a patient rights advocate system in hospitals.

P4   Becker, M.H., et al. "A Field Experiment To Evaluate Various Outcomes of Continuity of Physician Care." AMERICAN JOURNAL OF PUBLIC HEALTH 64 (1974): 1062-70.

This study examines outcomes from two organization methods for the provision of ambulatory health and medical care. Using a field experiment design, patient families and medical staff of a large children and youth project were assigned randomly to a conventional sequential clinic. Emphasis was placed on continuity of records and linking the patient with the first available physician or to four medical panels, each with one pediatrician and a permanent supportive staff, and on physician continuity by having the patient see only his or her assigned doctor on each return visit.

P5   Bernstein, L., et al. INTERVIEWING: A GUIDE FOR HEALTH PROFESSIONALS. 2d ed. New York: Appleton-Century-Crofts, 1974. xx, 197 p. Bibliog. Index.

The contents of this volume include: the relationship between the health professional and the patient; an overview of interviewing techniques; evaluative, hostile, reassuring, probing, and understanding responses; emotional reactions to illness and treatment; and death and dying.

P6   Beutler, L.E., et al. "Outcomes in Group Psychotherapy: Using Persuasion Theory To Increase Treatment Efficiency." JOURNAL OF CONSULTING AND CLINICAL PSYCHOLOGY 42 (1974): 547-53.

Some research suggests that improvement in psychotherapy is related to the degree that a patient adopts his or her therapist's evaluative attitudes. This article pursues the possibility of predicting the outcomes of group psychotherapy using attitude theory.

P7   Bowden, C.L., et al. PSYCHOSOCIAL BASIS OF MEDICAL PRACTICE --AN INTRODUCTION TO HUMAN BEHAVIOR. Baltimore: Williams & Wilkins Co., 1974. 229 p.

This book includes topics on the therapeutic relationship; the physician, the patient, and the diagnostic process; interview skills; patterns of defense and adaptation; the angry patient; the anxious, tearful, and depressed patient; the hypochondriacal patient; and the denying patient.

P8   Bracken, M.B., et al. "Abortion Counseling: An Experimental Study of Three Techniques." AMERICAN JOURNAL OF OBSTETRICS AND GYNECOLOGY 117 (1973): 10-20.

> One hundred and seventy-one abortion patients assigned randomly to group orientation, group process, and individual counseling procedures were matched to a specific counselor. Data collected included circumstances surrounding the pregnancy, reaction to the counseling session, and response to the abortion.

P9 Browning, P.L., et al. "Counseling Process with Mentally Retarded Clients: A Behavioral Exploration." AMERICAN JOURNAL OF MENTAL DEFICIENCY 79 (1974): 292-96.

> Behavioral aspects of the counseling process with mentally retarded clients are examined. Transcripts from an early and a late counseling session were rated according to the initial client statement, therapist response, and client continuation of the topic.

P10 Bryson, S., and Cody, J. "Relationship of Race and Level of Understanding Between Counselor and Client." JOURNAL OF COUNSELING PSYCHOLOGY 20 (1973): 495-98.

> This study examined the relationship between race and the level of understanding between counselor and client. Inter- and intraracial differences in understanding during an initial counseling interview also were investigated.

P11 Canfield, R.E. "The Physician as a Teacher of Patients." JOURNAL OF MEDICAL EDUCATION 48, no. 12, pt. 2 (1973): 79-87.

> This article advocates developing a new primary and secondary school curriculum in human biology; establishing models in medical schools for the teaching of patients, at least about specific disease; and exploring the potential of focusing the resources of the university on the synthesis of new programs for health consumers about common psychological problems and the disorders that they create.

P12 Carter, D.K., and Pappas, J.P. "Systematic Desensitization and Awareness Treatment for Reducing Counselor Anxiety." JOURNAL OF COUNSELING PSYCHOLOGY 22 (1975): 147-51.

> This study compares the effects on reducing counselor anxiety of systematic desensitization with an awareness treatment designed to increase awareness of interpersonal anxiety, and with no treatment.

P13 Corbus, H.F., and Connell, R.W. "The Patient's Needs--Does Anyone

Care?" HOSPITALS--JOURNAL OF THE AMERICAN HOSPITAL ASSOCIATION 48 (16 January 1974): 46-49.

> A review by a patient care committee indicated a need for a hospital to be concerned with the extra-clinical aspects of patient care. Hospital personnel were found to be concerned with the patient's welfare and to respond enthusiastically to opportunities to improve the patient's well being.

P14   D'Augelli, A.R. "Nonverbal Behavior of Helpers in Initial Helping Interactions." JOURNAL OF COUNSELING PSYCHOLOGY 21 (1974): 360-63.

> This study examines the importance of nonverbal behavior in effective helping. Several nonverbal behaviors of helpers in a small group were tallied and related to independent judgments of the helper made by observers and the person being helped. Helpee-rated understanding and warmth were correlated with frequency of helper nodding.

P15   Dorroh, T. BETWEEN PATIENT AND HEALTH WORKER. New York: McGraw-Hill, 1974. 262 p. Bibliog.

> This book is designed to improve patient care by promoting better interpersonal relationships between health workers and patients; by increasing the health worker's knowledge of and sensitivity to the feelings and attitudes of patients; and to help him or her become aware of his or her own feelings toward illness, the health care system, and other people. Topics include the needs of patients' families and patients' attitudes toward illness.

P16   Etzwiler, D.D. "The Contract for Health Care." JOURNAL OF THE AMERICAN MEDICAL ASSOCIATION 224 (1973): 1034.

> The advantages of formalized medical contracts are outlined. The author suggests that failure to provide patient education and include the patient as a member of the health care team may constitute malpractice.

P17   Fletcher, S.W., et al. "Management of Hypertension: Effect on Improving Patient Compliance for Follow-up Care." JOURNAL OF THE AMERICAN MEDICAL ASSOCIATION 233 (1975): 242-44.

> A randomized, controlled study was carried out to determine whether the addition of a follow-up clerk to the emergency room staff would improve compliance among emergency room patients requiring follow-up for nonurgent conditions. Analysis of compliance according to sociodemographic characteristics showed that this intervention was associated with higher compliance, regardless of age, race, sex, or marital or employment status.

Factors Reinforcing Patient Behavior

P18   Frankenberg, R. "Functionalism and After? Theory and Developments in Social Science Applied to the Health Field." INTERNATIONAL JOURNAL OF HEALTH SERVICES 4 (1974): 411-27.

        It is suggested that sociology could have a totalizing theoretical function in relation to medicine. The solution is seen in identification with patients and an honest acceptance of class conflict and contradiction.

P19   Fuller, D.S., and Quesada, G.M. "Communication in Medical Therapeutics." JOURNAL OF COMMUNICATION 23 (1973): 361-70.

        This paper discusses characteristics of effective doctor-patient communication, the process of progressive failure of communication, and the manner in which such a process may be reversed. The need for teaching communication principles in programs of medical education is discussed.

P20   Fulton, M., et al. "Helping Diabetics Adapt to Failing Vision." AMERICAN JOURNAL OF NURSING 74 (1974): 54-57.

        The responsibility of nurses in helping diabetic patients maintain independence and take responsibility for their own care is described.

P21   Glogow, E. "The Bad Patient Gets Better Quicker." SOCIAL POLICY 4 (November-December 1973): 72-76.

        The practice of thinking of the compliant patient as the good patient and the noncompliant patient as bad is questioned on the grounds that the compliant patient is expressing a feeling of powerlessness. The author points out that the desired outcome of the treatment process is the patient's recovery, and health workers should be sensitized to the fact that confusing compliance with being a good patient and noncompliance with being a bad patient may hinder the therapeutic process.

P22   Gump, L.R. "Counselor Self-Awareness and Counseling Effectiveness." COUNSELOR EDUCATION AND SUPERVISION 13 (1974): 263-66.

        This study was conducted to determine if there were differences in perceived effectiveness between counselors who said they related self-awareness to their counseling role and counselors who believed self-awareness was unrelated to their role.

P23   Haar, E., et al. "Factors Related to the Preference for a Female Gynecologist." MEDICAL CARE 13 (1975): 782-90.

        Female patients (n = 409) of male and female physicians completed a self-administered questionnaire exploring their attitudes and practices regarding gynecologists and gynecological examinations.

P24   Hendershot, G.E., and Grimm, J.W. "Abortion Attitudes Among Nurses and Social Workers." AMERICAN JOURNAL OF PUBLIC HEALTH 64 (1974): 438-41.

    This study examines attitudes toward abortion among samples of nurses and social workers in Tennessee.

P25   Hilton, B., and Callahan, D., eds. ETHICAL ISSUES IN HUMAN GENETICS: GENETIC COUNSELING AND THE USE OF GENETIC KNOWLEDGE. New York: Plenum, 1973. 460 p.

    This volume explores issues arising in the field of human genetics. The topics covered are the promise and potential danger of cloning and synthetic gene construction, the meaning of genetic disease for individuals and populations, possible genetic criteria for a right of life, privacy in genetic counseling, and control of application of genetic knowledge.

P26   Kayser, J.S., and Minnigerode, F.A. "Increasing Nursing Students' Interest in Working with Aged Patients." NURSING RESEARCH 24 (1975): 23-26.

    The Tuckman-Lorge Attitude questionnaire, which measures stereotypes and misconceptions about the aged, was administered to 311 baccalaureate nursing students. The nurses were also asked to indicate their relative preferences for various fields of specialization within nursing and their preferences for working with child, adult, and elderly patients.

P27   Kilty, K.M. "Attitudes Toward Alcohol and Alcoholism Among Professionals and Nonprofessionals." JOURNAL OF STUDIES ON ALCOHOL 36 (1975): 327-47.

    Analysis of the responses of graduate students, professional service agency workers, and community residents to scales measuring attitudes and beliefs about alcohol and alcoholism is described. Results are discussed in terms of the usefulness of the multidimensional model of attitude structure for comparing attitudes among various groups and the implications for professional and community education programs.

P28   Knox, W.J. "Attitudes of Psychology Graduate Students Toward Drug Abuse." PROFESSIONAL PSYCHOLOGY 5 (1974): 185-90.

    The attitudes of psychology trainees of the Veteran's Administration were assessed on topics such as working with drug offenders, hospital vs. prison as a treatment environment, financial benefits for drug abusers, use of methadone in treatment, and the general relevance of graduate education today.

P29   Kupst, M.J., et al. "Evaluation of Methods To Improve Communication in the Physician-Patient Relationship." AMERICAN JOURNAL OF ORTHOPSYCHIATRY 45 (1975): 420-29.

> Parents of children with congenital heart disease were participants in this experiment to assess effects of communication methods on retention, anxiety, and satisfaction in regard to information received from physicians.

P30 Lally, J.J., and Barber, B. "The Compassionate Physician: Frequency and Social Determinants of Physician-Investigator Concern for Human Subjects." SOCIAL FORCES 53 (1974): 289-96.

> The importance of the compassion of the physician investigator for human subjects and his or her professional expertise and other individual qualifications in protecting subjects' rights and welfare are discussed. Data from interviews with 337 research physicians provided a basis for refinement and operationalization of the concept "compassion." Frequency of these physician's general concern for the actual subjects of their investigations was estimated.

P31 Lester, D., et al. "Attitudes of Nursing Students and Nursing Faculty Toward Death." NURSING RESEARCH 23 (1974): 50-53.

> A questionnaire was used to investigate attitudes toward death and dying of 128 undergraduates, 66 graduate nursing students, and 62 nursing faculty at a university school of nursing.

P32 Lillie, D.C. "Doctor-Patient Interaction." GERONTOLOGIA CLINICA 16 (1974): 44-53.

> The author contends that there are no uncooperative patients but rather many patients whose cooperation the doctor fails to secure. Cultivating personal qualities and involvement with and compassion for patients are suggested for physicians.

P33 Linn, L.S., and Davis, M.S. "Occupational Orientation and Overt Behavior: The Pharmacist as Drug Adviser to Patients." AMERICAN JOURNAL OF PUBLIC HEALTH 63 (1973): 502-8.

> The study tested the hypothesis that business-oriented pharmacists, as compared with professionally oriented pharmacists, would be more likely to recommend medication to patients than to refer them to physicians. Characteristics of economic interest or social position did not seem relevant factors in the type of advice pharmacists rendered and did not support the hypothesis under investigation.

P34 McIntosh, J. "Processes of Communication, Information Seeking and Control Associated with Cancer: A Selective Review of the Literature."

SOCIAL SCIENCE AND MEDICINE 8 (1974): 167-87.

> This paper reviews the literature on communication processes associated with malignant disease in the hospital setting. The questions considered are: what do doctors tell cancer patients and why, how much do patients want to know, and how do they obtain information?

P35 McKinlay, J.B. "Who is Really Ignorant--Physician or Patient?" JOURNAL OF HEALTH AND SOCIAL BEHAVIOR 16 (1975): 3-11.

> An investigation is reported of actual and perceived comprehension of frequently used medical terms by lower working-class users and underusers of maternity services in Scotland. Findings are consistent with some earlier studies in the United States.

P36 Mazzullo, J.M., et al. "Variations in Interpretation of Prescription Instructions: The Need for Improved Prescribing Habits." JOURNAL OF THE AMERICAN MEDICAL ASSOCIATION 227 (1974): 929-31.

> Instructions on each of ten prescription labels were interpreted by sixty-seven patients. Misinterpretations illustrated the need for physicians to provide medication instructions consistent with the patient's daily activities and to review the instructions with the patient.

P37 Mitchell, J.A. "Public Perceptions of the Responsibility of Consumers and Providers to the Quality of Health Care Relative to Cancer." HEALTH EDUCATION MONOGRAPHS 1, no. 36 (1973): 34-39.

> The author discusses the diagnosis and treatment of cancer in relation to informed consent for treatment and physicians' witholding of diagnoses, options for treatment, and prognoses from cancer patients.

P38 Ort, R.S. "Effects of Psychological Factors on the Delivery and Utilization of Health Care Services." PROFESSIONAL PSYCHOLOGY 5 (1974): 91-94.

> This paper deals with the self-perceptions and attitudes of doctors toward their work and clientele and with the health attitudes and behavior of the poor.

P39 Pender, N.J. "Patient Identification of Health Information Received During Hospitalization." NURSING RESEARCH 23 (1974): 262-67.

> Interviews with 162 patients indicated amount, type, and sources of information received about their health problems during hospitalization. Patients reported a need for more information before discharge on how to care for themselves at

Factors Reinforcing Patient Behavior

home, the effect of illness on their daily living habits, possible complications of their present illness, and prevention of future illness.

P40 Ramsden, E.L. "The Patient's Right to Know: Implications for Interpersonal Communication Processes." PHYSICAL THERAPY 55 (1975): 133-38.

The right of a patient to have information about his or her diagnosis, treatment, and prognosis within the framework of communication processes is discussed from the perspectives of the professional, the patient, and the interaction process. The traditional role of the patient and changes in his or her role are described.

P41 Reader, G.G. "The Physician as Teacher." HEALTH EDUCATION MONOGRAPHS 2 (1974): 34-38.

The author advocates physician use of the hospitalization period as an opportunity to engage patients in a long-term program of health education. Methods of reinforcing health education in general as well as for specific illnesses and the use of a written plan for meeting patient needs after discharge are suggested.

P42 Richards, R.F. "Patients Are Learning." HEALTH EDUCATION MONOGRAPHS 2 (1974): 30-33.

Describing the change in predominance of hospital admissions from acute to long-term care patients and changes in hospital services, the author points up the need for encouragement of patients' involvement in their own care and recovery.

P43 Riggs, R.C. "Attitudes of Counselor Trainees Toward Three Client Groups." REHABILITATION COUNSELING BULLETIN 18 (1974): 78-82.

Biases of rehabilitation personnel were questioned in a sample of forty-one counselor trainees. Preferences among clients from three client groups were given: the culturally deprived black, the ex-mental patient, and the ex-convict, for each of twenty-five situations, on a forced-choice questionnaire.

P44 Rosen, R.A.H., et al. "Health Professionals' Attitudes Toward Abortion." PUBLIC OPINION QUARTERLY 38 (1974): 159-73.

Attitudes toward abortion, as obtained in a 1971 nationwide survey, are presented for students and faculty in nursing, medicine, and social work, and are compared to attitudes of the general population. Results reveal implications for the abortion-related services provided by health professionals.

P45 Rosenzweig, S.P., and Folman, R. "Patient and Therapist Variables Affecting Premature Termination in Group Psychotherapy." PSYCHOTHERAPY: THEORY, RESEARCH AND PRACTICE 11 (1974): 76-79.

> To understand premature termination in group psychotherapy, this study investigated three types of behavior hypothesized to influence the therapy relationship: patient personality variables as measured by psychological tests, patient demographic data, and therapists' judgments and attitudes. The relationship between therapists' attitudes towards patients and outcome of therapy was a particular focus.

P46 Schweer, S.F., and Dayani, E.C. "The Extended Role of Professional Nursing: Patient Education." INTERNATIONAL NURSING REVIEW 20 (1973): 174-75.

> This paper examines existing problems in the patient education provided by professional nurses and offers possibilities for clarifying the nurse's role in this area.

P47 Shaw, D.W., and Thoreson, C.E. "Effects of Modeling and Desensitization in Reducing Dentist Phobia." JOURNAL OF COUNSELING PSYCHOLOGY 21 (1974): 415-20.

> To eliminate dental avoidance behavior and reduce stress, this study explored the effects of systematic desensitization and social modeling treatments with placebo and assessment control groups. The importance of demonstrating behaviors coupled with covert practices (self-modeling) is discussed.

P48 Shusterman, L.R., and Sechrest, L. "Attitudes of Registered Nurses Toward Death in a General Hospital." PSYCHIATRY IN MEDICINE 4 (1973): 411-26.

> A death anxiety questionnaire for six conceptually distinct aspects of attitudes toward death was developed from the work of previous investigators. It was administered, along with other measures, to 188 hospital nurses.

P49 Sorenson, J.R. "Biomedical Innovation, Uncertainty, and Doctor-Patient Interaction." JOURNAL OF HEALTH AND SOCIAL BEHAVIOR 15 (1974): 366-74.

> The doctor-patient relationship is examined in relation to biomedical innovations and patients who are health decision-makers. One area of medicine, clinical genetics, is examined, with particular emphasis on types of doctor-patient exchanges and the social forces that condition these relationships.

P50 Sowa, P.A., and Cutter, H.S. "Attitudes of Hospital Staff Toward

Alcoholics and Drug Addicts." QUARTERLY JOURNAL OF STUDIES ON ALCOHOL 35 (1974): 210-14.

> A sample of staff members of the Washingtonian Center for Addictions, consisting of thirty-three low status (clerical), thirty middle status (nurses), and nineteen high status (psychiatrists) personnel, completed a questionnaire derived from the Gough Adjective Check List, choosing from adjectives with positive and negative connotations as they apply to alcoholics and drug addicts.

P51 Steward, M., and Regalbuto, G. "Do Doctors Know What Children Know?" AMERICAN JOURNAL OF ORTHOPSYCHIATRY 45 (1975): 146-49.

> Informed about cognitive development by Piaget's theory, children of preschool and elementary age were asked to use two common pediatric tools and to explain how they function, to test their understanding of the information given them.

P52 Wallston, K.A., and Wallston, B.S. "Nurses' Decisions to Listen to Patients." NURSING RESEARCH 24 (1975): 16-22.

> A methodology using role playing responses by nurses to simulate patient disclosures was tested in this investigation. Four simulated patients with diagnoses of diabetes mellitus, alcoholism with bleeding ulcer, ulcerative colitis, and cancer of the large intestine tape-recorded twenty- to thirty-second segments on twelve topics pertaining to their illness, both physical and psychological problems. Nurses' willingness to listen and to pass along information to the next nurse was tested.

P53 White, C. "Patient Characteristics and Supportive Behavior of Nursing Personnel in Nursing Homes." Dr. P.H. dissertation, Johns Hopkins University School of Hygiene and Public Health, 1973.

> This study represents a partial test of a conceptual model derived from previous research indicating that selected patient characteristics were associated with preferences of nursing personnel for certain patients. Findings indicate differences in the behavior of nursing persons and suggest that social characteristics of patients account for some of those differences in behavior, independent of the amount of time the nursing person spends with the patient.

Chapter 7
COMMUNICATIONS THEORY AND PRACTICE

The subject of communications is so broad as to be potentially encompassing of every form of social interaction. Our selection of representative works for annotation here is conditioned by our professional judgment as to the applicability and utility of the work for patient education. Much has been written on communications in the past decade by and for health professionals, but much more is written by and for others with unintended but undeniable relevance for patient education.

Communications theory and practice, as outlined by the selections here, are the core of patient education as viewed by most health providers and by patients. Indeed, many physicians and nurses would not regard the material covered in previous chapters as relevant or appropriate concerns of patient education. Some clinicians would regard the activities covered in the next chapters as administrative or academic concerns beyond the scope of patient education, leaving only the direct communication with patients as the proper focus of patient education.

This book is organized on the premise that communications with patients should be planned on the basis of a prior assessment of the health problem, the specific behaviors contributing to the health problem or required to alleviate or control the problem, and a careful educational diagnosis of the predisposing, enabling, and reinforcing factors influencing those behaviors. Only then can the appropriate communication content, method, and medium be selected with any certainty. Only then can a claim be made that the communication has the purpose and potential of improving health.

Furthermore, direct communication with patients is only part of a patient education program. There is a responsibility of the communicator to be concerned with the physical and social barriers confronting the behavior advocated in communications with the patient. Such barriers must be addressed through communication with people other than the patient--family members, employers, administrators of resources needed by the patient, and health professionals. These can be viewed as indirect channels of communications with patients insofar as these significant others ultimately will support or disparage the communications patients have received directly from their health care providers. Subsequent chapters will deal with these broader aspects of patient education.

Our review of the methods and content of sixty-seven published "evaluations" of patient education for the years 1974 to 1978 revealed that 23 percent of the programs defined success solely on the criterion of increase in patients' knowledge.* Examples of those using pre- and post-educational knowledge scores on written questionnaires are the following:

Black, L.F., and Mitchell, M.M. "Evaluation of a Patient Education Program for Chronic Obstructive Pulmonary Disease." MAYO CLINIC PROCEEDINGS 52 (1977): 106-11.

Caron, H.S., and Roth, H.P. "An Evaluation of a Program for Teaching Clinic Patients the Rationale of Their Peptic Ulcer Regimen." HEALTH EDUCATION MONOGRAPHS 5 (1977): 25-49.

Hassell, J., and Medved, E. "Group/Audiovisual Instruction for Patients with Diabetes: Learning Achievements and Time Economics." JOURNAL OF THE AMERICAN DIETETIC ASSOCIATION 66 (1975): 465-70.

Rahe, R.H., et al. "A Teaching Evaluation Questionnaire for Postmyocardial Infarction Patients." HEART AND LUNG 4 (1975): 759-66.

Teuscher, A., and Heidecker, B. "Evaluation of an Instruction Programme on Diabetes Diet by Means of a Teaching Machine." MEDICAL EDUCATION 10 (1976): 508-11.

Examples of patient education programs that have used a combination of knowledge gain and behavioral change as criteria of success are:

Kay, R.L., and Hammond, A.H. "Understanding Rheumatoid Arthritis: Evaluation of a Patient Education Program." JOURNAL OF THE AMERICAN MEDICAL ASSOCIATION 239 (1978): 2466-67.

Lawson, V.K., et al. "An Audio-Tutorial Aid for Dietary Instruction in Renal Dialysis." JOURNAL OF THE AMERICAN DIETETIC ASSOCIATION 69 (1976): 390-96.

Sly, R.M. "Evaluation of a Sound-Slide Program for Patient Education." ANNALS OF ALLERGY 34 (1975): 94-97.

Some have used a combination of knowledge gain and attitudinal change as criteria of success:

Alkhateeb, W., et al. "A Comparison of Three Educational Techniques Used in a Venereal Disease Clinic." PUBLIC HEALTH REPORTS 90 (1975): 159-64.

Laugharne, E., and Steiner, G. "Tri-Hospital Diabetes Education Centre: A Cost-Effective, Cooperative Venture." CANADIAN NURSE 73 (1977): 113-19.

*This review was supported by NIH research training grant HL07180 at Johns Hopkins University and is forthcoming in L.W. Green et al. "What Are Recent Evaluations of Patient Education Efforts Telling Us?" SECOND NATIONAL SYMPOSIUM ON PATIENT EDUCATION, PROCEEDINGS, edited by W. Squyres. New York: Springer, 1979 (in press).

Among the patient education evaluations, some were designed specifically to test the effectiveness of a specific medium or channel of communication, e.g., the studies by Sly and Alkhateeb, et al., above, and:

Bracken, M.B., et al. "Patient Education by Videotape After Myocardial Infarction: An Empirical Evaluation." ARCHIVES OF PHYSICAL MEDICINE REHABILITATION 58 (1977): 213-19.

Solfin, D., et al. "Development and Evaluation of an Individualized Patient Education Program About Digoxin." AMERICAN JOURNAL OF HOSPITAL PHARMACY 34 (1977): 367-71.

Vignos, P.J., et al. "Evaluation of a Clinic Education Program for Patients with Rheumatoid Arthritis." JOURNAL OF RHEUMATOLOGY 3 (1976): 155-65.

Communication strategies in patient education are designed sometimes to reduce anxiety or increase satisfaction with health care rendered. Examples of programs that have had these, among other objectives are:

Lazes, P.M. "Health Education Project Guides Outpatients to Active Self-Care." HOSPITALS: JOURNAL OF THE AMERICAN HOSPITAL ASSOCIATION 51 (15 February 1977): 81-86.

Pozen, M.W., et al. "A Nurse Rehabilitator's Impact on Patients with Myocardial Infarction." MEDICAL CARE 15 (1977): 830-37.

Roter, D.L. "Patient Participation in the Patient-Provider Interaction: The Effects of Patient Question Asking on the Quality of Interaction, Satisfaction and Compliance." HEALTH EDUCATION MONOGRAPHS 5 (1977): 281-315.

Salzer, J.E. "Classes To Improve Diabetic Self-Care." AMERICAN JOURNAL OF NURSING 75 (1975): 1324-26.

Wallace, N., and Wallace, D.C. "Group Education After Myocardial Infarction: Is It Effective?" MEDICAL JOURNAL OF AUSTRALIA 2 (1977): 245-47.

Examples of other evaluations that have given particular attention to behavioral and biomedical changes resulting from communication strategies in patient education programs are the following:

Altshuler, A., et al. "Even Children Can Learn To Do Clean Self-Catheterization." AMERICAN JOURNAL OF NURSING 77 (1977): 97-101.

Bryant, N.H., et al. "VD Hotline: An Evaluation." PUBLIC HEALTH REPORTS 91 (1976): 231-35.

D'Altroy, L., et al. "Patient Drug Self-Administration Improves Regimen Compliance." HOSPITALS: JOURNAL OF THE AMERICAN HOSPITAL ASSOCIATION 52 (1978): 131-36.

Flegle, J.M. "Teaching Self-Dialysis to Adults in a Hospital." AMERICAN JOURNAL OF NURSING 77 (1977): 270-72.

Flowers, R.V. EFFECTS OF SOCIAL SUPPORT ON ADHERENCE TO THERA-

PEUTIC REGIMENS. Ph.D. dissertation, University of Michigan of Ann Arbor, 1978.

Green, L.W., et al. "Clinical Trials of Health Education for Hypertensive Outpatients: Design and Baseline Data." PREVENTIVE MEDICINE 4 (1975): 417-25.

_____. "Research and Demonstration Issues in Self-Care: Measuring the Decline of Medicocentrism." In CONSUMER SELF-CARE IN HEALTH, edited by J. Gallicchio, pp. 20-26. Research Proceedings Series. DHEW Pub. No. (HRA) 77-3181. Washington, D.C.: National Center for Health Services Research, August 1977.

Haynes, R.B., et al. "Improvement of Medication Compliance in Uncontrolled Hypertension." LANCET 4 (1976): 1265-68.

Jesudasan, K., et al. "An Evaluation of the Self-Administration of DDS in Gudiyatham Taluk." LEPROSY INDIA 48 (1976): 668-76.

Johnston, B.L., et al. "Eight Steps to Inpatient Cardiac Rehabilitation: The Team Effort-Methodology and Preliminary Results." HEART AND LUNG 5 (1976): 97-111.

Jones, R.J., et al. "The Educational Diagnosis in Nutrition Counseling for Serum Cholesterol Reduction." In PROCEEDINGS OF THE NUTRITION-BEHAVIORAL RESEARCH CONFERENCE, edited by J. Tillotston, pp. 41-44. Bethesda: National Heart, Lung and Blood Institute, 1975.

Kirscht, J.P., et al. "Effects of Threatening Communication and Mothers Health Beliefs on Weight Change in Obese Children." JOURNAL OF BEHAVIORAL MEDICINE 1 (1978): 147-57.

Latos, D.L., et al. "Home Dialysis Program of the Nashville VA Hospital." SOUTHERN MEDICAL JOURNAL 70 (1977): 1431-35, 1439.

Levine, D.M., et al. "Health Education for Hypertensive Patients." JOURNAL OF THE AMERICAN MEDICAL ASSOCIATION 241 (1979): 1700-1703.

Maiman, L.A., et al. "Education for Self-Treatment by Adult Asthmatics." JOURNAL OF THE AMERICAN MEDICAL ASSOCIATION 241 (1979): 1919-22.

Radius, S.M., et al. "Factors Influencing Mothers' Compliance with a Medication Regimen for Asthmatic Children." JOURNAL OF ASTHMA RESEARCH 15 (1978): 133-47.

Steckel, S.B., and Swain, M. "Contracting with Patients To Improve Compliance." HOSPITALS: JOURNAL OF THE AMERICAN HOSPITAL ASSOCIATION 51 (1 December 1977): 81-84.

Witschi, J.D., et al. "Family Cooperation and Effectiveness in a Cholesterol-Lowering Diet." JOURNAL OF THE AMERICAN DIETETICS ASSOCIATION 72 (1978): 384-89.

Ziesat, H.A. "Behavior-Modification in Treatment of Hypertension." INTERNATIONAL JOURNAL OF PSYCHOLOGICAL MEDICINE 8 (1978): 257-65.

Communications Theory & Practice

These examples of patient education programs were listed here by the types of outcomes they featured. Their titles reflect the variety of communication strategies used. In the following four sections, additional works are annotated under the types of communication strategies used (interpersonal, small group, audiovisual or other educational technology, and behavioral modification).

The relative effectiveness of different communication methods appears to depend largely on the circumstances and the enthusiasm or novelty of their application. This leaves the appropriate selection of patient education methods dependent on considerations of cost, logistics, and feasibility. These considerations have been reviewed in the following recent works:

American Hospital Association. HOSPITAL INPATIENT EDUCATION: SURVEY FINDINGS AND ANALYSES. Atlanta: DHEW, Public Health Service, Center for Disease Control, 1977.

_____. "Research Capsule No. 7." HOSPITALS: JOURNAL OF THE AMERICAN HOSPITAL ASSOCIATION 46 (1 July 1977): 102.

American Public Health Association, Public Health Education Section, Committee on Educational Tasks in Chronic Illness. A MODEL FOR PLANNING PATIENT EDUCATION. Rockville, Md.: Health Resources Administration, 1972.

Bloom, J.R. "Hypertension Control Through Design of Targeted Delivery Models." PUBLIC HEALTH REPORTS 93 (1978): 35-40.

Fudge, R.P., and Vlasses, P.H. "Third-Party Reimbursement for Pharmacist Instruction About Anti-Hemophilic Factor." AMERICAN JOURNAL OF HOSPITAL PHARMACY 34 (1977): 831-34.

Gardner, M.E., and Trinca, C.E. "The Pharmacy Clinic: A New Approach to Ambulatory Care." AMERICAN JOURNAL OF HOSPITAL PHARMACY 35 (1978): 429-31.

Gillum, R.F., and Barsky, A.J. "Diagnosis and Management of Patient Non-Compliance." JOURNAL OF THE AMERICAN MEDICAL ASSOCIATION 228 (1974): 1563-67.

Gillum, R.F., et al. "Improving Hypertension Detection and Referral in an Ambulatory Setting." ARCHIVES OF INTERNAL MEDICINE 138 (1978): 700-703.

Green, L.W. "Determining the Impact and Effectiveness of Health Education as it Relates to Federal Policy." HEALTH EDUCATION MONOGRAPHS 6, Suppl. (1978): 28-66.

_____. "Educational Strategies To Improve Compliance with Therapeutic and Preventive Regimens: The Recent Evidence." In COMPLIANCE IN HEALTH CARE, edited by R.B. Haynes et al., pp. 157-73. Baltimore: Johns Hopkins University Press, 1979.

Green, L.W., et al. HEALTH EDUCATION PLANNING: A DIAGNOSTIC APPROACH. Palo Alto: Mayfield Publishers, 1979.

_____. "How Cost-Effective Are Smoking Cessation Methods?" WORLD SMOKING AND HEALTH 3 (1978): 33-40.

Harris, C.L. "Hospital-Based Patient Education Programs and the Role of the Hospital Librarian." BULLETIN OF THE MEDICAL LIBRARY ASSOCIATION 66 (1978): 210-17.

Hart, L.K., and Frantz, R.A. "Characteristics of Postoperative Patient Education Programs for Open Heart Surgery Patients in the U.S." HEART AND LUNG 6 (1977): 137-42.

Redman, B.K. "Patient Teaching." NURSING DIGEST 6 (Spring 1978): entire issue.

_____. THE PROCESS OF PATIENT TEACHING IN NURSING. 2d ed. St. Louis: C.V. Mosby Co., 1976.

Rosenberg, S., and Judkins, B.A. "Federal Programs Make Education an Integral Part of Patient Care." HOSPITALS: JOURNAL OF THE AMERICAN HOSPITAL ASSOCIATION (1 May 1976): 62-65.

Ruley, E.J. "Compliance in Young Hypertensive Patients." PEDIATRIC CLINICS OF NORTH AMERICA 25 (1978): 175-82.

Simmons, J.J., ed. "Making Health Education Work." AMERICAN JOURNAL OF PUBLIC HEALTH 65, Suppl. (October 1975): vi, 1-49.

Slepcevitch, E., comp. RX: EDUCATION FOR THE PATIENT: WHO, WHAT, WHERE, WHY. . . AND AT WHAT COST? Conference proceedings, June 25-26, 1974. Carbondale: Southern Illinois University, Department of Health Education, April 1975. 129 pp. Append.

Werlin, S.H., and Schauffler, H.H. "Structuring Policy Development for Consumer Health Education." AMERICAN JOURNAL OF PUBLIC HEALTH 68 (1978): 596-97.

Young, M.A.C. "Reviews of Research and Studies on Health Education (1961-66): Patient Education." HEALTH EDUCATION MONOGRAPHS 1, no. 26 (1968): 1-64.

Q. THEORIES AND METHODS OF INTERPERSONAL COMMUNICATIONS

Q1   Arnold, W.E. "The Effect of Nonverbal Cues on Source Credibility." CENTRAL STATES SPEECH JOURNAL 24 (1973): 227-30.

> The purpose of this study was to measure the effects of nonverbal cues on the credibility of a source and to determine the influence of variations in a speaker's physical appearance on his or her credibility.

Q2   Beebe, S.A. "Eye Contact: A Nonverbal Determinant of Speaker Credibility." SPEECH TEACHER 23 (1974): 21-25.

> This study investigates the importance of direct eye contact to a speaker's perceived credibility in a live public-speaking situation. Some support for the theory that eye contact is an

important delivery characteristic in establishing speaker credibility is provided.

Q3 Britton, M. "Should Relatives Be Informed that Autopsy Is Intended? Opinions of Relatives with Recent Experience." SCANDANAVIAN JOURNAL OF SOCIAL MEDICINE 1, no. 1 (1974): 81-90.

Relatives of deceased patients were informed that autopsy was intended, and 278 relatives subsequently were interviewed to investigate whether the notification procedure had positive consequences for them or whether they felt it could be omitted. Suggestions are offered concerning the procedure for informing relatives of an intended autopsy.

Q4 Broll, L., and Gross, A.E. "Effects of Offered and Requested Help on Help Seeking and Reactions To Being Helped." JOURNAL OF APPLIED SOCIAL PSYCHOLOGY 4 (1974): 244-58.

Subjects in this study could receive assistance on a difficult logic problem by requesting help or by accepting an offer of aid from a helper. Negative consequences of requesting help are interpreted in terms of attribution theory, and implications of the findings for help-delivery systems are discussed.

Q5 Burleson, G. "Modeling: An Effective Change Technique for Teaching Blind Persons." THE NEW OUTLOOK 10 (1973): 433-69.

Bandura's modeling techniques were applied in a case study of a mentally retarded, blind, thirty-eight-year-old institutionalized, severely anxious male. Apparent significant decreases in anxiety and increases in social and interpersonal interactions resulted.

Q6 Carlson, K.W. "Increasing Verbal Empathy as a Function of Feedback and Instruction." COUNSELOR EDUCATION AND SUPERVISION 13 (1974): 208.

Twenty-four counselor-trainees were assigned randomly to three experimental groups and one control group: immediate feedback, feedback and instructions, equipment present, and control. An FM-radio system enabled supervisors to communicate to the counselor-trainees in the experimental groups. Audio-tapes were rated on empathy before and after the treatment.

Q7 Cialdini, R.B., et al. "Reciprocal Concessions Procedure for Inducing Compliance: The Door-in-the-Face Technique." JOURNAL OF PERSONALITY AND SOCIAL PSYCHOLOGY 31 (1975): 206-15.

Three experiments were conducted to test the effectiveness of a rejection-then-moderation procedure for inducing compliance

Communications Theory & Practice

with a request for a favor. The three included a condition in which a requester first asked an extreme favor (which was refused) and then for a smaller favor. In each instance, this procedure produced more compliance with the smaller favor than a procedure in which the requester asked solely for the smaller favor.

Q8 Cuskey, W.R., and Premkumar, T. "A Differential Counselor Role Model for the Treatment of Drug Addicts." HEALTH SERVICES REPORTS 88 (1973): 663-68.

A model is presented for counseling drug addicts during differing treatment phases. Cost benefit estimates of matching counselor skill to patient need are given.

Q9 Drumheller, S.J. "Some Sticky Wickets in the Behavioral Objective Game--Looking at Learning Curves." JOURNAL OF EDUCATIONAL TECHNOLOGY SYSTEMS 2 (Summer 1973): 39-47.

The relationship between behavioral objectives and learning curves is examined. Havinghurst's development tasks are presented as a model for providing better learning retention, and the science of psychometrics is discussed briefly.

Q10 Fletcher, C.M. COMMUNICATION IN MEDICINE. London: Nuffield Provincial Hospitals Trust, 1973. ix, 121 p.

Contents of part 1 of this book include communication between individuals in the National Health Service, communication with patients, communication in hospitals, and communication between the three divisions of the National Health Service. Part 2 discusses communication with the public about medicine.

Q11 Gregory, I.D. "A New Look at the Lecture Method." BRITISH JOURNAL OF EDUCATIONAL TECHNOLOGY 6 (January 1975): 55-61.

A definition of the lecture method, a list of uses, the method's strengths and weaknesses, and suggestions for effective lecturing are presented.

Q12 Gross, A.E., et al. "Beneficiary Attractiveness and Cost as Determinants of Responses to Routine Requests for Help." SOCIOMETRY 38 (1975): 131-40.

An experiment to determine effects of beneficiary attractiveness and cost on responses to a routine request for help is described. The results are contrasted with previous studies that failed to find a positive relationship between interpersonal attraction and helping.

Q13 Hammerman, S.R. COMMUNICATIONS IN REHABILITATION. REPORT OF THE INTERNATIONAL EXPERTS MEETINGS ON COMMUNICATIONS IN REHABILITATION. New York: Rehabilitation International, 1974. 163 p.

> A meeting of experts in the communications field and their counterparts in the field of rehabilitation focused on three areas of rehabilitation activity: treatment and counseling of disabled persons, training of rehabilitation personnel, and public education toward supporting rehabilitation programs from private and governmental sources.

Q14 Insko, C.A., et al. "Facilitative and Inhibiting Effects of Distraction on Attitude Change." SOCIOMETRY 37 (1974): 508-28.

> Subjects in this study listened to a persuasive communication while working on a task and were told either to attend primarily to the communication or to the task. There was also a message-only condition. On the basis of the results, it is argued that different mediational models are operating for message-set distraction and task-set distraction.

Q15 Kaplan, M.F., and Anderson, N.H. "Information Integration Theory and Reinforcement Theory as Approaches to Interpersonal Attraction." JOURNAL OF PERSONALITY AND SOCIAL PSYCHOLOGY 28 (1973): 301-12.

> Two theoretical approaches to interpersonal attraction are compared: Anderson's information integration theory and Byrne's reinforcement theory.

Q16 Kinsella, N.A. "Some Psychological Dimensions of the Trusting Attitude." HUMANITAS 9 (1973): 253-71.

> Experimental studies on trust and on personality theory and trust are reviewed.

Q17 Kopel, S.A., and Arkowitz, H.S. "Role Playing as a Source of Self-Observation and Behavior Change." JOURNAL OF PERSONALITY AND SOCIAL PSYCHOLOGY 29 (1974): 677-86.

> In this study, role playing was used as the vehicle for self-perception to test the hypothesis that, within a perceived-choice paradigm, role playing an upset reaction to electric shocks would lead to subsequent decreases in pain and tolerance thresholds, whereas role playing a calm reaction to electric shocks would lead to subsequent increases. Results are discussed in terms of self-perception theory and clinical applications.

Q18 Kriss, M., et al. "Message Type and Status of Interactants as Determinants of Telephone Helping Behavior." JOURNAL OF PERSONALITY AND SOCIAL PSYCHOLOGY 30 (1974): 856-59.

> The wrong telephone number technique was used to investigate the emotional tone or impact of the help-seeking message as a variable mediating the help seeker and potential benefactor interaction.

Q19 Labov, W. LANGUAGE IN THE INNER CITY: STUDIES IN THE BLACK ENGLISH VERNACULAR. Pennsylvania Paperback Series, 51. Philadelphia: University of Pennsylvania Press, 1972.

> Studying not only the normal processes of communication in the inner city but also such art forms as the ritual insult and ritualized narrative, Labov advances arguments for the existence of the black vernacular as a separate and independent dialect of English, with its own internal logic and grammar. His analysis of this vernacular could be helpful in patient communications.

Q20 McGarry, J., and Hendrick, C. "Communicator Credibility and Persuasion." MEMORY AND COGNITION 2 (1974): 82-86.

> This paper reports an experiment designed to test the hypothesis that the perceived vested interest of a speaker, the position the speaker advocates, and the social similarity between audience and speaker will influence attributions of credibility and affect the speaker's persuasiveness.

Q21 McGinnies, E. "Initial Attitude, Source Credibility, and Involvement as Factors in Persuasion." JOURNAL OF EXPERIMENTAL AND SOCIAL PSYCHOLOGY 9 (1973): 285-96.

> Source credibility was manipulated in factorial combination with the measured variables of initial attitude, issue involvement, and sex, using Japanese university students as subjects. Results indicate greater resistance of highly involved subjects to persuasion under extreme rather than under moderate communicator-recipient discrepancy.

Q22 McLachlan, J.F.C. "Therapy Strategies, Personality Orientation and Recovery from Alcoholism." CANADIAN PSYCHIATRIC ASSOCIATION JOURNAL 19 (1974): 25-30.

> The conceptual level matching model to predict the response of alcoholics to group psychotherapy was tested in a follow-up study of drinking behavior. Patients "matched" to their therapists showed higher recovery rates than "mismatched" patients.

Q23 Marks, S.E., et al. "Cognitive Flexibility and Communication of

Therapeutic Conditions." PSYCHOLOGICAL REPORTS 34 (1974): 486.

This study investigated the relationship of cognitive flexibility to ability to communicate the therapeutic conditions of accurate empathy, nonpossessive warmth, and genuineness.

Q24 Mettlin, C., and Woelfel, J. "Interpersonal Influence and Symptoms of Stress." JOURNAL OF HEALTH AND SOCIAL BEHAVIOR 15 (1974): 311-19.

Three fundamental features of the influence process believed pertinent to stress are delineated: discrepancy among influences, level of influence, and number of influence sources. Implications for future studies of stress and the influence process are noted.

Q25 Michener, H.A., and Burt, M.R. "Components of 'Authority' as Determinants of Compliance." JOURNAL OF PERSONALITY AND SOCIAL PSYCHOLOGY 31 (1975): 606-14.

This study investigates factors affecting compliance to orders from a formal authority.

Q26 Miller, G.R., and Burgoon, M. NEW TECHNIQUES OF PERSUASION. New York: Harper & Row, 1973. 120 p.

Three concepts in the field of persuasion theory are presented: inducing resistance to persuasion, role playing, and counter-attitudinal advocacy. The application of the techniques and ethical implications are discussed.

Q27 Missett, M.A. "The Effect of Specific Educational Input by the Care Provider on the Antepartal Patients' Perception of Themselves as Co-Participants in Their Care." Master of Public Health thesis, Johns Hopkins University School of Hygiene and Public Health, 1974. 100 p.

This research measured the extent to which a specific health education activity, the care provider's explanation of certain management procedures for the antepartal care of pregnant women during the clinic process, altered the patient's attitude of nonparticipation.

Q28 Nash, H. "Perception of Vocal Expression of Emotion by Hospital Staff and Patients." GENETIC PSYCHOLOGY MONOGRAPHS 89 (1974): 25-87.

Using responses of patients on admission to medical and psychiatric wards of a general hospital and responses of hospital staff, this study describes the ability to perceive emotional communication from tone of voice independent of content. Patient's improvement in this skill was measured at hospital discharge.

Q29 Norton, R.W., and Miller, L.D. "Dyadic Perception of Communication Style." COMMUNICATION RESEARCH 2 (1975): 50-67.

> This experiment was designed to assess variations in perception of communication style by persons with differing Communication Style Measure scores.

Q30 Olson, D.R. "What Is Worth Knowing and What Can Be Taught." SCHOOL REVIEW 82 (November 1973): 27-43.

> This essay discusses the role that experience, both direct and vicarious, plays in education and the communication process and the kind of bias it puts upon both the means and ends of education.

Q31 Orkow, B.M., and Ross, J.L. "Weight Reduction Through Nutrition Education and Personal Counseling." JOURNAL OF NUTRITION EDUCATION 7 (1975): 65-67.

> A nutritionist and a social worker who led weight reduction groups for women for one year assess the value of having nutrition education and counseling techniques taught in tandem.

Q32 Packwood, W.T. "Loudness as a Variable in Persuasion." JOURNAL OF COUNSELING PSYCHOLOGY 21 (1974): 1-2.

> A persuasion scale that assesses counselor conviction and client agreement was used to rate 900 counselor statements in determining that loudness is a characteristic of persuasion.

Q33 Pearce, W.B. "Trust in Interpersonal Communication." SPEECH MONOGRAPHS 41 (1974): 236-44.

> This paper reviews three relevant literatures on the concept of trust and presents a conceptualization of trust that draws from and extends them. A bias for explaining similar and dissimilar communication behaviors is provided.

Q34 Pliner, P., et al. "Compliance Without Pressure: Some Further Data on the Foot-in-the-Door Technique." JOURNAL OF EXPERIMENTAL AND SOCIAL PSYCHOLOGY 10 (1974): 17-22.

> A replication of the Freedman and Fraser (1966): "foot-in-the-door" technique was attempted in which subjects were exposed to one of two prior requests and then asked to comply with a larger request. The mechanism by which the technique operates is discussed.

Q35 Reddy, W.B. "The Impact of Sensitivity Training on Self-Actualization: A One-Year Follow-up." SMALL GROUP BEHAVIOR 4 (1973): 407.

This investigation explores two areas of sensitivity training: the stability of changes in self-actualization over time and the role of anxiety as it relates to these changes.

Q36 Rees, S. "No More than Contact: An Outcome of Social Work." BRITISH JOURNAL OF SOCIAL WORK 4 (1974): 255-79.

This study of the experiences of eight sets of clients with their respective social workers illustrates how the outcome of each case is a reflection of each party's interpretations of crucial aspects of their meetings.

Q37 Reiss, D., et al. "Assimilating the Patient Stranger." JOURNAL OF NERVOUS AND MENTAL DISEASE 158 (1974): 118-41.

This paper reports an experimental study of interaction between psychiatric staff and patients. The influence of the professional value orientation and the personality of the staff members on assimilating the patient into a working relationship was studied.

Q38 Roberts, T.B. "Transpersonal: The New Educational Psychology." PHI DELTA KAPPAN 56 (1974): 191-93.

The author discusses his theory that transpersonal psychology that deals with altered states of consciousness, man's impulse to higher states of being, psychic phenomena, biofeedback, and voluntary control of internal states will lead to new educational understandings and practices.

Q39 Rosen, S., et al. "Interactive Effects of New Valence and Attraction on Communicator Behavior." JOURNAL OF PERSONALITY AND SOCIAL PSYCHOLOGY 28 (1973): 298-300.

The relationship between attractiveness of the recipient and a communicator's behavior in transmitting good or bad news is explored.

Q40 Samaan, M.K., and Parker, C.A. "Effects of Behavioral (Reinforcement and Advice-Giving Counseling on Information-Seeking Behavior." JOURNAL OF COUNSELING PSYCHOLOGY 20 (1973): 193-201.

This article compares the differential effectiveness of reinforcement and persuasive advice-giving to increase information seeking. Behavioral reinforcement counseling was found to be superior to advice giving.

Q41 Snyder, M., and Cunningham, M.R. "To Comply or not Comply: Testing the Self-Perception Explanation of the 'Foot-in-the-Door' Phenomenon." JOURNAL OF PERSONALITY AND SOCIAL PSYCHOLOGY 31 (1975): 64-67.

This field experiment tested the self-perception explanation of the "foot-in-the-door" phenomenon of increased compliance with a substantial request after prior compliance with a smaller demand.

Q42 Steele, C.M., and Ostrom, T.M. "Perspective Mediated Attitude Change: When Is Indirect Persuasion More Effective than Direct Persuasion?" JOURNAL OF PERSONALITY AND SOCIAL PSYCHOLOGY 29 (1974): 737-41.

The prediction that a persuasion-induced shift for one attitude issue will mediate indirect attitude change toward sharing a comparable reference scale was supported in two experiments. Consistency models of indirect influence are discussed as explanations of indirect change in excess of direct change.

Q43 Stokes, S.J., and Bickman, L. "The Effect of the Physical Attractiveness and Role of the Helper on Help Seeking." JOURNAL OF APPLIED PSYCHOLOGY 4 (1974): 286-94.

The effects of a helper's physical attractiveness and role on help-seeking behavior were investigated.

Q44 Stone, V.A., and Hoyt, J.L. "The Emergence of Source-Message Orientation as a Communication Variable." COMMUNICATION RESEARCH 1 (1974): 89-109.

In two experiments, persons were classified as source-oriented by a bipolar test to support the prediction that source likability would increase attitude change in relation to health truisms (toothbrushing and chest X-rays) more for source-oriented than for message-oriented persons.

Q45 Tagliacozzo, D.M., et al. "Nurse Intervention and Patient Behavior: An Experimental Study." AMERICAN JOURNAL OF PUBLIC HEALTH 64 (1974): 596-603.

This study reports the consequences of one type of nurse intervention for selected attitudes and behaviors of black clinic outpatients. It examines the effects of nurse teaching on patient compliance as measured by prolonged clinic attendance, regularity of attending scheduled visits, compliance with the request to take medication, and compliance with requests for laboratory tests and visits to other clinics.

Q46 Tedeschi, J.T., et al. CONFLICT, POWER, AND GAMES: THE EXPERIMENTAL STUDY OF INTERPERSONAL RELATIONS. Chicago: Aldine, 1973. 280 p.

This book gives an account of power and influence in interpersonal situations of conflict and examines the use of experimental games as a research strategy. The authors clarify a wide range of concerns in experimental social psychology-exchange theory, persuasive communication, trust, leadership, coalition theory, modeling and imitation, proxemics, and other topics. Recent literature on research and theory is reviewed.

Q47 Van Rooijen, L. "Talking About the Bright Side: Pleasantness of the Referent as a Determinant of Communication Accuracy." EUROPEAN JOURNAL OF SOCIAL PSYCHOLOGY 3 (1973): 473-78.

The importance of the pleasant-unpleasant dimension (of facial expressions) was investigated in a referential communication setting. Special attention was paid to possible differential performance between male and female respondents.

Q48 Veninga, R. "The Management of Conflict." JOURNAL OF NURSING ADMINISTRATION 3 (July-August 1973): 12-16.

Suggestions are given for analyzing interpersonal conflict, diminishing defensiveness, confronting problems, and determining courses of action to achieve more effective working relationships.

Q49 Ward, C.D., and McGinnies, E. "Perception of a Communicator's Credibility as a Function of When He Is Identified." PSYCHOLOGICAL RECORD 23 (Fall 1973): 561-62.

This article gives possible explanations for the sequence effect, the tendency for a communicator to be judged as more credible when he or she is identified after the persuasive communication rather than before.

Q50 Ward, C.D., and McGinnies, E. "Persuasive Effects of Early and Late Mention of Credible and Noncredible Sources." JOURNAL OF PSYCHOLOGY 86 (1974): 17-23.

A standard persuasive communication was presented to 248 subjects in an attitude-change experiment. The credibility of the communication's source was varied (low versus high) along with the sequence in which the credibility information was presented (before the persuasive communication versus after). The implications of these results for understanding the effects of source credibility are discussed.

Q51 Weiss, R.L., and Swearingen, R.V. CHAIRSIDE PSYCHOLOGY IN PATIENT EDUCATION: A SELF-INSTRUCTION COURSE. San Francisco: U.S. Department of Health, Education, and Welfare, National Institutes of Health, Dental Health Center, 1973. 200 p.

This course, prepared specifically for the practicing dentist, discusses some general principles of education and the significance of each in the educational situation in private dental practice. The educational process is explained generally and step-by-step, and factors that influence the educational process are discussed.

Q52 Whittaker, J.K. SOCIAL TREATMENT: AN APPROACH TO INTER-

PERSONAL HELPING. Chicago: Aldine, 1974. 280 p.

> This book provides an introduction to interpersonal helping in the context of social work practice. It develops a conceptual framework for interpersonal helping called social treatment, to enable the social worker and members of other helping professions to use effectively the various methods and strategies currently practiced. A conceptual framework for practice that allows for systematic eclecticism in theory and technique is developed.

Q53 Young, M.E. FAMILY COUNSELING (A BIBLIOGRAPHY WITH ABSTRACTS). Springfield, Va.: National Technical Information Service, 1978. 80 p.

> This bibliography contains eighty abstracts on advising and counseling families of physically or socially handicapped persons, training of counselors, and counseling effectiveness.

R. SMALL GROUP COMMUNICATIONS IN PATIENT SETTINGS

R1 Astrachan, B.M. "Learning Theory and a Social Systems Perspective." JOURNAL OF APPLIED BEHAVIORAL SCIENCE 10 (1974): 175-79.

> This article comments on two papers dealing with learning theory models in group therapy. Whether the art of therapy can be examined scientifically and what behaviors ought to be studied are issues raised.

R2 Balch, P., and Ross, A.W. "A Behaviorally Oriented Didactic-Group Treatment of Obesity: An Exploratory Study." JOURNAL OF BEHAVIOR THERAPY AND EXPERIMENTAL PSYCHIATRY 5 (1974): 239-43.

> This exploratory study reports on the application of a behaviorally oriented didactic-group approach to the treatment of obesity with a diverse population of employees and patients in a medical school setting. The applicability of this program to the increasing numbers of obese and the possibility of paraprofessionals leading such programs are discussed.

R3 Bednar, R.L., et al. "Empirical Guidelines for Group Therapy: Pretraining, Cohesion, and Modeling." JOURNAL OF APPLIED BEHAVIORAL SCIENCE 10 (1974): 149-79.

> This paper suggests that group therapy is reaching a level of empirical sophistication that permits the development of clinical models from empirical data. Pretherapy training, cohesion, and modeling are suggested as significant parameters of effective group treatment. Evidence defining the properties and effects of those variables are reviewed and their theoretical and practical implications are discussed.

R4   Clark, R.D., III. "Risk Taking in Groups: A Social Psychological Analysis." JOURNAL OF RISK AND INSURANCE 41 (1974): 75-92.

    This article provides a general summary of risk-taking in groups, evaluates prevalent hypotheses, and discusses implications for group decision making in managerial settings.

R5   Feldman, R.A. "Power Distribution, Integration, and Conformity in Small Groups." AMERICAN JOURNAL OF SOCIOLOGY 79 (1973): 639-64.

    Using data from sixty-one children's groups, this study examines interrelationships among peer power distribution, conformity behavior, and three modes of social integration: interpersonal, functional, and normative.

R6   Goodacre, D. "Experience of Group Work in a Rehabilitation Unit." GERONTOLOGIA CLINICA 15 (1973): 352-56.

    A hospital chaplain's (the author's) work with groups of patients within a rehabilitation unit is set against a background of a theory of groups. Illustrations from an actual group, a summary of the kind of subjects discussed over a year, and an assessment of staff attitudes and the value of group work are presented.

R7   Held, J.P., et al. "Sexual Attitude Reassessment Workshops: Effect on Spinal Cord Injured Adults, Their Partners and Rehabilitation Professionals." ARCHIVES OF PHYSICAL MEDICINE AND REHABILITATION 56 (1975): 14-18.

    Five workshops, focused on the sexuality of adults with acquired spinal cord injuries, were offered for rehabilitation professionals and spinal cord injured adults. The objectives were to assist the professional to be more helpful with others and the disabled to be more helpful to themselves. Preworkshop, postworkshop, and follow-up questionnaires evaluated the effects of the seminars upon all participants: disabled, able-bodied, professional, and nonprofessional.

R8   Jacobs, A., et al. "Anonymous Feedback: Credibility and Desirability of Structured Emotional and Behavioral Feedback Delivered in Groups." JOURNAL OF COUNSELING PSYCHOLOGY 21 (1974): 106-11.

    This article reports on the credibility of personal information selected by members of small groups to describe each other's characteristics. The information was delivered to the members by the group leader without naming the source.

R9   Karpen, M.L., and Lipke, L.A. "Sex Education as Part of an Agency's Four Week Summer Workshop for Visually Impaired Young People." NEW

OUTLOOK FOR THE BLIND 68 (1974): 260-67.

> A sex education program developed by the staff of a school for the blind, parents, and resource persons is described. Curriculum content included the physical, emotional, and social aspects of sexual maturation, grooming and hygiene, cosmetics, and drug use and abuse, and was based on expressed interests of the students. Teaching strategies were student oriented and included group discussion, group activities, problem solving, and demonstration/participation.

R10 Kilmann, P.R. "Direct and Nondirect Marathon Group Therapy and Internal-External Control." JOURNAL OF COUNSELING PSYCHOLOGY 21 (1974): 380-84.

> This study investigated whether direct and nondirect therapist techniques within a twenty-three-hour marathon format would differentially induce client shifts in locus of control.

R11 Lubin, B., and Lubin, A.W. "The Group Psychotherapy Literature: 1972." INTERNATIONAL JOURNAL OF GROUP PSYCHOTHERAPY 23 (1973): 474-513.

> A summary of the 1972 group therapy literature is organized under the headings: (1) "Group Psychotherapy," (2) "Client Populations," (3) "Intensive Small Group Experiences," and (4) "Research."

R12 McCall, R.J. "Group Therapy with Obese Women of Varying MMPI Profiles." JOURNAL OF CLINICAL PSYCHOLOGY 30 (1974): 466-70.

> This article compares the Minnesota Multiphasic Personality Inventory profiles of women who either are unable to lose weight or to maintain their weight loss for six months or more with obese women who lost weight successfully and maintained the loss. Use of the profile for predicting success in group therapy is examined.

R13 McLeish, J., et al. THE PSYCHOLOGY OF THE LEARNING GROUP. London: Hutchinson & Co., 1973. 221 p.

> Theoretical approaches to small group research are reviewed and selected empirical evidence is evaluated. Three schools of small group research, the psychoanalytic, behavioristic, and interactionist approaches, are described; and processes of teaching and learning in groups are reviewed.

R14 Ohlmeier, D., et al. "Psycho-Analytic Group Interview and Short-Term Group Psychotherapy with Post-Myocardial Infarction Patients." PSYCHIATRICA CLINICA 6 (1973): 240-49.

Group-analytical methods were used in a research project on psychodynamic personality factors and psychotherapeutic possibilities in the after-treatment of postmyocardial infarction patients.

R15 Oradei, D.M., and Waite, N.S. "Group Psychotherapy with Stroke Patients During the Immediate Recovery Phase." AMERICAN JOURNAL OF ORTHOPSYCHIATRY 44 (1974): 386-95.

This paper discusses psychosocial issues presented by members of daily group therapy sessions with hospital patients recovering from strokes. The impact of the sessions on patients, staff, and ward milieu is described.

R16 Paden, R.C., et al. "Videotape vs. Verbal Feedback in the Modification of Meal Behavior of Chronic Mental Patients." JOURNAL OF CONSULTING AND CLINICAL PSYCHOLOGY 42 (1974): 623.

This study investigated the comparative effectiveness of videotaped versus verbal feedback in therapy groups for improving deficit meal behaviors of chronic mental patients within a token economy.

R17 Paulson, M.J., et al. "Parents of the Battered Child: A Multidisciplinary Group Therapy Approach to Life-Threatening Behavior." LIFE-THREATENING BEHAVIOR 4 (1974): 18-31.

This paper reports the demographic findings and the experience of a three-year, multidisciplinary group psychotherapy program with thirty-one child-abusing families.

R18 Power, L. "New Approaches to the Old Problem of Diabetes Education." JOURNAL OF NUTRITION EDUCATION 5 (1973): 230-32.

This article suggests the use of trained paraprofessionals with professional backup for managing uncomplicated diabetic problems, instituting preventive measures, and teaching at visiting times with the diabetic patient. Several patient instruction methods used in a general hospital are described.

R19 Ribner, N.G. "Effects of an Explicit Group Contract on Self-Disclosure and Group Cohesiveness." JOURNAL OF COUNSELING PSYCHOLOGY 21 (1974): 116-20.

The effects of an explicit group contract on self-disclosure and group cohesiveness were investigated. The contract served to increase significantly both the frequency and depth of self-disclosure and to enhance the cohesiveness of the groups (i.e., attraction to the group). It had, however, the opposite effect on members' mutual liking.

R20 Rowe, W., et al. "The Relationship of Counselor Characteristics and Counseling Effectiveness." REVIEW OF EDUCATIONAL RESEARCH 45 (1975): 231-46.

> This paper examines the relationship between counselor characteristics and counseling effectiveness through a review of research since 1960. The authors comment on various approaches and present conclusions supported by available evidence.

R21 Sheridan, M.S. "Talk Time for Hospitalized Children." SOCIAL WORK 20 (1975): 40-44.

> A "Talk Time" program in a university hospital focused on discussion and encouraged children to share their fears and fantasies of the hospital experience. It also helped hospital personnel to learn about pediatric patients and to listen to their needs.

R22 Strupp, H.H., and Bloxom, A.L. "Preparing Lower-Class Patients for Group Psychotherapy: Development and Evaluation of a Role-Induction Film." JOURNAL OF CONSULTING AND CLINICAL PSYCHOLOGY 41 (1973): 373-84.

> A role-induction film was developed to improve the performance of lower-class patients in psychotherapy. Its effects were studied systematically under field conditions.

R23 Tavormina, J.B. "Relative Effectiveness of Behavioral and Reflective Group Counseling with Parents of Mentally Retarded Children." JOURNAL OF CONSULTING AND CLINICAL PSYCHOLOGY 43 (1975): 22-31.

> This study evaluated the relative effectiveness of behavioral and reflective group parent counseling with fifty-one mothers of mentally retarded children. Subjects were assigned to behavioral, reflective, or waiting-list control groups. Six success criteria, including direct observations, attitudinal scale, maternal reports, and frequency counts, were used to measure outcome.

R24 Wilson, C.J., et al. "Time-Limited Group Counseling for Chronic Home Hemodialysis Patients." JOURNAL OF COUNSELING PSYCHOLOGY 21 (1974): 376-79.

> The effects of six sessions of group counseling on nine chronic, home hemodialysis patients are compared with a treatment control group. Comparisons between the experimental and control groups were made for Rotter's locus of control scale and for selected California Personality Inventory scales.

S. EDUCATIONAL MEDIA AND TECHNOLOGY

S1 Abelle, B.E. "The Teaching-Learning Implications of Educational Technology." JOURNAL OF RISK AND INSURANCE 40 (1973): 607-15.

> Instructional technology is described as a system incorporating communication media (hardware such as radio and videotape), learning theory, human skills, and various instructional settings. The systems approach considers learner objectives, availability of communications media, instructional design, and cost effectiveness.

S2 Bashshur, R.L., et al., eds. TELEMEDICINE: EXPLORATIONS IN THE USE OF TELECOMMUNICATIONS IN HEALTH CARE. Springfield, Ill.: Charles C Thomas, 1975. 376 p. Illus., Tables.

> Published and unpublished materials on experience with telemedicine are reviewed. Items discussed include communication aspects, economic implications, patient attitudes, and the impact of telemedicine on accessibility.

S3 Carroll, J.B., and Freedle, R.O., eds. LANGUAGE COMPREHENSION AND THE ACQUISITION OF KNOWLEDGE. Washington, D.C.: Hemisphere Publishing Corp., 1974. 390 p.

> This volume explores the concept of comprehension from recent theoretical viewpoints and fundamental practical applications such as controlling and measuring the degree of comprehension in both spoken and written discourse.

S4 Chalmers, D.K., and Rosenbaum, M.E. "Learning by Observing Versus Learning by Doing." JOURNAL OF EDUCATIONAL PSYCHOLOGY 66 (1974): 216-24.

> Differences between performers and observers were investigated in a concept-transfer task. The method assessed the transfer due to reversal, nonreversal, and irrelevant or control shifts as a function of either performing or observing during an initial training task.

S5 Coke, E.U. "The Effects of Readability on Oral and Silent Reading Rates." JOURNAL OF EDUCATIONAL PSYCHOLOGY 66 (1974): 406-9.

> The usefulness of reading rate as a measure of reading difficulty was evaluated in two studies relating reading rate to two text-derived measures of readability.

S6 Deeds, S.G. A GUIDEBOOK FOR FAMILY PLANNING EDUCATION.

Columbia, Md.: Westinghouse Health Systems Division, 1973. 93 p.

This book describes the state of the art of educational technology in several related fields. Developments and future possibilities are defined in the process of exploring potentials for family planning patient education.

S7 Feurzeig, W., et al. OPTIMIZING THE EFFICIENCY OF COMPUTER AIDED INSTRUCTION. Cambridge, Mass.: Bolt, Beranek and Newman, 1975. 52 p.

This report describes the development and testing of an adaptive training model for optimizing path sequencing in computer-aided instruction (CAI) to minimize training time. Computer program implementation of the model and CAI course materials was done through use of the PLATO IV CAI system.

S8 Fisher, D.F., et al. SHORT-TERM MEMORY (1958-1973): AN ANNOTATED BIBLIOGRAPHY. Aberdeen Proving Ground, Md.: Human Engineering Lab., 1974. 404 p.

This annotated bibliography of 1,393 references deals with short-term memory. An alphabetical index of pertinent parameters and special topics of interest is provided.

S9 Goodstadt, M., ed. RESEARCH ON METHODS AND PROGRAMMES OF DRUG EDUCATION: INTERNATIONAL SYMPOSIA ON ALCOHOL AND DRUG PROBLEMS, PROCEEDINGS, NO. 3. Toronto: Addiction Research Foundation, 1974. 190 p.

The papers in these proceedings discuss the social and psychological dynamics involved in drug use and drug use modification.

S10 Graves, V. "A Medical Tape-Slide Library." PROGRAMMED LEARNING AND EDUCATIONAL TECHNOLOGY 11 (1974): 253-57.

This article describes a medical lending library service using tape-slide programs that are supplied to doctors, nurses, and a wide variety of paramedical workers. The tape-slide programs are suitable for individual study in libraries or for small discussion groups.

S11 Hartley, J. "Programmed Instruction 1954-1974: A Review." PROGRAMMED LEARNING AND EDUCATIONAL TECHNOLOGY 11 (1974): 278-91.

This article examines the research on programmed instruction in light of B.F. Skinner's 1954 call to educationalists. The widening scope of programmed instruction and some areas of controversy are discussed.

S12 Hearnshaw, T., and Roach, D.K. "A Self-Instructional Course in Audio-Visual Techniques." BRITISH JOURNAL OF EDUCATIONAL TECHNOLOGY 3 (October 1974): 60-71.

    The authors describe the design, organization, application, and implications of a practical, self-instructional course to teach audiovisual media and methods to teachers and student-teachers. Extracts of the printed course texts illustrate the format and content of the course.

S13 Hurst, J.C., et al. "Encountertapes: Evaluation of a Leaderless Group Procedure." SMALL GROUP BEHAVIOR 4 (1973): 476-85.

    An innovation in group therapy, the Encountertape program, is described. Eight prerecorded audiotapes are designed for ninety-minute encounter group sessions, with no leader required.

S14 Ivey, A.E. "Media Therapy: Educational Change Planning for Psychiatric Patients." JOURNAL OF COUNSELING PSYCHOLOGY 20 (1973): 338-43.

    Media therapy, a systematic video program in behavior change, was used with psychiatric patients. Patients engaged in short videotaped interactions with a consultant-facilitator. Case illustrations are presented, implications for an educational treatment program for psychiatric patients are discussed, and the role of the therapist as change agent is examined.

S15 Ivey, A.E. "Microcounseling and Media Therapy: State of the Art." COUNSELOR EDUCATION AND SUPERVISION 13 (1974): 172-83.

    The history and method of microcounseling and media therapy, video methods of teaching single behavioral skills of counseling, and communications are detailed. Recent research on the effectiveness of these techniques is summarized.

S16 Klement, J.J., et al. "The Learning Effects of Three Amounts of Reinforcement in Computer Assisted Instruction." AUSTRALIAN JOURNAL OF ADULT EDUCATION 13 (1973): 131-33.

    The purpose of this study was to provide an empirical test of the hypothesis that increases in reinforcement, in addition to the knowledge of results, will cause corresponding increases in the amount individuals learn with computer-assisted instruction.

S17 Koch, J.H. "Riding the Behavioral Objective Bandwagon." SCHOOL COUNSELOR 21 (1974): 196-202.

    The author suggests that to prevent the misuse of educational tools, administrators, counselors, and teachers should look critically into the problem before jumping aboard the behavioral objective bandwagon. Some guidelines are given.

S18 Kundu, M.R. "What a Low-Cost Closed-Circuit Television System Can Do." EDUCATIONAL TECHNOLOGY 15 (February 1975): 57-59.

This article describes uses of a low-cost, closed-circuit television system; the equipment and operation involved; and the development of such a system with limited funds.

S19 Lawson, T.E. "Effects of Instructional Objectives on Learning and Retention." INSTRUCTIONAL SCIENCE 3 (1974): 1-22.

A synthesis is presented of the various rationales that predict the facilitative influence of instructional objectives on learning and retention.

S20 Lippey, G. "Computer Managed Instruction: Some Strategic Considerations." EDUCATIONAL TECHNOLOGY 15 (January 1975): 9-13.

Three strategic considerations in developing computer-based instructional systems are proposed: a classification of applications, an admonition to supporters of individualized systems to become familiar with the history of such approaches, and three practical tests that may be applied to new applications to judge their viability.

S21 McGrane, H.F. "Tape Recorded Evaluation: A Method of Teaching." JOURNAL OF NURSING EDUCATION 14 (1975): 11-17.

This paper describes the use of individual tape-recorded cassettes as a supplemental evaluation method for two groups of senior students in a baccalaureate nursing program. The method is intended to promote educational growth and is relevant in hospital and community settings.

S22 Malo-Juvera, D. "Seeing Is Believing." NURSING OUTLOOK 21 (1973): 583-85.

Audiovisual technology, combined with role playing, was used to help nursing students learn to teach the subject of human sexuality to patients or parents and to develop sensitivity to the attitudes of others.

S23 Meichenbaum, D. "Self-Instructional Strategy Training: A Cognitive Prothesis for the Aged." HUMAN DEVELOPMENT 17 (1974): 273-80.

A self-instructional strategy training procedure, designed to explicitly teach the use of heuristic processes and mediational devices, is proposed for use in compensating for age-associated deficits such as poor problem solving. The format for the self-instructional procedure is derived from the developmental research of Soviet psychologists Vygotsky and Luria.

S24 Messner, E., and Schmidt, D.D. "Videotape in the Training of Medical Students in Psychiatric Aspects of Family Medicine." INTERNATIONAL JOURNAL OF PSYCHIATRY IN MEDICINE 5 (1974): 269-73.

This paper describes a course in family medicine using the preceptorship method and including patient-centered psychiatric conferences recorded on videotape. Considerations of confidentiality and informed consent are emphasized.

S25 Metzner, R.J., and Bittker, T.E. "Videotape Production by Medical Educators: Some Practical Considerations." JOURNAL OF MEDICAL EDUCATION 48 (1973): 743-51.

The authors explore practical difficulties and make recommendations for the medical educator creating videotaped teaching materials.

S26 Milgram, G.G. "A Descriptive Analysis of Alcohol Education Materials." JOURNAL OF STUDIES ON ALCOHOL 36 (1975): 416-21.

In this article, 832 alcohol education materials (books, pamphlets, and leaflets) are reviewed.

S27 O'Connor, R.J.J. "Integration of Programmed Instruction with Instructional Television in a Health Education Program." Ph.D. dissertation, Louisana State University and Agricultural and Mechanical College, 1974. viii, 96 p.

The effectiveness of instructional television in a hospital specializing in the treatment of leprosy was studied. This study specifically attempted to learn if concepts of programmed instruction could be successfully integrated into locally produced televised educational materials.

S28 Pearson, K.M., Jr., and Bloch, A.D. "Dial Access Libraries: Their Use and Utility." JOURNAL OF MEDICAL EDUCATION 49 (1974): 882-96.

The utility and efficiency of dial access service were studied by reviewing the literature, analyzing available operational statistical data, and visiting ten dial access libraries throughout the country.

S29 Persons, R.W., and Persons, M.K. "Psychotherapy Through Media." PSYCHOTHERAPY: THEORY, RESEARCH AND PRACTICE 10 (1973): 234-35.

The authors describe the development of a college course that presented fifty-two films and five tapes of psychotherapists both working with patients and discussing theoretical models.

S30   Pope, D. "A Course in Educational Technology." PROGRAMMED LEARNING AND EDUCATIONAL TECHNOLOGY 11 (1974): 236-39.

This article outlines the development of a course in educational technology and details steps taken to assist students in learning to use equipment in an educational setting.

S31   Rossiter, C.M., and Luecke, J.R. "The Use of Videotape Recordings in Teaching Interpersonal Communication." SPEECH TEACHER 23 (1974): 59-60.

Classroom techniques to maximize educational benefits from viewing videotape recordings of television programs are described.

S32   Rubin, M.L., et al. EVALUATION OF THE EXPERIMENTAL CAI NETWORK (1973-1975) OF THE LISTER HILL NATIONAL CENTER FOR BIOMEDICAL COMMUNICATIONS, NATIONAL LIBRARY OF MEDICINE. Alexandria, Va.: Human Resources Research Organization, 1975. 85 p.

This report describes an evaluation of the biomedical Computer Aided Instruction Network Experiment, established to test the feasibility of sharing CAI learning materials through a national computer network. Data sources included case studies; user reports; and interviews with hospital and medical school administrators, faculty, librarians, computer laboratory staff, and students. Network use by program type, by user institution, and by class of user is analyzed.

S33   Rulin, M.C., and Chez, R.A. "Interdepartmental Sharing of Audiovisual Teaching Aids: A Valid Concept?" OBSTETRICS AND GYNECOLOGY 43 (1974): 461-65.

An audiovisual tutorial course in obstetrics and gynecology for third-year medical students was shared with fifty-five medical schools. The role of this material in departmental teaching programs is assessed, and appropriateness of content is analyzed.

S34   Sinnett, E.R., et al. "Credibility of Sources of Information About Drugs." PSYCHOLOGICAL REPORTS 36 (1975): 299-309.

Several samples of youthful drug users and nonusers were asked to rate the credibility of a variety of sources of information about drugs. Implications of the findings for drug education and drug counseling are discussed.

Communications Theory & Practice

S35 Skiff, A.W. "Experiences with Methods for Patient Teaching from a Public Health Service Hospital." HEALTH EDUCATION MONOGRAPHS 2 (1974): 48-52.

> Methods employed in carrying out an educational program in a hospital are described. The author considers the approach and methods that are useful in a hospital not appreciably different from those used in other educational settings.

S36 Solomon, L. "CAI: A Study of Efficiency and Effectiveness." EDUCATIONAL TECHNOLOGY 14 (October 1974): 39-41.

> This article describes the time-savings element of a computer-assisted instruction tutorial experiment.

S37 Stritter, F.T., et al. "Documentation of the Effectiveness of Self-Instructional Materials." JOURNAL OF MEDICAL EDUCATION 48 (1973): 1129-32.

> Comparison of self-instructional materials having programmatic characteristics with other instructional approaches indicates that learning through self-instructional materials is superior to learning through other types of instruction.

S38 Teather, D.C.B., and Marchant, H. "Learning from Film with Particular Reference to the Effects of Cueing, Questioning, and Knowledge of Results." PROGRAMMED LEARNING AND EDUCATIONAL TECHNOLOGY 11 (1974): 317-27.

> Methods are reviewed for improving the effectiveness of instructional films through devices to encourage learner participation during brief intervals in the film showing. The article also describes an experiment in which the main variables, cueing, questioning, and providing knowledge of results, have been incorporated into a single research design.

S39 U.S. Department of Health, Education, and Welfare. NATIONAL MEDICAL AUDIOVISUAL CENTER CATALOG, 1977: AUDIOVISUALS FOR THE HEALTH SCIENTIST. DHEW Publication no. NIH 77-506. Atlanta: National Medical Audiovisual Center, 1977. 319 p.

> This catalog contains information on videotape duplication programs; instructions for ordering films; foreign loans; purchasing films; series, subject, and motion picture listing; as well as lists by title and fields of study. Current information can be found in the NATIONAL LIBRARY OF MEDICINE AUDIOVISUAL CATALOG, 1977 to date.

S40 Young, M.E. AUDIOVISUAL EDUCATION (A BIBLIOGRAPHY WITH ABSTRACTS). 2 vols. Springfield, Va.: National Technical Information Service, vol. 1, 1964-November 1976. 227 p. Vol. 2, November 1976-December 1978. 170 p.

> This bibliography of 397 citations covers audiovisual aids, visual aids, and other supports useful for instruction and training.

S41 _____. COMPUTER AIDED INSTRUCTION (A BIBLIOGRAPHY WITH ABSTRACTS). 3 vols. Springfield, Va.: National Technical Information Service, vol. 1, 1970-73. 222 p. Vol. 2, 1974-77. 273 p. Vol. 3, January-July 1978. 77 p.

> Reports on the use of computers in education are abstracted in this bibliography that includes studies on motivation, technical training, learning factors, and human factors engineering.

T. BEHAVIOR MODIFICATION IN PATIENT CARE

T1 Arkowitz, H. "Desensitization as a Self-Control Procedure: A Case Report." PSYCHOTHERAPY--THEORY, RESEARCH AND PRACTICE 11 (1974): 172-74.

> This report illustrates the potential of self-desensitization as a self-control procedure. Self-desensitization is defined as a process of learning of general anxiety-reducing skill. A case history is presented.

T2 Azrin, N.H., et al. "Eliminating Self-Injurious Behavior by Educative Procedures." BEHAVIOUR RESEARCH AND THERAPY 13 (1975): 101-11.

> An alternative treatment to pain-shock punishment for preventing self-injury in autistic and severely retarded persons is described. The method includes positive reinforcement, a period of required relaxation or incompatible postures upon each occurrence of a self-injurious episode, and a hand-awareness training procedure.

T3 Bandura, A., et al. "Efficacy of Participant Modeling as a Function of Response Induction Aids." JOURNAL OF ABNORMAL PSYCHOLOGY 83 (1974): 56-65.

> This experiment tested the efficacy of participant modeling, as a function of the amount of response induction aids employed, in changing the behavior and attitudes of adult phobics.

T4 Bardo, H.R., et al. "Black Concern with Behavior Modification."

PERSONNEL AND GUIDANCE JOURNAL 53 (1974): 20-25.

> In this article two black psychologists and a white educator explain why the black community should be concerned with and actively involved in behavior modification practices. Concerns specified were that blacks have been and are being used as subjects in behavior modification studies with and without their consent and that social value judgments are being made concerning behaviors of blacks without their input.

T5 Barrish, I.J. "Ethical Issues and Answers to Behavior Modification." CORRECTIVE AND SOCIAL PSYCHIATRY AND JOURNAL OF APPLIED BEHAVIOR THERAPY 20, no. 2 (1974): 30-37.

> This paper deals with objections and concerns regarding the nature of and uses of behavior modification in a variety of applied settings. Responses in the literature to the objections are reviewed.

T6 Bell, C.E., et al. "Communicating Dental Hygiene Practices to Chronically, Emotionally Ill, Hospitalized Patients." AMERICAN JOURNAL OF PUBLIC HEALTH 63 (1973): 778-81.

> An empirical analysis of the effectiveness of a dental hygiene program for hospitalized, psychotic patients was undertaken. Persuasive and fear-arousing communications were used to promote changes in attitudes and practices of dental hygiene, with generally negative results.

T7 Best, J.A. "Tailoring Smoking Withdrawal Procedures to Personality and Motivational Differences." JOURNAL OF CONSULTING AND CLINICAL PSYCHOLOGY 43 (1975): 1-8.

> Concentrated cigarette smoke was used as an aversive agent in treating eighty-nine habitual cigarette smokers. A factorial design assessed the incremental efficacy of three procedures: treatment focus, punishment, and timing of attitude change. Findings are interpreted as a support for the principle of tailoring therapeutic procedures to individual differences.

T8 Best, J.A., and Steffy, R.A. "Smoking Modification Procedures for Internal and External Locus of Control Clients." CANADIAN JOURNAL OF BEHAVIORAL SCIENCE 7 (1975): 155-65.

> Relationships between locus of control and effectiveness of two treatment procedures were studied in a smoking clinic to develop treatments of choice for internal and external locus of control clients. Guidelines are suggested for matching internally and externally focused treatment to the respective internal and external orientations of clients.

Communications Theory & Practice

T9 Bindra, D. "A Motivational View of Learning, Performance, and Behavior Modification." PSYCHOLOGICAL REVIEW 81 (1974): 199-213.

    The theoretical formulation advanced in this study discards the response-reinforcement principle. It attributes learned behavior modifications to the building of central representations of contingencies between situational stimuli and incentive stimuli.

T10 Blanchard, E.B., and Young, L.D. "Clinical Applications of Biofeedback Training: A Review of Evidence." ARCHIVES OF GENERAL PSYCHIATRY 30 (1974): 573-89.

    Published reports on clinical applications of biofeedback training are summarized and reviewed.

T11 Blanchard, E.B., et al. "Differential Effects of Feedback and Reinforcement in Voluntary Acceleration of Human Heart Rate." PERCEPTUAL AND MOTOR SKILLS 38 (1974): 683-91.

    Six single-subject experiments were conducted in an attempt to differentiate the roles of feedback and reinforcement in changing heart rate.

T12 Blanchard, E.B., et al. "The Effects of Feedback Signal Information Content on a Long-Term Self-Control of Heart Rate." JOURNAL OF GENERAL PSYCHOLOGY 91 (1974): 175-87.

    Groups of subjects were trained to raise and to lower heart rate with the use of either binary visual feedback (n = 12), proportional visual feedback (n = 12) or no feedback (n = 6).

T13 Blanchard, E.B., et al. "A Simple Feedback System for the Treatment of Elevated Blood Pressure." BEHAVIOR THERAPY 6 (1975): 241-45.

    An open-loop feedback system for teaching patients with elevated blood pressure to lower their blood pressure is described. Four single subject design experiments are described in which the elevated systolic blood pressures of four patients suffering from essential hypertension or borderline hypertension were lowered to the normal range.

T14 Bleecker, E.R., and Engel, B.T. "Learned Control of Cardiac Rate and Cardiac Conduction in the Wolff-Parkinson-White Syndrome." NEW ENGLAND JOURNAL OF MEDICINE 288 (1973): 560-62.

    The purpose of this study was to determine whether a patient with intermittent Wolff-Parkinson-White syndrome can learn to control heart rate and modify the pathway of cardiac conduction.

T15 _____. "Learned Control of Ventricular Rate in Patients with Atrial Fibrillation." PSYCHOSOMATIC MEDICINE 35 (1973): 161-75.

> Patients (n = 6) with chronic atrial fibrillation and rheumatic heart disease on stable digitalis regimens were trained to slow and to speed ventricular rate.

T16 Boisvert, M.J. "Behavior Shaping as an Alternative to Psychotherapy." SOCIAL CASEWORK 55 (1974): 43-47.

> This article describes simultaneous use of a time-out procedure and positive and negative reinforcement to modify the deviant behavior of a seven-year-old boy.

T17 Brady, J.P., et al. "Blood Pressure Reduction in Patients with Essential Hypertension Through Metronome-Conditioned Relaxation: A Preliminary Report." BEHAVIOR THERAPY 5 (1974): 203-9.

> Four male patients with essential hypertension were subjects of an experiment on the effects of a behavioral procedure, metronome-conditioned relaxation, on blood pressure.

T18 Brown, B.S., et al. BEHAVIOR MODIFICATION: PERSPECTIVE ON A CURRENT ISSUE. DHEW Publication no. ADM 75-202. Rockville, Md.: U.S. Department of Health, Education, and Welfare, National Institute of Mental Health, 1975. 26 p. Bibliog.

> This report describes the variety of approaches to the management of mental and behavioral disorders that has evolved from early experimental studies of the principles of learning. Illustrations of diverse behavior modification techniques are given, with explanations of the psychological theories underlying their application. The effectiveness of behavior modification approaches is reviewed, and contemporary concerns for the ethical use of behavior modification techniques in institutional and public settings are outlined. Included are descriptions of standards established by professional organizations and the federal government and implications of recent legal rulings pertaining to behavior modification.

T19 Budzynski, T.H. "Biofeedback Procedures in the Clinic." SEMINARS IN PSYCHIATRY 5 (1973): 537-47.

> Techniques for clinical applications of biofeedback training are presented for patients with headaches, migraine headaches, anxiety, insomnia, and certain phobias. The use of biofeedback training as a preventive technique for coping with stress is suggested.

T20 Budzynski, T.H., et al. "EMG Biofeedback and Tension Headache:

A Controlled Outcome Study." PSYCHOSOMATIC MEDICINE 35 (1973): 484-96.

> Patients were trained in the relaxation of the forehead musculature through EMG biofeedback during sixteen semiweekly twenty-minute EMG feedback sessions, plus daily home practice.

T21 Davis, M.S. "The Responsibility of Caring for an Unmotivated Population." HEALTH EDUCATION MONOGRAPHS 2, Suppl. 1 (1974): 26-33.

> The problem of motivation of the patient/public and the institutional supports of the sick role are discussed. Some behavior modification proposals are presented.

T22 Elder, S.T., et al. "The Role of Systolic-Versus Diastolic-Contingent Feedback in Blood Pressure Conditioning." PSYCHOLOGICAL RECORD 24 (Spring 1974): 171-76.

> An experiment designed to compare systolic-contingent with diastolic-contingent feedback is reported.

T23 Engel, B.T. "Clinical Applications of Operant Conditioning Techniques in the Control of the Cardiac Arrhythmias." SEMINARS IN PSYCHIATRY 5 (1973): 433-38.

> The clinical applications of biofeedback technology in the control of cardiac arrhythmias were evaluated to determine whether patients can learn to control their cardiac arrhythmias and whether such learning may be therapeutic.

T24 Ewing, J.A. "Behavioral Approaches for Problems with Alcohol." INTERNATIONAL JOURNAL OF THE ADDICTIONS 9 (1974): 389-99.

> This paper argues for a holistic approach to the alcoholic, using all or any methods of therapy that seem appropriate. It describes behavioral approaches, techniques aimed at developing aversion or indifference to alcohol, coupled with a planned future of total abstinence; techniques aimed at replacing alcoholic drinking patterns with those of controlled drinking; and aversive approaches in total abstinence programs.

T25 Flannery, R.B., Jr. "Behavior Modification of Geriatric Grief: A Transactional Perspective." INTERNATIONAL JOURNAL OF AGING AND HUMAN DEVELOPMENT 5 (1974): 197-203.

> A behavior modification program for treating grief in a geriatric patient is presented. The program was carried out in a community mental health setting and is discussed in relation to transactional analysis.

T26   Franks, C.M., and Wilson, G.T., eds. ANNUAL REVIEW OF BEHAVIOR THERAPY: THEORY & PRACTICE, 1973. New York: Brunner/Mazel, 1974. 848 p.

   The inaugural volume of this annual series provides a distillation of the literature in the field of behavior therapy, with appropriate commentaries to place the many developments in perspective. The editors analyze the developments and a wide variety of articles in both the theory and practice of behavior therapy.

T27   Gatchel, R.J. "Frequency of Feedback and Learned Heart Rate Control." JOURNAL OF EXPERIMENTAL PSYCHOLOGY 103 (1974): 274-83.

   The effects of varying frequency of feedback information on learning to accelerate and decelerate heart rate were investigated.

T28   Gaylin, W. "On the Borders of Persuasion: A Psychoanalytic Look at Coercion." PSYCHIATRY 37 (1974): 1-9.

   The author discusses developments in the biological sciences in the last twenty years in terms of ethical and value issues. The issue of freedom and coercion is examined: society's right to coerce versus the individual's right to freedom, what distinguishes coercion from a free act, and the true determinants of behavior.

T29   Geiger, O.G., and Johnson, L.A. "Positive Education for Elderly Persons: Correct Eating Through Reinforcement." GERONTOLOGIST 14 (1974): 432-36.

   A positive continuous reinforcement procedure implemented for six geriatric patients with severely low rates of correct eating is described. Applications of the technique in professional practice are suggested.

T30   Gentry, D.L. "Directive Therapy Techniques in the Treatment of Migraine Headaches: A Case Study." PSYCHOTHERAPY--THEORY, RESEARCH AND PRACTICE 10 (1973): 308-11.

   This article describes a migraine patient treated with directive therapy, a communication-oriented approach to the treatment of psychiatric symptoms postulating that psychopathology in a person is produced by that person's attempt to gain control of an interpersonal relationship.

T31   Goldstein, G.S. "Behavior Modification: Some Cultural Factors." PSYCHOLOGICAL RECORD 24 (Winter 1974): 89-91.

Some precautions for applying behavior modification programs to culturally unique populations, specifically American Indian children, are identified. The practitioner must apply contingencies that are not reflections of his or her own cultural values.

T32 Gordon, S.B., and Hall, L.A. "Therapy Determined by Assessment in the Modification of Smoking: A Case Study." JOURNAL OF BEHAVIOR THERAPY AND EXPERIMENTAL PSYCHIATRY 4 (1973): 379-82.

> This study demonstrates that effectiveness of techniques to reduce smoking depends on matching the treatment to the client.

T33 Gottfried, A.W., and Verdicchio, F.G. "Modifications of Hygienic Behaviors Using Reinforcement Therapy." AMERICAN JOURNAL OF PSYCHOTHERAPY 28 (1974): 122-28.

> This paper reports the results of a reinforcement therapy intervention program aimed at eliciting and maintaining the hygienic habits of chronic psychotic patients.

T34 Gygi, C., et al. "Self-Confrontation and Weight Reduction: A Controlled Experiment." PSYCHOTHERAPY: THEORY, RESEARCH AND PRACTICE 10 (1973): 315-20.

> This study was designed to test whether self-confrontation, a planned time-out where one can experience troubling feelings without the opportunity to act on these feelings, is a useful tool in weight reduction.

T35 Hagen, R.L. "Group Therapy Versus Bibliotherapy in Weight Reduction." BEHAVIOR THERAPY 5 (1974): 222-34.

> Comparison of weight loss was made in eighty-nine coeds treated for obesity under three conditions: group therapy, use of a written manual (bibliotherapy), and group therapy and bibliotherapy combined.

T36 Hall, S.M., and Hall, R.G. "Outcome and Methodological Considerations in Behavioral Treatment of Obesity." BEHAVIOR THERAPY 5 (1974): 352-64.

> Studies determining the efficacy of behavioral treatment of obesity are divided into self and experimenter managed categories and are reviewed as to outcome and adequacy of design. Methodological factors are considered, and conclusions with regard to outcome and suggestions for research are offered.

T37 Harris, M.B., and Hallbauer, E.S. "Self-Directed Weight Control Through Eating and Exercise." BEHAVIOUR RESEARCH AND THERAPY 11 (1973): 523-29.

> A weight control program using a written contract and other self-control behavior modification techniques for changing eating habits was compared with a program concentrating on both eating and exercise behavior and with an attention-placebo control condition.

T38 Jacobs, S.H. ALCOHOL ABUSE AND ALCOHOLISM PREVENTION MODEL LEARNING SYSTEMS PRELIMINARY DESIGNS. Los Angeles: Sutherland Learning Associates, 1974. 256 p.

> This report on a project to develop a model learning system for alcohol abuse and alcoholism prevention contains format details of four specific programs. Each program is geared to give maximum reinforcement of responsible behavior, to change learner behavior, and to insure effective implementation in a variety of institutional settings.

T39 Jeffrey, D.B. "A Comparison of the Effects of External Control and Self-Control on the Modification and Maintenance of Weight." JOURNAL OF ABNORMAL PSYCHOLOGY 83 (1974): 404-10.

> In an experimental study of weight reduction, sixty-two obese men and women were administered a pretreatment questionnaire and assigned randomly to three experimental treatment groups: an external control group with a nonrefundable contingency, a self-control group with a refundable contingency, and a self-control group with a nonrefundable contingency.

T40 Kapche, R. "Aversion-Relief Therapy: A Review of Current Procedures and the Clinical and Experimental Evidence." PSYCHOTHERAPY: THEORY, RESEARCH AND PRACTICE 11 (1974): 156-62.

> This paper reviews the clinical and experimental literature relevant to aversion-relief therapy procedures. It is restricted to those studies that have used physical aversive stimuli; symbolic relief is not considered.

T41 Kazdin, A.E. BEHAVIOR MODIFICATION IN APPLIED SETTINGS. Homewood, Ill.: Dorsey Press, 1975. 292 p.

> This book is intended as a college text in applied operant techniques. Diverse techniques across varied treatment populations and settings are discussed. Relevant research, limitations, and salient issues are reviewed for each procedure.

T42 Kessler, S. "Treatment of Overweight." JOURNAL OF COUNSELING

PSYCHOLOGY 21 (1974): 395-98.

> Methods of treating overweight were compared. Participants (n = 18) were divided randomly into three groups. Subjects in both Treatment 1, which applied learning theory plus group therapy, and in Treatment 2, involving the same therapy as Treatment 1 plus mutual help principles, lost significantly more weight over a seven-week period than did the control group. Little difference existed, however, between the two treatment groups.

T43 Khan, A.U., et al. "Role of Counter-Conditioning in the Treatment of Asthma." JOURNAL OF ASTHMA RESEARCH 11, no. 2 (1973): 57-62.

> Twenty asthmatic children aged eight to sixteen were given a counter-conditioning treatment, involving the instigation of bronchial constriction, followed by training in bronchial dilation through biofeedback reinforcement. Results of a follow-up for one year indicated improvement in the experimental group with regard to the frequency of asthmatic attack, emergency room visits, and the amount of medication during that period.

T44 Kirscht, J.P., and Haefner, D.P. "Effects of Repeated Threatening Health Communications." INTERNATIONAL JOURNAL OF HEALTH EDUCATION 16 (1973): 268-77.

> This study investigated belief and behavior change in response to level of threat and message repetition, utilizing high and low threat films of similar content on the topic of coronary heart disease. Groups of adult subjects (n = 120) were exposed to a film one, two, or three times a day with a one-day interval between, with questionnaires administered after the films and again after eight months.

T45 Kroll, H.W. "Bibliography on Behavioral Approaches to Modification of Smoking: January 1964 Through December 1973." PSYCHOLOGICAL REPORTS 35 (1974): 435-40.

> This bibliography of 126 references on behavioral approaches to smoking modification was compiled from a review of PSYCHOLOGICAL ABSTRACTS; the National Clearinghouse for Smoking and Health's BIBLIOGRAPHY ON SMOKING AND HEALTH; and EDUCATIONAL RESOURCES INFORMATION CENTER, and from reference sections of relevant journal articles.

T46 Lazarus, R.S. "A Cognitively Oriented Psychologist Looks at Biofeedback." AMERICAN PSYCHOLOGIST 30 (1975): 553-61.

The three interrelated themes in this article are: that the somatic reactions with which biofeedback deals are part of a broader set of issues, those related to the stress emotions and their role in human adaptation; emotional processes and their self-regulation are products of mediating cognitive appraisals about the significance of an event for a person's well being; and the control of somatic processes is an integral aspect of emotional states and their self-regulation.

T47   Le Bow, M.D. "Behavior Modification Process for Parent-Child Therapy." FAMILY COORDINATOR (July 1973): 313-19.

   This paper discusses the importance of incorporating the parents of children with behavior problems into the treatment process by teaching the parents behavior modification principles.

T48   Leibowitz, J.M., and Holcer, P. "Building and Maintaining Self-Feeding Skills in a Retarded Child." AMERICAN JOURNAL OF OCCUPATIONAL THERAPY 28 (1974): 545-48.

   Through the use of operant conditioning techniques, an echolalic child with severe behavior problems was taught to accept an increasing variety and texture of foods while concurrently developing appropriate self-feeding skills. Progress in other self-help areas accompanied the acquisition of self-feeding skills, and the skills were still present five months after termination of treatment.

T49   Levin, H.S., and Benton, A.L. "Age Effects in Proprioceptive Feedback Performance." GERONTOLOGIA CLINICA 15 (1973): 161-69.

   Age effects in performance on a proprioceptive feedback task were examined in neurologically intact hospitalized patients.

T50   Levine, B.A. "Effectiveness of Contingent and Non-Contingent Electric Shock in Reducing Cigarette Smoking." PSYCHOLOGICAL REPORTS 34 (1974): 223-26.

   The effects of negative practice with both contingent and non-contingent electric shock upon the cigarette smoking of fifteen college students are examined.

T51   MacDonald, M.L., and Butler, A.K. "Reversal of Helplessness: Producing Walking Behavior in Nursing Home Wheelchair Residents Using Behavior Modification Procedures." JOURNAL OF GERONTOLOGY 29 (1974): 97-101.

   This article describes a study in which two nursing home residents who had been transported by wheelchair for several months were effectively encouraged to walk through environmental programming. The implications of these results for the institutional treatment of the aged are discussed.

T52 Mahoney, M.J. COGNITION AND BEHAVIOR MODIFICATION. Cambridge, Mass.: Ballinger Publishing Co., 1974. 389 p. Bibliog. Figs. Tables.

Conceptual and philosophical issues regarding inference and cognitive-symbolic processes are examined, and an empirical evaluation is made of two mediational models, covert conditioning and cognitive learning. Clinical implications of cognitive behavior modification and a general paradigm for therapeutic science are suggested.

T53 _____. "Self Reward and Self-Monitoring Techniques for Weight Control." BEHAVIOR THERAPY 5 (1974): 48-57.

Obese adult volunteers (n = 49) were assigned randomly to one of four treatment groups: self-reward for weight loss, self-reward for habit improvement, self-monitoring, and delayed treatment control. Improvements were more pronounced when subjects rewarded themselves for habit change rather than weight loss. Clinical implications and contemporary research issues are discussed briefly.

T54 Mash, E.J. "Has Behaviour Modification Lost Its Identity?" CANADIAN PSYCHOLOGIST 15 (1974): 271-80.

This paper considers the current identity for behavior modification from a recent historical perspective. An examination is made of the identity changes occurring during the evolution of behavior modification from academic operant psychology to a treatment approach concerned with socially important behaviors.

T55 Melin, G.L., and Gotestam, K.G. "A Contingency Management Program on a Drug-Free Unit for Intravenous Amphetamine Addicts." JOURNAL OF BEHAVIOR THERAPY AND EXPERIMENTAL PSYCHIATRY 4 (1973): 331-37.

In a contingency management program, high frequency behaviors were used as reinforcers for low frequency behaviors on a ward for intravenous amphetamine addicts. It was possible to change behaviors on the ward; patient activity increased; and there was increased contact between personnel and patients.

T56 Miller, P.M., et al. "Electrical Aversion Therapy (Vs. Control Conditioning and Group Therapy) with Alcoholics: An Analogue Study." BEHAVIOUR RESEARCH AND THERAPY 11 (1973): 491-97.

No statistically significant differences in reduced alcohol consumption or attitudes toward alcohol were found among groups of chronic alcoholic patients matched on age, education, and length of problem drinking. Patients were assigned to one of three treatment conditions: electrical aversion conditioning, control conditioning, and group therapy.

T57 _____. "Relative Effectiveness of Instructions, Agreements, and Reinforcement in Behavioral Contracts with Alcoholics." JOURNAL OF ABNORMAL PSYCHOLOGY 83 (1974): 548-53.

This study evaluated experimentally the effects of the components of behavioral contracting on drinking in forty chronic alcoholics exposed to one of the following conditions: verbal instructions to limit alcohol consumption, signed written agreement to limit consumption, verbal instructions plus reinforcement for compliance, and signed agreement plus reinforcement.

T58 Nicassio, P., and Bootzin, R. "A Comparison of Progressive Relaxation and Autogenic Training as Treatments for Insomnia." JOURNAL OF ABNORMAL PSYCHOLOGY 83 (1974): 253-60.

Two relaxation techniques, progressive relaxation and autogenic training, were evaluated as treatments for insomnia. No-treatment, a baseline control group, and a self-relaxation group designed to control for nonspecific therapeutic elements were employed. Subjects were thirty adult insomniacs who had chronic and severe difficulties in falling asleep.

T59 Nimmer, W.H., and Kapp, R.A. "A Multiple Impact Program for the Treatment of Injection Phobias." JOURNAL OF BEHAVIOR THERAPY AND EXPERIMENTAL PSYCHIATRY 5 (1974): 257-58.

Three female college students with long-standing histories of aversive reactions to injections were treated in a program consisting of prolonged presentation of hierarchical stimuli, in vivo work, modeling, and "homework." An average of five sessions completed the therapy, and six-month follow-up contacts were made.

T60 Redmond, D.P., et al. "Blood Pressure and Heart-Rate Response to Verbal Instruction and Relaxation in Hypertension." PSYCHOSOMATIC MEDICINE 36 (1974): 285-97.

Six hypertensive patients were instructed alternately to raise and lower their blood pressure by concentrating on changing "heart rate, force of contraction, and blood vessel resistance to flow." This study adds to other data on the potential for nonspecific or "placebo" effects to be operative in conditioning studies.

T61   Repp, A.C., et al. "Reducing Stereotypic Responding of Retarded Persons by the Differential Reinforcement of Other Behavior." AMERICAN JOURNAL OF MENTAL DEFICIENCY 79 (1974): 279-84.

> Stereotypic responding of three retarded persons was reduced when reinforcement was delivered for specific periods of time in which the behavior did not occur. Differentially reinforcing behavior other than a specified one is discussed as an alternative to extinction.

T62   Rinn, R.C., et al. "Behavior Modification with Outpatients in a Community Mental Health Center." JOURNAL OF BEHAVIOR THERAPY AND EXPERIMENTAL PSYCHIATRY 4 (1973): 243-47.

> This paper describes the use of behavior modification with outpatients in a community mental health center. The structure of the outpatient service of the center is described along with the functions of the center's various divisions.

T63   Romanczyk, R.G., et al. "Behavioral Techniques in the Treatment of Obesity: A Comparative Analysis." BEHAVIOUR RESEARCH AND THERAPY 11 (1973): 629-40.

> Using obese adult volunteers, the relative efficacy of the major techniques typically used in behavioral treatment programs for weight reduction was investigated.

T64   Schmidt, M.P.W., and Duncan, B.A.B. "Modifying Eating Behavior in Anorexia Nervosa." AMERICAN JOURNAL OF NURSING 74 (1974): 1646-48.

> Operant conditioning was used to treat physical manifestations, and psychotherapy was employed to treat the underlying emotional pathology.

T65   Schwartz, G.E., and Shapiro, D. "Biofeedback and Essential Hypertension: Current Findings and Theoretical Concerns." SEMINARS IN PSYCHIATRY 5 (1973): 493-503.

> Biofeedback research on self-regulation of systolic and diastolic pressure, heart rate, and patterns of these functions are reviewed. The authors analyze applications of these techniques to the control of systolic and diastolic pressure in patients with essential hypertension.

T66   Sherwood, G.G., and Gray, J.E. "Two Classic Behavior Modification Patients: A Decade Later." CANADIAN JOURNAL OF BEHAVIORAL SCIENCE 6 (1974): 420-27.

> Follow-up was done on two "classic" patients treated by Dr. Teodoro Ayllon in his original behavior modification unit.

Two of the four behaviors specifically eliminated were still absent a decade later.

T67 Sieg, K.W. "Applying the Behavioral Model to the Occupational Therapy Model." AMERICAN JOURNAL OF OCCUPATIONAL THERAPY 28 (1974): 421-28.

> The basic systems of applied behavioral analysis and the procedures for designing a patient behavior modification project are presented in the context of their application in occupational therapy.

T68 Silverstein, S.J., et al. "Blood Alcohol Level Estimation and Controlled Drinking by Chronic Alcoholics." BEHAVIOR THERAPY 5 (1974): 1-15.

> Four chronic alcoholic subjects were trained over a ten-day period by feedback, social reinforcement, and token reinforcement methods to estimate their own blood alcohol levels.

T69 Sipich, J.F., et al. "A Comparison of Covert Sensitization and Nonspecific Treatment in the Modification of Smoking Behavior." JOURNAL OF BEHAVIOR THERAPY AND EXPERIMENTAL PSYCHIATRY 5 (1974): 201-3.

> Forty-nine volunteer participants in a study to reduce cigarette smoking were assigned randomly to one of five experimental conditions: covert sensitization, attention-placebo, self-control, monitoring control, or no-contact control. Subjects in the first three treatment conditions significantly decreased smoking behavior and were still smoking significantly fewer cigarettes than baseline at the end of six months.

T70 Sirota, A.D., and Mahoney, M.J. "Relaxing on Cue: The Self-Regulation of Asthma." JOURNAL OF BEHAVIOR THERAPY AND EXPERIMENTAL PSYCHIATRY 5 (1974): 65-66.

> A forty-one year-old woman with severe asthmatic difficulties was given brief training in muscular relaxation as a means of avoiding and reducing bronchospasm. A portable timer was used to cue naturalistic self-monitoring of muscle tension and self-relaxation.

T71 Sirota, A.D., et al. "Voluntary Control of Human Heart Rate: Effect on Reaction to Aversive Stimulation." JOURNAL OF ABNORMAL PSYCHOLOGY 83 (1974): 261-67.

> In anticipation of receiving painful stimuli, twenty female subjects learned to control their heart rate when provided with external feedback and reward for criterion heart rate changes.

T72 Sobell, M.B., and Sobell, L.C. "Alcoholics Treated by Individualized Behavior Therapy: One Year Treatment Outcome." BEHAVIOUR RESEARCH AND THERAPY 11 (1973): 599-618.

> One-year treatment outcome results are reported for seventy male alcoholics who, while hospitalized, served as subjects in an experiment evaluating Individualized Behavior Therapy techniques. The difficulties of designing and applying sophisticated follow-up procedures and measures are discussed.

T73 _____. "Evidence of Controlled Drinking by Former Alcoholics: A Second Year Evaluation of Individualized Behavior Therapy." Paper read at the 81st Annual Convention of the American Psychological Association, Montreal, 31 August 1973. Mimeographed.

> This presentation dealing with research conducted at a state hospital in California is a preliminary evaluation of second-year treatment outcomes for alcoholics.

T74 Stainback, S., et al. "The Behavioral Orientation: Misunderstandings and Ethical Issues." EDUCATIONAL TECHNOLOGY 7 (January 1974): 49-50.

> This article discusses the major criticisms of the underlying behavioral principles upon which recent technological advances in education are based.

T75 Stroebel, C.F., and Glueck, B.C. "Biofeedback Treatment in Medicine and Psychiatry: An Ultimate Placebo?" SEMINARS IN PSYCHIATRY 5 (1973): 379-92.

> This paper explores the possibility that biofeedback procedures may prove to be an ultimate placebo by placing both the placebo effect and the patients themselves in a position of importance in the prevention and treatment of illness.

T76 Stunkard, A.J. "From Explanation to Action in Psychosomatic Medicine: The Case of Obesity." PSYCHOSOMATIC MEDICINE 37 (1975): 195-236.

> Psychosomatic medicine as a social movement within medicine and its changing emphasis from understanding to action are discussed in relation to obesity. Planned social intervention, particularly through behavior modification, as a more effective treatment of obesity is described. The question of treatment delivery and the relative merits of medical auspices, patient self-help groups, and commercial enterprises are discussed.

T77 Tanner, B.A., and Zeiler, M. "Punishment of Self-Injurious Behavior Using Aromatic Ammonia as the Aversive Stimulus." JOURNAL OF

APPLIED BEHAVIOR ANALYSIS 8 (1975): 53-57.

> Punishment with aromatic ammonia was used to eliminate self-injurious behavior of an autistic woman during experimental sessions.

T78 Thomas, E.J., et al. "Repertoires of Behavioral and Nonbehavioral Treatment Methods Used in Social Work." SOCIAL SERVICE REVIEW 48 (1975): 107-14.

> This report discussed behavioral treatment methods used in social work practice, along with their nonbehavioral counterparts. Information was obtained from questionnaire responses of 140 practicing social workers. Implications for specialized and polarized use of treatment methods are discussed briefly.

T79 Veterans Administration. Cooperation Study Group on Antihypertensive Agents. "Return of Elevated Blood Pressure after Withdrawal of Antihypertensive Drugs." CIRCULATION 51 (1975): 1107-13.

> The rate at which arterial pressure rises after discontinuing active treatment was investigated in a group of eighty-six hypertensive patients who received treatment with hydrochlorothiazide, reserpine, and hydralazine for two years or longer and whose diastolic pressures averaged below 96 mm Hg for the last year of treatment.

T80 Walter, G.A. "Effects of Video Tape Feedback and Modeling on the Behaviors of Task Group Members." HUMAN RELATIONS 28 (1975): 121-38.

> This study tested the relative and combined merits of two behavior modification inputs, video-tape feedback and modeling, for effecting predictable and productive task group behavior change.

T81 Weiss, T., and Engel, B.T. "Operant Conditioning of Heart Rate in Patients with Premature Ventricular Contractions." PSYCHOSOMATIC MEDICINE 33 (1971): 301-21.

> Findings in this study suggest that some aspects of cardiac ventricular function can be brought under voluntary control and that such control can mediate clinically significant changes in cardiac function.

T82 Welgan, P.R. "Learned Control of Gastric Acid Secretions in Ulcer Patients." PSYCHOMATIC MEDICINE 36 (1974): 411-19.

> Attempts to train increases in the pH of gastric acid secretions in peptic ulcer patients were made. This preliminary work

suggests that gastric acid secretions may be altered and controlled with appropriate feedback.

T83 Whitman, H.H., and Lukes, S.J. "Behavior Modification for Terminally Ill Patients." AMERICAN JOURNAL OF NURSING 75 (1975): 98-101.

The authors advocate the use of behavior modification techniques to help preterminal and terminally ill patients deal with problems, alter behavior, begin realistic problem-solving, and communicate more effectively with families and hospital staff.

T84 Wickramaskera, I.E. "Temperature Feedback for the Control of Migraine." JOURNAL OF BEHAVIOR THERAPY AND EXPERIMENTAL PSYCHIATRY 4 (1973): 343-44.

Case studies describe the treatment of two patients treated with EMG feedback training without positive response, who were later treated with the "temperature trainer" with positive response.

T85 Williams, J.L., and Adkins, J.R. "Voluntary Control of Heart Rate During Anxiety and Oxygen Deprivation." PSYCHOLOGICAL RECORD 24 (Winter 1974): 3-16.

Three experiments on biofeedback-trained subjects tested acceleration and deceleration of their heart rates. Theoretical and clinical implications of biofeedback training in counteracting anxiety and psychological stress are discussed.

T86 Young, L.D., and Blanchard, E.B. "Effects of Auditory Feedback of Varying Information Content on the Self-Control of Heart Rate." JOURNAL OF GENERAL PSYCHOLOGY 91 (1974): 61-68.

The relative efficacy of auditory feedback, varying in the amount of information contained in the feedback signal, for the self-control of heart rate was determined by comparing groups of ten subjects. Subjects received continuous proportional feedback, discontinuous proportional feedback, binary feedback, heart sounds, or no feedback.

T87 Zlutnick, S., et al. "Modification of Seizure Disorders: The Interruption of Behavioral Chains." JOURNAL OF APPLIED BEHAVIOR ANALYSIS 8 (1975): 1-12.

This study investigated the effects of interruption and differential reinforcement on seizure in children, with a strategy aimed at identifying and modifying behaviors that reliably preceded the seizure climax. Parents and school personnel were used as change agents.

Chapter 8
ORGANIZATIONAL THEORY AND METHODS
FOR PATIENT EDUCATION

In this and the next chapter, a variety of administrative issues, problems, and solutions for patient education programs are addressed. Chapter 8 addresses macro-administrative issues at an institutional or organizational and interorganizational level. Chapter 9 reduces administrative problems to the level of intraorganizational relations, including staff development, training, and consultation.

Patient education does not function in isolation from other activities and goals of the medical institution, office, clinic, or agency in which it occurs. It must be integrated with other services, compete for resources with other demands, and link to other organizations, groups, and services in the community. These considerations force the patient educator to maintain a broader perspective than a strictly clinical one.

The broader perspective that prevails among health education specialists is from the literature of community organization and organizational development. This literature is applied to health education in the following general works:

Beckhard, R. ORGANIZATION DEVELOPMENT: STRATEGIES AND MODELS. Reading, Mass.: Addison-Wesley, 1966.

D'Onofrio, C.A. REACHING OUR HARD-TO-REACH--THE UNIMMUNIZED. Berkeley: California State Department of Public Health, 1966.

Griffiths, W. "Achieving Change in Health Practices." HEALTH EDUCATION MONOGRAPHS 1, no. 20 (1965): 27-41.

Hepner, J.O., and Hepner, D.M. THE HEALTH STRATEGY GAME: A CHALLENGE FOR REORGANIZATION AND MANAGEMENT. St. Louis: C.V. Mosby Co., 1973.

Julian, J. "Organizational Involvement and Social Control." SOCIAL FORCES 47 (September 1968): 12-16.

Levine, S., et al. "Community Interorganizational Problems in Providing Medical Care and Social Services." AMERICAN JOURNAL OF PUBLIC HEALTH 53 (1963): 1183-95.

Organizational Theory and Methods

Mico, P.R., and Ross, H.S. HEALTH EDUCATION AND BEHAVIORAL SCIENCE. Oakland, Calif.: Third Party Associates, 1975. 207 p.

Mullen, P.D., et al. "Health Education in Health Maintenance Organizations." In HANDBOOK OF HEALTH EDUCATION, edited by M.M. Lazes, pp. 53-76. Germantown, Md.: Aspen Systems Corp., 1979.

Richards, N.D. "Methods and Effectiveness of Health Education." SOCIAL SCIENCES AND MEDICINE 9 (1975): 141-56.

Ross, H.S., et al. "Health Education Discussion Groups for 'Worried Well' Patients in an Ambulatory Setting." HEALTH EDUCATION MONOGRAPHS 5 (1977): 51-61.

Additional texts on the broader organizational and community context of patient education, but without specific reference to patient education or health education, are the following general works:

Bennis, W.G. CHANGING ORGANIZATION. New York: McGraw-Hill Book Co., 1966.

Bennis, W.G., et al. THE PLANNING OF CHANGE. 3d ed. New York: Holt, Rinehart and Winston, 1976.

Burke, W.W., and Hornstein, H.A., eds. THE SOCIAL TECHNOLOGY OF ORGANIZATION DEVELOPMENT. Fairfax, Va.: NTL Learning Resources Corp., 1972.

Freidson, E. "The Organization of Medical Practice." In HANDBOOK OF MEDICAL SOCIOLOGY. 2d ed., edited by H.E. Freeman et al., pp. 299-319. Englewood Cliffs, N.J.: Prentice-Hall, 1972.

French, W.L., and Bell, C.H. ORGANIZATION DEVELOPMENT: BEHAVIORAL SCIENCE INTERVENTIONS FOR ORGANIZATION IMPROVEMENT. Englewood Cliffs, N.J.: Prentice Hall, 1973.

Perrow, C. ORGANIZATIONAL ANALYSIS: A SOCIOLOGICAL VIEW. Belmont, Calif.: Wadsworth Publishing Co., 1970.

Torrens, P.R. THE AMERICAN HEALTH CARE SYSTEM: ISSUES AND PROBLEMS. St. Louis: C.V. Mosby Co., 1978.

In patient education, more than in most clinical activities, the linkages with resources outside the clinical setting may be decisive in determining success or failure. This has been so partly because the resources for patient education have been underdeveloped in most clinical settings, leaving health educators with no choice but to mobilize external resources on behalf of patients.

Even with resources, however, the clinically based health educator is well advised to link patient education with the settings and groups to which patients will be discharged. Without the continued support, encouragement, and reinforcement from such external environments, discharged patients and ambulatory patients will have difficulty sustaining any prescribed behavior over time. Without the understanding and support of family, neighbors, employers, and

Organizational Theory and Methods

friends, the ambulatory or homebound patient will be compelled to deny his or her condition, to hide its symptoms, and to act as though there were no need for caution, care, dietary discretion, physical limitation, or continued monitoring. Medical devices and prescriptions will be disguised at first, then neglected, and eventually abandoned.

The literature on community resources has been evolving and changing character with every new legislative initiative at the federal or state level that attempts to stimulate greater capacity at the local level to manage health problems outside the expensive medical institutions. Neighborhood and community centers and self-help groups, halfway houses, residential treatment centers, home health care services, meals on wheels, voluntary health associations, recovery and convalescent programs, and health maintenance organizations all have flourished in recent years as governments and consumers strive to reverse the trends of increasingly costly facilities and technological dependency that have escalated health care costs.

Some major reference works, and illustrative resources, in addition to those annotated in the following two sections, in chapter 5 on enabling factors, and in chapter 6 on reinforcing factors, are the following:

Almond, R. THE HEALING COMMUNITY: DYNAMICS OF THE THERAPEUTIC MILIEU. New York: Aronson, 1974.

American Public Health Association. CONSUMER HEALTH EDUCATION: A DIRECTORY. Rev. ed. DHEW Publication no. (HRA) 77-607. Rockville, Md.: Office of Health Resources Opportunity, Public Health Service, October 1976.

BRIEF GUIDE TO ALCOHOLICS ANONYMOUS. New York: Alcoholics Anonymous World Services, 1975.

Cox, F.M., et al. TACTICS AND TECHNIQUES OF COMMUNITY PRACTICE. Itasca, Ill.: F.E. Peacock Publishers, 1977.

Gartner, A., and Riessman, F. "The Consumer: A Hidden Resource for Improving Productivity." SOCIAL POLICY 9 (September-October 1978): 54-55.

_____. THE SERVICE SOCIETY AND THE CONSUMER VANGUARD. New York: Harper & Row, 1974.

Grant, R.H. "Family and Self-Help Education in Isolated Rural Communities." HEALTH EDUCATION MONOGRAPHS 5 (1977): 145-160.

Green, L.W. "Constructive Consumerism and Health Economics." HEALTH EDUCATION 2 (1975): 3-6. Also in DISEASE CONTROL AND HEALTH EDUCATION AND PROMOTION. Congressional Hearings Before the Subcommittee on Health of the Committee on Labor and Public Welfare, U.S. Senate (7-8 May 1975): 966-79.

_____. "The Potential of Health Education Includes Cost-Effectiveness." HOSPITALS: JOURNAL OF THE AMERICAN HOSPITAL ASSOCIATION 50 (1 May 1976): 57-61.

Harris, C.L. "Hospital-Based Patient Education: Programs and Role of the Hospital Librarian." BULLETIN OF THE MEDICAL LIBRARY ASSOCIATION 66 (1978): 210-17.

Howell, M.C. HELPING OURSELVES: FAMILIES AND THE HUMAN NETWORK. Boston: Beacon, 1975.

Hurvitz, N. "The Origins of the Peer Self-Help Psychotherapy Group Movement." JOURNAL OF APPLIED BEHAVIORAL SCIENCES 12 (1976): 283-94.

THE INTERNATIONAL ASSOCIATION OF LARYNGECTOMEES DIRECTORY, 1976. New York: International Association of Laryngitis, 1976.

Jencks, S.F. "Problems in Participatory Health Care." In SELF-HELP AND HEALTH: A REPORT, pp. 86-98. New York: Graduate and University Center, New Human Services Institute, 1976.

Levin, L.S. "Forces and Issues in the Revival of Interests in Self-Care: Impetus for Redirection in Health." HEALTH EDUCATION MONOGRAPHS 5 (1977): 115-20.

Lewis, J.A., and Lewis, M.A. COMMUNITY COUNSELING: A HUMAN SERVICES APPROACH. New York: John Wiley and Sons, 1976.

Lohr, W. "System Characteristics that Influence Behavior." In CANCER: THE BEHAVIORAL DIMENSIONS, edited by J.W. Cullen et al., pp. 125-36. New York: Raven Press, 1976.

Martin, J. "Health Promotion: Importance of Demedicalized and Active Approaches." AMERICAN JOURNAL OF PUBLIC HEALTH 68 (1978): 686-87.

Milio, N. "Self-Care in Urban Settings." HEALTH EDUCATION MONOGRAPHS 5 (1977): 136-44.

Morris, D., and Hess, K. NEIGHBORHOOD POWER. Boston: Beacon Press, 1975.

Parsell, S., and Tagliareni, E.M. "Cancer Patients Help Each Other." AMERICAN JOURNAL OF NURSING 74 (1974): 650-57.

Sehnert, K. HOW TO BE YOUR OWN DOCTOR (SOMETIMES). New York: Grossett and Dunlap, 1975.

Simmons, J.J. "Models of Patient Education: What Is Feasible in Different Settings?" In PROCEEDINGS: NATIONAL SYMPOSIUM ON PATIENT EDUCATION, 1977, pp. 10-23. Atlanta, Ga.: Bureau of Health Education, Center for Disease Control, Public Health Service, 1978.

Smith, D.H. "A Psychological Model of Individual Participation in Formal Voluntary Associations." AMERICAN JOURNAL OF SOCIOLOGY 73 (1967): 235-44.

Somers, A.R. "Consumer Health Education: Where Are We? Where Are We Going? CANADIAN JOURNAL OF PUBLIC HEALTH 68 (1977): 362-68.

Stoeckle, J.D. "Prevention In and Out of Practice: Reorganization and Regulation for It." In APPLYING BEHAVIORAL SCIENCE TO CARDIOVASCULAR RISK, edited by A.J. Enelow and J.B. Henderson, pp. 33-36. New York: American Heart Association, 1975.

Organizational Theory and Methods

Strauss, A. "Chronic Illness." SOCIETY 10 (1973): 26-36.

Stunkard, A.J. "The Success of TOPS, a Self-Help Group." POSTGRADUATE MEDICINE 18 (1972): 143-47.

Tamney, J.B. SOLIDARITY IN A SLUM. New York: John Wiley and Sons, 1975.

Vickery, D.M., and Fries, J.F. TAKE CARE OF YOURSELF: A CONSUMER'S GUIDE TO MEDICAL CARE. Reading, Mass.: Addison-Wesley, 1976.

Wang, V.L. "Using Cooperative Extension Programs for Health Education." AMERICAN JOURNAL OF PUBLIC HEALTH 64 (1974): 107-11.

U. MOBILIZING INSTITUTIONAL AND VOLUNTEER RESOURCES

U1   Abernathy, W.J., et al., eds. THE MANAGEMENT OF HEALTH CARE: A TECHNOLOGY PERSPECTIVE. Cambridge, Mass.: Ballinger Publishing Co., 1975. 192 p. Append. Bibliog. Figs. Tables.

>   The authors suggest a new policy perspective on the management of health care organizations. They discuss issues and techniques for the application and integration of technology into the typical health care organization and the management of the evolving organizations. Papers and selected cases are used to define problems.

U2   Bartow, J.C. "Volunteer Services." HOSPITALS: JOURNAL OF THE AMERICAN HOSPITAL ASSOCIATION 48 (1 April 1974): 181-85.

>   This review article reports on the state of volunteer services in the United States.

U3   Berman, H.J., and Weeks, L.E. THE FINANCIAL MANAGEMENT OF HOSPITALS. 2d ed. Ann Arbor, Mich.: Health Administration Press, 1974. 400 p.

>   This book provides background on and clarification of the financial environment in which hospitals exist. Sources of operating revenue, including third-party relationships and payment guidelines, the mechanics of rate setting, and future trends in payment arrangements are discussed.

U4   Bracht, N.F. "Health Maintenance Organizations: Legislative and Training Implications." JOURNAL OF EDUCATION FOR SOCIAL WORK 11 (Winter 1975): 36-44.

>   This article reviews and analyzes major features of the Health Maintenance Organization Act. Expanded roles for social work practitioners are discussed and suggestions are made for strengthening social work training programs to prepare social

Organizational Theory and Methods

work professionals for practice in HMOs and the health care field.

U5 Branscomb, A.B., and Branscomb, E.W. "Sharing: A Death Research Information Exchange." OMEGA--JOURNAL OF DEATH AND DYING 4 (1973): 243-49.

> A proposal is outlined for the establishment of a computer-supported system of direct data sharing between death researchers.

U6 Clipson, C.W., and Wehrer, J.J. PLANNING FOR CARDIAC CARE: A GUIDE TO THE PLANNING AND DESIGN OF CARDIAC CARE FACILITIES. Ann Arbor, Mich.: Health Administration Press, 1974. 400 p. Illus.

> This book is concerned with developing cardiac care facilities. It is illustrated with floor plans, photographs, and drawings.

U7 Collen, F.B., et al. "The Educational Adjunct to Multiphasic Health Testing." PREVENTIVE MEDICINE 2 (1973): 247-60.

> This paper describes a patient education approach in a multiphasic health testing system.

U8 Corning, M.E. "National Library of Medicine: International Cooperation for Biomedical Communications." BULLETIN OF THE MEDICAL LIBRARY ASSOCIATION 63 (1975): 14-22.

> Policy and operational aspects of the international programs of the National Library of Medicine are detailed. All have the objective of improved medical research, education, and practice.

U9 Counte, M.A., and Kimberly, J.R. "Organizational Innovation in a Professionally Dominated System: Responses of Physicians to a New Program in Medical Education." JOURNAL OF HEALTH AND SOCIAL BEHAVIOR 15 (1974): 188-98.

> Variability in responses of practicing physicians to an organizational innovation is examined in relation to hypotheses derived from the literature on adoption of innovation.

U10 Crawford, S., and Dandurand, G.L. "Health Science Libraries in the United States: A Five Year Perspective." BULLETIN OF THE MEDICAL LIBRARY ASSOCIATION 63 (1975): 7-13.

> Two surveys of health science libraries in the United States were completed by the Medical Library Association's Committee on Surveys and Statistics over a five-year period. Sum-

Organizational Theory and Methods

mary data for 1969 and 1973 are compared for distribution of libraries, resources, personnel, and salaries.

U11 Distefano, M.K., Jr., and Pryer, M.W. "Work Behavior Dimensions of Psychiatric Attendants and Aides." JOURNAL OF APPLIED PSYCHOLOGY 60 (1975): 140-42.

Factor analysis of responses of 136 psychiatric attendants and aides to an 80-item self-report job questionnaire was used to conceptualize the work behavior of these paraprofessionals and to differentiate work activities associated with various types of treatment programs.

U12 Dorken, H., and Whiting, J.F. "Psychologists as Health-Service Providers." PROFESSIONAL PSYCHOLOGY 5 (1974): 309-19.

A national sample study of fee-for-service practice of psychologists provided data for this study on manpower distribution, fees, type of practice, extent of health insurance coverage, and carrier experience.

U13 Fink, D.J. "The Cancer Control Program." CANCER 35 (1975): 72-75.

Program objectives and specific projects of the Cancer Control Program are described. This program aims at disseminating existing knowledge concerning cancer in the areas of prevention, detection, diagnosis, treatment, and rehabilitation.

U14 Fiori, F., et al. "Health Education in a Hospital Setting: Report of a Public Health Service Project in Newark, New Jersey." HEALTH EDUCATION MONOGRAPHS 2 (1974): 11-29.

The purpose of the Presbyterian Hospital project was the systematic exploration of educational needs of patients and their families and ways in which health education skills might be used in a hospital setting to meet these needs.

U15 Gardner, J.E. PARAPROFESSIONAL WORK WITH TROUBLED CHILDREN. New York: Halstead Press, 1975. 208 p.

This account of the use of paraprofessionals in remedial reading, child therapy, parent counseling, and preventive kindergarten programs considers the evolution and development of paraprofessional programs in a mental health educational setting and provides a perspective to the treatment of troubled children.

U16 Gold, R.A., et al. "The Health Information Specialist: A New Re-

Organizational Theory and Methods

source for Hospital Library Services and Education Programs." BULLETIN OF THE MEDICAL LIBRARY ASSOCIATION 62 (1974): 266-72.

> Roles for a community hospital librarian are suggested. A one-week training program for librarians and special orientation for hospital educators and administrators, with follow-up field consultation for all three is described. The program is proposed to help hospital librarians expand their role into that of health information specialists.

U17 Gordon, D.W. "Health Maintenance Service: Ambulatory Patient Care in the General Medical Clinic." MEDICAL CARE 12 (1974): 648-58.

> This study compares the health care of two groups of ambulatory patients, those treated primarily by a nurse-clinician under physician supervision with those treated by attending physicians, in the general medical clinic of a large university teaching hospital.

U18 Harper, R., and Balch, P. "Some Economic Arguments in Favor of Primary Prevention." PROFESSIONAL PSYCHOLOGY 6 (1975): 17-25.

> This economic review of tertiary-secondary and primary prevention suggests that primary prevention services, if utilized optimally, would be the most cost effective for mental health services.

U19 Howe, R., et al. "A New Role for the Hygienist: The Preventive Prescription." JOURNAL OF DENTAL EDUCATION 38 (1974): 403-5.

> A health care delivery system in which the dental hygienist has many nonconventional functions is described. Two functions, the preventive prescription and supervision of the disease control program, are detailed.

U20 Kaluzny, A.D., et al. "Innovation of Health Services: A Comparative Study of Hospitals and Health Departments." MILBANK MEMORIAL FUND QUARTERLY 52 (1974): 51-82.

> The authors investigate the differential contribution of various organizational variables that affect the innovation of high-risk versus low-risk health service programs in hospitals and health departments.

U21 Katkin, S., et al. "Using Volunteer Therapists to Reduce Hospital Readmissions." HOSPITAL AND COMMUNITY PSYCHIATRY 26 (1975): 151-53.

> Two aftercare programs using volunteers as therapists for former inpatients are described as effective in reducing hospital read-

Organizational Theory and Methods

missions. The volunteer therapists ensured that patients took medications, evaluated them for decompensation, helped them find housing and jobs, and gave them supportive counseling.

U22 Kraegel, J.M., et al. PATIENT CARE SYSTEMS. Philadelphia: J.B. Lippincott Co., 1974. 219 p.

An introduction to the organizational aspects of patient care, this book focuses on patients' needs and nurses' responsibilities for meeting those needs. Nursing procedures for admitting, assessing, and classifying patients; making planned rounds; reporting patients' conditions; giving shift reports; and charting are detailed.

U23 Kress, J.R., and Singer, J. HMO HANDBOOK. Rockville, Md.: Aspen Systems Corporation, 1975. 250 p. Bibliog. Charts. Tables.

This handbook is a step-by-step guide through every phase of the development of a health maintenance organization.

U24 Lamb, D., and Clack, R.J. "The Effect of Professional vs. Paraprofessional Approaches to Orientation on Subsequent Contacts with a Counseling Center." JOURNAL OF COUNSELING PYSCHOLOGY 21 (1974): 61-65.

Incoming college freshmen were exposed to one of three types of orientation procedures designed to acquaint them with available counseling services: one conducted by an undergraduate paraprofessional, one by a professional staff member, and one by a combination of professional staff and videotaped presentation. Students in all presentations expressed equally positive immediate reactions. A disproportionate number who made subsequent counseling contacts were in the orientation procedure conducted by a fellow student (paraprofessional). Implications regarding the use of paraprofessionals in counseling service orientation programs are discussed.

U25 Lee, E.A. "Health Education." HOSPITALS: JOURNAL OF THE AMERICAN HOSPITAL ASSOCIATION 48 (1 April 1974): 133-35, 138-39.

This article is a review of the literature concerning recent developments in health education, especially in relation to the need for increasing federal interest; the hospital's role; reimbursement of health education expenses; planning and implementation of programs; examples of current programs; evaluation; and health education needs.

U26 Lewis, A.B., Jr. "Brief Psychotherapy in the Hospital Setting: Techniques and Goals." PSYCHIATRIC QUARTERLY 47 (1973): 341-52.

Organizational Theory and Methods

The author argues that brief hospitalization should provide considerable psychotherapy for acute psychoses to prepare patients for long-term outpatient care.

U27 Mechanic, D. "Ideology, Medical Technology, and Health Care Organization in Modern Nations." AMERICAN JOURNAL OF PUBLIC HEALTH 65 (1975): 241-47.

The author discusses the convergence in medical care organization in modern nations, resulting from the growth and elaboration of medical technology and the rising aspirations and expectations of populations for accessible, comprehensive, and effective medical care.

U28 _____. POLITICS, MEDICINE, AND SOCIAL SCIENCE. New York: Wiley-Interscience, 1974.

This book explores the relationships between medicine, social science research, politics, and society. It examines health care policy, focusing on the importance of medical care as a supportive service, and stresses the importance of the reactions of patients and health workers to new health policies before and after they are implemented.

U29 Oakes, C.G. THE WALKING PATIENT AND THE HEALTH CRISIS. Columbia, S.C.: University of South Carolina Press, 1973. xxii, 432 p. Bibliog. Index.

Contents include the nature of policy for ambulatory care; planning ambulatory care programs; community and regional organization of ambulatory services; the organization of ambulatory care programs; medical records, evaluation, and research; continuity of care; and the Delphic Oracle's miscellany.

U30 Petersen, H.M., et al. "An Approach to Citizen Involvement in Education and Health Care Delivery." JOURNAL OF MEDICAL EDUCATION 49 (1974): 189.

This paper presents a planning model and methodology for bringing citizens (including consumers and providers) together so that the providers' delivery of health services and the consumers' subsequent evaluation of these services are congruent.

U31 Raske, K.E. "The Components of Inflation: An Analysis of the Causes of Increases in Hospital Costs." HOSPITALS: JOURNAL OF THE AMERICAN HOSPITAL ASSOCIATION 48 (1 July 1974): 67-70.

The meaning of hospital costs, ways in which hospital costs have increased, and reasons for the increases are discussed.

Organizational Theory and Methods

U32 Reid, R.A. "A Work Sampling Study of Midlevel Health Professionals in a Rural Medical Clinic." MEDICAL CARE 13 (1975): 241-49.

    A work sampling study provided a description of tasks performed by mid-level health personnel in the rural component of an experimental medical care delivery system. The investigation determined the proportion of time spent on various activities by the staff members of a rural clinic that is linked to supervisory physicians in a distant urban medical center. Findings and proposed changes in administrative and medical policy provide a basis for development of cost-reducing alternative staffing configurations.

U33 Roach, A.A., and Addington, W.W. "The Effects of an Information Specialist on Patient Care and Medical Education." JOURNAL OF MEDICAL EDUCATION 50 (1975): 176-80.

    A medical librarian joined a hospital's pulmonary medicine health care team to respond to information needs regarding patient care, graduate medical education, continuing education, and research. Regular attendance at rounds and conferences enabled the librarian to initiate immediate literature searches in response to both clinical problems and educational requirements.

U34 Robertson, L.S., et al. CHANGING THE MEDICAL CARE SYSTEM: A CONTROLLED EXPERIMENT IN COMPREHENSIVE CARE. New York: Praeger Publishers, 1974. xviii, 182 p. Bibliog. Index. Tables.

    Chapter contents of this book include: medical care--system or disorder?; methodology and experimental treatment; engagement in comprehensive care; utilization, attitudes, and costs; the problems of morbidity; race and care; the issues of social pathology; middle-class physicians and lower-class patients; and the medical care system reconsidered.

U35 Rubenstein, A.H., et al. "Behavioral Factors Influencing the Adoption of an Experimental Information System." HOSPITAL ADMINISTRATION 18 (Winter 1973): 27-43.

    The aim of this study was to improve understanding of the information-seeking styles of medical researchers, clinicians, and medical staff supervisors and thus to increase the ability to predict (within limits) their reactions to the introduction of a new information system into their environment. The nature and design of this field experiment can indicate to administrators the type of research they can perform in their own hospital as a source of information system design data.

U36 Rushmer, R.F. HUMANIZING HEALTH CARE: ALTERNATIVE FUTURES

Organizational Theory and Methods

FOR MEDICINE. Cambridge: MIT Press, 1975. 210 p.

This book identifies the problems and requirements of the health care system and proposes solutions. Topics covered include projections of future developments in health care delivery, creating desirable futures for health care; concepts of cost-benefit and value-added applied to health care; health-personnel requirements, meeting health-facilities requirements, home-based health care--alternative modes of medical management, and patient participation--partnerships for health.

U37 Schoenrich, E.H. "The Potential of Health Education in Health Services Delivery." HEALTH SERVICES REPORTS 89 (1974): 3-7.

This paper defines health education and patient education, and discusses the organization and personnel required for patient education programs. The need for and the potential of patient education, as well as constraints placed on health education are detailed.

U38 Schulberg, H.C., and Baker, F. THE MENTAL HOSPITAL AND HUMAN SERVICES. New York: Behavioral Publications, 1975. 385p.

The contents of this text include the prospective role of the mental hospital, clinical and administrative organizational patterns, utilizing the mental hospital, staff ideology, professional roles and relations, the mental hospital and its environments, and clinical programs and patient outcome.

U39 Shochet, B.R. "The Role of the Mental Health Counselor in the Psychiatric Liaison Service of the General Hospital." INTERNATIONAL JOURNAL OF PSYCHIATRY IN MEDICINE 5 (1974): 1-16.

Functions of mental health counselors in the medical and surgical units of general hospitals are identified.

U40 Simonds, S.K. "President's Committee on Health Education." HOSPITALS: JOURNAL OF THE AMERICAN HOSPITAL ASSOCIATION 47 (1 March 1973): 55-59.

Charges to the President's Committee on Health Education are detailed, and the role and responsibilities of hospitals in participating in and benefitting from expanded health education activities are discussed.

U41 Skiff, A. "Hospital Office Promotes Health Education." HOSPITALS: JOURNAL OF THE AMERICAN HOSPITAL ASSOCIATION 47 (1 March 1973): 117-24.

A six-year-old health education project in a general hospital is

Organizational Theory and Methods

described. The goal of the education program is to help both well and sick individuals assume responsibility for their health maintenance. The activities of the office of education are categorized into nine areas and described in this article.

U42 Slepcevitch, E., comp. RX: EDUCATION FOR THE PATIENT: WHO, WHAT, WHERE, WHY . . . AND AT WHAT COST? Conference Proceedings, June 25-26, 1974. Carbondale: Southern Illinois University, Department of Health Education, April 1975. 129 p. Append.

These proceedings report ten presentations made at a two-day institute for health professionals. The purposes of the institute were to review the state-of-the-art in patient education; discuss the concept of patient education; develop guidelines and evaluation methods for patient education programs; outline differences between information and education in patient compliance; and discuss reimbursement for and cost-effectiveness of patient education. "A Patient's Bill of Rights" is appended to the proceedings.

U43 Somers, A.R. "Adapting Institutions to Changing Needs." HOSPITALS: JOURNAL OF THE AMERICAN HOSPITAL ASSOCIATION 48 (1 May 1974): 41-44.

The author discusses briefly four significant characteristics of contemporary U.S. society and their implications for future health needs: affluence, the sexual and marital revolution, mobility and individual "aloneness," and rising death rates for men. Implications of these characteristics for the planning of health care delivery are discussed, and some guidelines are proposed.

U44 Steen, J. "Liaison Nurse: Ombudsman for the Chronically III." AMERICAN JOURNAL OF NURSING 73 (1973): 2102-4.

The use of a liaison nurse in a general hospital was found to improve the quality and continuity of care for patients with chronic illness.

U45 Stein, G.H. "The Use of a Nurse Practitioner in the Management of Patients with Diabetes Mellitus." MEDICAL CARE 12 (1974): 885-90.

The medical management of twenty-three female patients with maturity-onset diabetes mellitus was supervised for six months by a nurse practitioner trained in the management of diabetes mellitus. Eleven diabetes patients were managed in the traditional fashion by a clinical physician. Standard measurements of the course of the disease were followed during the six-month interval; and comparisons were made. This study supports the view that the nurse practitioner role may be extended to encompass more comprehensive treatment programs in patients with chronic disease.

U46 Stoeckle, J.D., and Twaddle, A.C. "Non-Physician Health Workers: Some Problems and Prospects." SOCIAL SCIENCE AND MEDICINE 8 (1974): 71-76.

> This paper discusses changes in the clinical organization of medical practice, the rising demand for care and the introduction of technology and public interest in both services and costs. The problems and potentials of using nonphysician health workers in the reorganization and reform of medical practice are reviewed.

U47 Stuart, B., and Stockton, R. "Control over the Utilization of Medical Services." MILBANK MEMORIAL FUND QUARTERLY 5 (1973): 341-93.

> The issues and problems involved in the control of medical utilization are considered in this article.

U48 Task Force on Patient Education for the President's Commission on Health Education. "Concept of Planned, Hospital-Based Patient Education Programs." HEALTH EDUCATION MONOGRAPHS 2 (1974): 1-10.

> This paper deals with hospital-based inpatient and ambulatory patient education programs. The Task Force's recommendations include hospitals and health professionals' acceptance of the patient's "right to know," encouragement and provision for patients' participation in their own care, establishment of educational centers in hospitals, employment of professional education specialists, and education as a part of health care inservice programs. Other components in patient education programs are recommended and enumerated.

U49 Weingarten, V. "Report of Findings and Recommendations of the President's Committee on Health Education." HEALTH EDUCATION MONOGRAPHS 2, Suppl. 1 (1974): 11-19.

> Findings of the President's Committee on Health Education are presented in the areas of health information versus health education, expenditures for health education, expenditures against health education, patient education, and school health education. The central recommendation, the creation of a National Center for Health Education, and a list of five functions for the proposed center are discussed.

U50 Wolf, J.N. "Decision-Making in a Prepaid Group Practice: The Congruence of Attitudes between Decision-Makers and Members." Sc.D. dissertation, Johns Hopkins University School of Hygiene and Public Health, 1973.

> The idea of consumer participation in decision-making for health care services would lead one to believe that consumer participation is the deus ex machina for health care crisis. Recent federal legislation has mandated the implementation of broad-based participation. This study investigated the potential consequences of implementing member participation in a prepaid group practice.

V. REFERRAL PLANS AND COMMUNITY RELATIONS

V1 Blum, H.L. PLANNING FOR HEALTH: DEVELOPMENT AND APPLICATION OF SOCIAL CHANGE THEORY. New York: Behavioral Publications, 1974. 662 p.

> This is a systems-oriented book presenting an analysis of health planning within the context of social change theory. The logic, nature, and choice of approaches taken in different systems are analyzed, with a look at the political, social, and economic situations in which the planning is to take place.

V2 Cauffman, J.G., et al. "Health Information and Referral Services Within Los Angeles County." AMERICAN JOURNAL OF PUBLIC HEALTH 63 (1973): 872-77.

> Findings of a survey of health information and referral services in Los Angeles County are discussed as part of a comprehensive and integrated data system project.

V3 _____. "Study of Health Referral Patterns." AMERICAN JOURNAL OF PUBLIC HEALTH 64 (1974): 331-56.

> To assist in evaluating a computerized network of health education and referral centers, a method for assessing the effectiveness of referrals was developed. This report focuses on the construction of a conceptual model for measuring referral outcomes, a system for classifying health problems and services, a system for tracking referrals, and factors related to outcomes.

V4 Daggett, D.R., et al. "Mental Health Consultation Improves Care of the Aged in Community Facilities." HOSPITAL AND COMMUNITY PSYCHIATRY 25 (1974): 170-73.

> A hospital-based community mental health center developed a consultation and training program for nursing home staff mem-

bers to improve the care and treatment of aged residents. The program was evaluated by the project staff, nursing home staff, and an independent group.

V5 Hoge, A.F., and Humphrey, G.B. "Towards Control of Breast Cancer in Oklahoma." JOURNAL OF OKLAHOMA STATE MEDICAL ASSOCIATION 68 (1975): 8-12.

> The Oklahoma Medical Research Foundation's statewide demonstration network for disseminating information to hospitals is described. Patient management plans for controlling breast cancer are a focus of the network.

V6 Jackson, M. "Integration of Psychosomatic Medicine in a Teaching Hospital: Experience with a Discussion Seminar." PSYCHOTHERAPY AND PSYCHOSOMATICS 22 (1973): 205-18.

> This paper describes an ongoing discussion seminar with psychiatrists training in a large teaching hospital. Evaluation and treatment of psychosomatic referrals within the hospital were discussion topics.

V7 Meloan, J.B. "Total Health Care--Our Responsibility." JOURNAL OF THE AMERICAN OPTOMETRIC ASSOCIATION 46 (1975): 133-41.

> Procedures to enable optometrists in recognizing patients who need referral are detailed. Various ocular and systemic conditions are reviewed, and the appropriate consultant is suggested. The nonclinical aspects of this proposed type of practice are also discussed.

V8 Metsch, J.M., and Veney, J.E. "A Model of the Adaptive Behavior of Hospital Administrators to the Mandate to Implement Consumer Participation." MEDICAL CARE 12 (1974): 338-50.

> This study develops a model of the adaptive behavior of hospital administrators to the mandate to implement consumer participation. The model consists of contextual variables describing the economic and social milieu of the hospitals, implementation variables describing structural aspects of the advisory boards, and outcome variables related to the study program objective of facilitating both delivery and community utilization of ambulatory services.

V9 Metsch, J.M., et al. "The Impact of Training on Consumer Participation in the Delivery of Health Services." HEALTH EDUCATION MONOGRAPHS 3 (1975): 251-61.

> A retrospective evaluation of training provided to six hospital consumer advisory groups is described. The training program

was designed to provide information and decision-making skills to consumers of services and community representatives serving on voluntary hospital advisory boards.

V10 Nagi, S.Z. "Gate-Keeping Decisions in Service Organizations: When Validity Fails." HUMAN ORGANIZATION 33 (1974): 47-58.

This paper presents a framework for the analysis of gatekeeping decisions in service organizations and the role of professionals in these decisions. Some implications of the patterns of gatekeeping decisions for the clients' relations to organizations and professions are also explored.

V11 Rogers, K.D. "Screening in Pediatric Practice: Review and Commentary." PEDIATRIC CLINICS OF NORTH AMERICA 21 (1974): 167-74.

This review discusses some of the rationale for screening, some of its limitations, and some developments needed to make it more effective.

V12 Stacey, M. "Consumer Complaints Procedures in the British National Health Service." SOCIAL SCIENCE AND MEDICINE 8 (1974): 429-35.

Procedures for dealing with consumer complaints in the British National Health Service are discussed, and their effectiveness for various purposes are assessed. The variation between the local authority, general practitioner, and hospital procedures is noted, and four inhibitions to the effective release and redress for consumer complaints are indicated.

V13 Tancredi, L.R. ETHICS OF HEALTH CARE. Washington, D.C.: National Academy of Sciences, 1974. 316 p.

These papers, written by authorities in such fields as law, clinical and administrative medicine, ethics and religion, economics, and psychiatry, focus on ethical implications of a variety of health-related decisions affecting individuals and the consequences of societal actions in the health field.

V14 White, P.E., and Vlasak, G.J. INTER-ORGANIZATIONAL RESEARCH IN HEALTH: BIBLIOGRAPHY (1960-1970). DHEW Publication no. (HSM) 72-3028. Rockville, Md.: U.S. Department of Health, Education, and Welfare, 1972. 167 p.

This bibliography encompasses research undertakings that focused on interorganizational relationships and dealt with health services, as reported in the periodical literature from 1960 to 1970.

Chapter 9
STAFF DEVELOPMENT AND ADMINISTRATION

The range of topics covered by this and the next chapter might be characterized as quality control for patient education. The administration strategies of continuing education, in-service training, supervision, and consultation covered in this chapter are directed primarily at improving the quality of patient education practice. The methods of evaluation covered in chapter 10 are designed to provide quality assurance or accountability for patient education practices. This perspective on education, training, and evaluation for patient education is developed more fully in the following works:

Green, L.W. "Impressions of an Overviewer." In PREPARATION AND PRACTICE OF COMMUNITY, PATIENT AND SCHOOL HEALTH EDUCATORS: PROCEEDINGS ON COMMONALITIES AND DIFFERENCES, pp. 48-54. Bethesda, Md.: Bureau of Health Manpower, Public Health Service, U.S. Department of Health, Education, and Welfare, 1978.

_____. "Suggested Procedures for Moving from Programmatic Accreditation to Peer Review Under Broader Institutional Accreditation." HEALTH EDUCATION MONOGRAPHS 4 (1976): 278-84.

_____. "Toward National Policy for Health Education." In ALCOHOL, YOUTH AND SOCIAL POLICY, edited by H. Blane and M.E. Chafetz, pp. 283-305. New York: Plenum Publishers, 1979.

_____. "What is Quality in Patient Education and How Do We Assess it?" In PROCEEDINGS: NATIONAL PATIENT EDUCATION SYMPOSIUM, 1977, edited by W. Squyres, pp. 51-66. Atlanta: Bureau of Health Education, Center for Disease Control, Public Health Service, U.S. Department of Health, Education, and Welfare, 1978.

Green, L.W., and Brooks-Bertram, P.A. "Peer Review and Quality Control in Health Education." HEALTH VALUES: ACHIEVING HIGH LEVEL WELLNESS 2 (1978): 191-97.

Institute of Medicine. ASSESSING QUALITY IN HEALTH CARE: AN EVALUATION. Washington, D.C.: National Academy of Sciences, 1976.

Ireton, H.R., and Casata, D.M. "A Psychological Systems Review." JOURNAL OF FAMILY PRACTICE 3 (1976): 155-59.

Jencks, S.F., and Green, L.W. "Developing a Hospital-Based Patient Education Program." QUALITY REVIEW BULLETIN 4 (October 1978): 8-11.

Staff Development & Administration

The remaining administrative concerns of this chapter have been covered comprehensively by volume 1 in the Health Affairs Information Guide Series:

Morris, D.A., and Morris, L.D. HEALTH CARE ADMINISTRATION: A GUIDE TO INFORMATION SOURCES. Health Affairs Information Guide Series, Vol. 1. Detroit: Gale Research Co., 1978.

Current developments in patient education for the continued development of quality in programs and practice are reported in the journals listed in the earlier chapters and in:

PATIENT EDUCATION IS IMPORTANT. Chicago: American Hospital Association, 1978.

PHYSICIAN'S PATIENT EDUCATION NEWSLETTER. Baltimore, Md.: Division of Health Education, School of Hygiene and Public Health, Johns Hopkins University (1978-- . Monthly).

The former is a package program with eighty color slides, an audio-cassette sound track, and a script for discussion leaders, designed to educate hospital staff about the need for patient education. The American Hospital Association also provides a RESOURCE CATALOG on patient education, free of charge.

W. IN-SERVICE TRAINING AND CONTINUING EDUCATION

W1 Abrams, R.D. NOT ALONE WITH CANCER: A GUIDE FOR THOSE WHO CARE; WHAT TO EXPECT; WHAT TO DO. Springfield, Ill.: Charles C Thomas, 1976. 120 p.

> This manual demonstrates that the clinical stage of cancer creates particular actions and reactions on the part of the patient that alter communication patterns between the patient and the caregivers. Written for physicians and other professionals in the health field, the book should also help the families and friends of patients with cancer to recognize and cope with the changing attitudes, needs, and wishes of the patient.

W2 Adelson, R. "The Role of the Hospital in Continuing Education." JOURNAL OF DENTAL EDUCATION 38 (1974): 487-90.

> The hospital is suggested as the appropriate setting for continuing education in dentistry, because of its existing facilities, faculty, and patients, and because it would be a reminder that dental care is an integral part of the patient's comprehensive care.

W3 Anthony, W.A. "Human Relations Training and Rehabilitation Counseling: Further Implications." REHABILITATION COUNSELING BULLETIN 17 (1974): 171-75.

Staff Development & Administration

This article responds to misperceptions that the author believes exist over the issue of systematic human relations training for rehabilitation counselors. Directions for rehabilitation counseling training from the human resource development model of Carkhuff are presented.

W4 Back, K.W. BEYOND WORDS: THE STORY OF SENSITIVITY TRAINING AND THE ENCOUNTER MOVEMENT. New York: Russell Sage Foundation, 1972.

The history of sensitivity training--T-groups; encounter groups, and the background of the movement's successes, varieties, and failures are presented. Sensitivity training is discussed as a social phenomenon.

W5 Barber, W.H., and Lurie, H.J. "Designing an Experientially Based Continuing Education Program." AMERICAN JOURNAL OF PSYCHIATRY 130 (1973): 1148-50.

The authors describe techniques used in sensitivity groups and present formats for continuing education workshops for mental health workers.

W6 Bassin, A. "Taming the Wild Paraprofessional." JOURNAL OF DRUG ISSUES 4 (1973): 333-40.

This paper examines issues and problems of paraprofessionals performing tasks and assuming responsibilities that were formerly the domain of highly trained psychiatrists, psychologists, and social workers.

W7 Bertcher, H. "Developing Group Leadership Skills." MANPOWER 6, no. 2 (1974): 10-13.

A self-instructional, self-administered workshop series on group leadership techniques is described. It is designed for counselors as well as for leaders of orientation groups, staff meetings, staff development programs, and similar activities.

W8 Blanton, G.W., and Heestand, D.E. "Use of Two-Way Television in Physical Therapy." PHYSICAL THERAPY 53 (1973): 867-68.

A closed circuit television system is described as a continuing education program tool for physical therapists in rural areas in Nebraska. The system also facilitates communications among health professionals and assists in patient transfers to other hospitals.

W9 Blumenfeld, W.S., and Crane, D.P. "Opinions of Training Effectiveness. How Good?" TRAINING DEVELOPMENT JOURNAL 27 (December 1973): 42-51.

Staff Development & Administration

The empirical research study reported in this paper concerns management perceptions of the effectiveness of various training techniques and the extent to which these perceptions are based on quality evidence.

W10 Boller, J.D. "Differential Effects of Two T-Group Styles." COUNSELOR EDUCATION AND SUPERVISION 14 (1974): 117-23.

A rationale for examining effects of the T-group on introverts and extroverts is presented. Two T-group styles are examined: a sensory awareness group and a verbal cognitive group.

W11 Braby, R., et al. A TECHNIQUE FOR CHOOSING COST-EFFECTIVE INSTRUCTIONAL DELIVERY SYSTEMS. Orlando, Fla.: Naval Training Equipment Center, Training Analysis and Evaluation Group, April, 1975. 117 p.

This report describes the Training Effectiveness, Cost Effectiveness Prediction technique. This technique is an approach for training system designers to use in making delivery system choices during the conceptual design phase. A three-step procedure is described in which training objectives are classified and organized into groups; appropriate learning strategies are defined; media are identified; and costs of alternative forms of training are projected.

W12 Calia, V.F. "Systematic Human Relations Training: Appraisal and Status." COUNSELOR EDUCATION AND SUPERVISION 14 (1974): 85-94.

The impact of the Systematic Human Relations Training model, the rudiments of the model, and its historical antecedents are described, and various technical and philosophical issues are identified.

W13 Canada, R.M. "Immediate Reinforcement versus Delayed Reinforcement in Teaching a Basic Interview Technique." JOURNAL OF COUNSELING PSYCHOLOGY 20 (1973): 395-98.

Thirty state employment service interviewers in a two-year paraprofessional training program were subjects in an exercise to test immediate versus delayed reinforcement in teaching counselors-in-training a basic interview skill.

W14 Canfield, E. "Pregnancy and Birth Control Counseling." JOURNAL OF SOCIAL ISSUES 30, no. 1 (1974): 87-96.

Clinical illustrations of pregnancy counseling are given, and points that should be covered in counseling are outlined. The selection and training of counselors and screening procedures are covered, with "how to" procedures included.

Staff Development & Administration

W15   Canter, L., and Paulson, T. "A College Credit Model of In-School Consultation: A Functional Behavioral Training Program." COMMUNITY MENTAL HEALTH JOURNAL 10 (1974): 268-75.

        This pilot study combined the didactic structure and credit of a college seminar with the immediate relevance, support, and training of a traditional in-school consultation. Teachers were trained in the implementation of functional-behavioral intervention skills. Evaluative data were collected and reported.

W16   Caplan, R.M. "Measuring the Effectiveness of Continuing Medical Education." JOURNAL OF MEDICAL EDUCATION 48 (1973): 1150-52.

        An example is given of behaviorally oriented evaluation research in continuing medical education. Aspects of the value of conventional learning formats for physicians and their patients are discussed.

W17   Chambers, D.W., and Hamilton, D.L. "Continuing Dental Education: Reasonable Answers to Unreasonable Questions." JOURNAL OF THE AMERICAN DENTAL ASSOCIATION 90 (1975): 116-20.

        Arguments from educational research and evaluation theory are used to show potential dangers involved in compulsory, mass-media continuing dental education with pretest and posttest measures of success.

W18   Clark, T.L., et al. PATIENT SERVICES TRAINING MANUAL. San Francisco: Planned Parenthood/World Population, 1973.

        This manual, written to aid volunteers and staff in Planned Parenthood/World Population contraceptive centers in the San Francisco-Alameda area, covers human reproduction, birth control, psychological aspects of contraceptive use, venereal disease, sterilization, teen services, and pregnancy counseling. Guidelines for family planning class leaders and medical aides are provided.

W19   Danish, S.J., and Brock, G.W. "The Current Status of Paraprofessional Training." PERSONNEL AND GUIDANCE JOURNAL 53 (1974): 299-303.

        This paper discusses several unresolved issues concerning the utilization of paraprofessionals and how these issues have affected their training. Four training programs are briefly examined, with an elaboration on the implementation of one of them focusing on the objectives and processes of the training.

Staff Development & Administration

W20 D'Augelli, A.R., et al. "The Effect of Group Composition and Duration on Sensitivity Training." SMALL GROUP BEHAVIOR 5 (1974): 56-64.

> This research investigated the relationship of group composition and duration to the verbal behavior of sensitivity training groups. This study provides information about the process of sensitivity training and its assessment.

W21 Distefano, M.K., Jr., and Pryer, M.W. "Effect of Brief Training on Mental Health Knowledge and Attitudes of Nurses and Nurses' Aides in a General Hospital." NURSING RESEARCH 24 (1975): 40-42.

> A six-week psychiatric nursing training program for professional nurses and nurses' aides in a general hospital setting is described. The Opinions about Mental Illness scale was used to measure change in the attitudes of the trainees toward mental health.

W22 Dowd, E.T., and Blocher, D.H. "Effects of Immediate Reinforcement and Awareness on Beginning Counselor Behavior." COUNSELOR EDUCATION AND SUPERVISION 13 (1974): 190-97.

> This investigation studied the effects of immediate reinforcement and awareness of response on the acquisition of a complex counselor behavior by beginning counselors. Implications for counselor education and for the "learning without awareness" controversy are discussed.

W23 Dyer, E.D., et al. "Increasing the Quality of Patient Care Through Performance Counseling and Written Goal Setting." NURSING RESEARCH 24 (1975): 138-44.

> In seven Veterans Administration hospitals, staff nurses who used a performance counseling protocol that required written goal setting, coupled with head nurse modeling and support to achieve goals were given "patient care scores." This group's scores was higher on five patient care scales when compared with scores of nurses who continued with usual practice. Staff nurses chose their own goals: to improve patient care and to improve staff nurse professional competence. Relationships of three instruments that measure quality of patient care are presented.

W24 Eiben, R., and Clack, R.J. "Impact of a Participatory Group Experience on Counselors in Training." SMALL GROUP BEHAVIOR 4 (1973): 486.

> The relevance of group experiences in training counselors and the appropriateness of the Personal Orientation Inventory as a sole measure of interpersonal change are examined.

Staff Development & Administration

W25 Farmer, E.D. "The Delivery of Continuing Dental Education." INTERNATIONAL DENTAL JOURNAL 24 (1974): 427-34.

>This article describes means of sponsoring and systems of delivering continuing education in selected countries.

W26 Fitzhugh, Z.A., et al. "A Patient-Centered Orientation Program." HOSPITALS: JOURNAL OF THE AMERICAN HOSPITAL ASSOCIATION 48 (16 January 1974): 72-78.

>A patient-centered approach that focused on an individual patient, following the course of his hospitalization, was used to orient new nurses at an urban university hospital.

W27 Flomenhaft, K., and Carter, R.E. "Family Therapy Training: A Statewide Program for Mental Health Centers." HOSPITAL AND COMMUNITY PSYCHIATRY 25 (1974): 789-91.

>This paper describes staff training programs in family therapy techniques. The programs are conducted one day a week for five months at mental health centers throughout Pennsylvania.

W28 Fry, L.J., and Miller, J.P. "The Impact of Interdisciplinary Teams on Organizational Relationships." SOCIOLOGICAL QUARTERLY 15 (1974): 417-31.

>Based upon research in an alcoholism treatment organization, this study explores the impact of interdisciplinary team treatment on organizational participants and structure.

W29 Gilandas, A.J. "Implications of the Problem-Oriented Record for Utilization Review and Continuing Education." HOSPITAL AND COMMUNITY PSYCHIATRY 25 (1974): 22-24.

>This article describes the application of the problem-oriented approach to recordkeeping to utilization review and continuing education in a hospital treatment unit. Deficiencies in patient management revealed in the audit provide a basis for constructing relevant continuing education programs for professional staff.

W30 Gluckstern, N.B., and Ivey, A.E. BASIC ATTENDING SKILLS: SYSTEMATIC VIDEOTRAINING FOR BEGINNING HELPERS AND PARAPROFESSIONALS. Rev. ed. North Amherst, Mass.: Microtraining Associates, 1975. 7 p.

>Designed as an introductory program for beginning helpers, the Basic Attending Skills program consists of microtraining video materials, a step-by-step trainer's guide, and participant manuals. This paper provides basic details about the program, its origins, and its effectiveness.

Staff Development & Administration

W31 Grotjahn, M. "Teaching the Teacher to Teach." INTERNATIONAL JOURNAL OF EPIDEMIOLOGY 2 (1973): 435-38.

    Participants, individually and in groups, undertook exercises in developing educational objectives, designing experimental teaching units, and selecting appropriate mechanisms for evaluation. Hypothetical undergraduate curricula were drafted.

W32 Hiemstra, R., and Long, R. "A Survey of 'Felt' Versus 'Real' Needs of Physical Therapists." ADULT EDUCATION (Washington) 24 (1974): 270-79.

    This research determined educational needs of physical therapists in order to plan continuing education. A questionnaire administered to seventy-seven physical therapists determined and compared felt needs (personal perceptions symptomatic of problems) and real needs (actual knowledge or skill weaknesses).

W33 Hirsch, E.O. "Utilization Review as a Means of Continuing Education." MEDICAL CARE 12 (1974): 358-62.

    Utilization review was used to improve the efficiency of medical care and to achieve optimal appropriate outcome with reduced use of resources. Eight months after an educational effort, decreases in selected problems indicated that greater efficiency had taken place.

W34 Hollister, W.G., and Edgerton, J.W. "Teaching Relationship Building Skills." AMERICAN JOURNAL OF PUBLIC HEALTH 64 (1974): 41-46.

    This paper reports a training sequence, field-tested and refined over the last twenty years, to develop trustful, cooperative working relationships.

W35 Inui, T.S., et al. "Improved Outcomes in Hypertension and Physician Tutorials: A Controlled Trial. ANNALS OF INTERNAL MEDICINE 84 (1976): 646-51.

    This study dealt with postgraduate physician education and the care of hypertensive patients in the setting of a general medical outpatient clinic. Its purpose was to determine whether an experimental tutorial effort directed at clinic physicians could produce demonstrable change in process and outcomes in the care of patients with hypertension.

W36 Johnson, E.M. NURSING HOME EVALUATION CONTRACT: REPORT FOR AN EVALUATION OF ELEVEN TRAINING PROGRAMS FOR NURSING HOME PERSONNEL. Chicago: Johnson and Associates, August, 1973. 137 p.

Staff Development & Administration

This report is a comparative study of the two educational approaches used by the Health Resources Administration in its Long-Term Care Education Strategy to train staff who provide services to patients in nursing homes. Eleven projects were selected for this study of the multidisciplinary and unidisciplinary training approaches.

W37 Jongsma, E.A., and Gaines, W.G. "The Effectiveness of an In-Service Training Session Pertaining to Instructional Objectives." SOUTHERN JOURNAL OF EDUCATIONAL RESEARCH 7 (1973): 148-56.

This study assessed knowledge of and attitudes toward instructional objectives of eighteen participants in an in-service training session, using a one-group, pretest-posttest design.

W38 Kaye, J.D. "Group Interaction and Interpersonal Learning." SMALL GROUP BEHAVIOR 4 (1973): 424-48.

Two T-groups were rated and compared in terms of the interpersonal learning that might be expected to occur as a result of a member-centered and emotionally-involving interactional climate within the groups.

W39 Kohle, K., et al. "Training of a Nursing Staff in Psychomatic Medicine in a Medical Clinic." PSYCHOSOMATICS 14 (1973): 336-40.

A group was formed to give nurses the opportunity to talk about emotional problems experienced in their association with patients.

W40 Levin, E.M., and Kurtz, R.R. "Structured and Nonstructured Human Relations Training." JOURNAL OF COUNSELING PSYCHOLOGY 21 (1974): 526-31.

Participant perceptions following structured and nonstructured human relations training are investigated.

W41 Lindsay, C.A., et al. "Professional Obsolescence: Implications for Continuing Professional Education." ADULT EDUCATION (Washington) 25 (1974): 3-22.

This paper describes the utility of the Content-Based Group-Assessment Model for developing programs of continuing education. The model provides a procedure for assessing educational needs of a defined group of practitioners and for translating identified needs into knowledge areas to form the basis for developing continuing education programs.

W42 McLaughlin, F.E., et al. "Modes of Interpersonal Feedback and Leadership Structure in Six Small Groups." NURSING RESEARCH 23 (1974): 307-18.

Staff Development & Administration

Forty-three graduate nursing students participated in a study of six small-group formats. Participants were tested on six instruments: Gough's Adjective Check List, the Semantic Differential Rating Scale, the Training Group Rating Scale, Whalen's Group Rating Category System, the Fundamental Interpersonal Relations Orientation--Behavior Scale, and a videotape test of cognitive-affective competence in knowledge of group process.

W43 Martin, R.D., and Shepel, L.F. "Locus of Control and Discrimination Ability with Lay Counselors." JOURNAL OF CONSULTING AND CLINICAL PSYCHOLOGY 42 (1974): 741.

The research reported examines the effects of brief training in helping relations on the variables of discrimination ability and locus of control. It tested the hypothesis that structured training would increase the ability to discriminate helpful counseling conditions and cause a shift toward the internality dimension of locus of control with associated increases in trust, insight, and self-confidence. Subjects were twenty-one senior female nurses from urban hospitals.

W44 Mims, F., et al. "Effectiveness of an Interdisciplinary Course in Human Sexuality." NURSING RESEARCH 23 (1974): 248-53.

A five-day human sexuality program was given to seventy medical students, thirty-seven nursing students, and other university students. Goals were to help participants increase sexual knowledge, to help desensitize against stressful and anxiety reactions to sexual stimuli, and to resensitize for a broader understanding of sexuality of self and others. A four-part Sexual Knowledge and Attitudes Test was used to measure participants' pre- and postcourse knowledge and attitudes.

W45 Montgomery, L.J. "The Sensitivity Movement: Questions to Be Researched." SMALL GROUP BEHAVIOR 4 (1973): 387-406.

Questions concerning the applicability of sensitivity group work are considered with reference to overall educational goals and to the particular practices of professional counselors.

W46 Moore, M. "Training Professionals to Work with Paraprofessionals." PERSONNEL AND GUIDANCE JOURNAL 53 (1974): 308-12.

This article presents a model for systematically training mental health professionals to work effectively with paraprofessionals.

W47 Nellis, D.H., and Burgess, J.H. "An Administrative Technique for Personnel Management and Training in Mental Health." PUBLIC AD-

Staff Development & Administration

MINISTRATION REVIEW 34 (1974): 496-99.

> The authors present a department of mental health's model of job classifications designed to compensate for manpower deficiencies through selection and training.

W48 Passons, W.R., and Garrett, L.D. "An Inservice Workshop on Group Counseling." PERSONNEL AND GUIDANCE JOURNAL 52 (1974): 482-86.

> This article reports on the contents, processes, and outcomes of a two-day workshop for training counselors in group counseling.

W49 Paul, G.L., and McInnis, T.L. "Attitudinal Changes Associated with Two Approaches to Training Mental Health Technicians in Milieu and Social-Learning Programs." JOURNAL OF CONSULTING AND CLINICAL PSYCHOLOGY 42 (1974): 21-33.

> Training programs are compared for two groups of nonprofessionals (n = twenty-one each) selected from a high unemployment area. Both groups received specific job-related training in behavioral principles and procedures of both milieu and social-learning treatment programs for chronic mental patients. The group receiving classroom instruction followed by on-the-job training had a better academic performance than the group receiving classroom instruction integrated with clinical observation.

W50 Powell, B.J., et al. "Attitude Changes of General Hospital Personnel Following an Alcoholism Training Program." PSYCHOLOGICAL REPORTS 34 (1974): 461-62.

> Improved attitudes of twenty-five nursing personnel toward a concept about alcoholics and their treatment potential, were observed after a three-day training program. The purpose of the study was to determine if these changes in concepts would remain stable and influence the effectiveness of the participants in their subsequent work with alcoholics.

W51 Prentice, E.D., and Metcalf, W.K. "A Teaching Workshop for Medical Educators." JOURNAL OF MEDICAL EDUCATION 49 (1974): 1031-34.

> The Teaching Workshop for Medical Educators, a course designed to aid participants in developing teaching skills, is described. Workshop members participate in topic discussions, develop self-instructional units, design courses in the medical sciences, and present lectures and teaching demonstrations that are videotaped and critiqued.

Staff Development & Administration

W52 Quilitch, H.R., et al. "Teaching Personnel To Implement Behavioral Programs." EDUCATIONAL TECHNOLOGY 15 (January 1975): 27-31.

> This article outlines a program to train behavior modification skills, which specifies student terminal behaviors and criteria for judging these behaviors. A professional model demonstrates skills and supervises the students, providing them with feedback and reinforcement for conducting individual behavior modification projects.

W53 ROLE OF COMMUNITY HOSPITALS IN CONTINUING EDUCATION OF HEALTH PROFESSIONALS. Arlington, Va.: Association for Hospital Medical Education, 1975. 83 p.

> This study determined the functioning of consortia of community hospitals in continuing medical education.

W54 Sawyer, W.A., and Mangiaracina, J. "Area Health Education Centers as the Foundation for a Statewide Biomedical Communications Network." BULLETIN OF THE MEDICAL LIBRARY ASSOCIATION 62 (1974): 343-47.

> Four Area Health Education Centers in South Carolina, based in community hospitals, provide residency programs and clinical instruction for medical, dental, nursing, pharmacy, and allied health students and continuing educational programs for health professionals. A plan for delivering library and other information services to the designated Area Health Education Centers is described.

W55 Schlesinger, E.R., et al. "A Controlled Test of the Use of Registered Nurses for Prenatal Care." HEALTH SERVICES REPORTS 88 (1973): 400-404.

> A Prenatal Care Personnel Utilization Project, initiated in 1967 at the University of Pittsburgh, focused on new approaches to prenatal care in an outpatient clinic serving a low socioeconomic, predominantly black population. The project was designed to develop, implement, and test, in a controlled situation, the use of registered nurses for the bulk of prenatal care traditionally reserved to more highly trained persons.

W56 Schwartz, A.N. "Staff Development and Morale Building in Nursing Homes." GERONTOLOGIST 14 (1974): 50-53.

> The high rate of staff turnover in long-term care facilities is briefly discussed in relation to costs and staff morale. Procedures for the development and training of staff in nursing homes are outlined.

W57 Shapiro, J.L., and Gust, T. "Counselor Training for Facilitative

Staff Development & Administration

Human Relationships." COUNSELOR EDUCATION AND SUPERVISION 13 (1974): 198-206.

>This study evaluated the effects of the prepracticum stage of a three-stage counselor training practicum designed to merge cognitive and experiential learning in a systematic manner, using professional trainers.

W58 Shaw, D.G. MEETING HEALTH MANPOWER NEEDS THROUGH MORE EFFECTIVE USE OF ALLIED HEALTH WORKERS. Manpower Research Monograph, No. 25. Washington, D.C.: U.S. Department of Labor, 1973. v, 23 p.

>This monograph summarizes the findings and discusses the implications of several research and development studies dealing with varied problems of the supply and training of allied health manpower.

W59 Stearns, N.S., et al. "Impact of Program Development Consultation on Continuing Medical Education in Hospitals." JOURNAL OF MEDICAL EDUCATION 49 (1974): 1158-65.

>A three-year study to assess the impact of educational consultation on the development of continuing medical education activities in community hospitals is reported. Characteristics of the consultation process, the hospital, and the medical staff were examined to determine whether facilitating or inhibiting relationships to consultation outcomes existed.

W60 Stephenson, R.W., and Burkett, J.R. AN ACTION ORIENTED REVIEW OF THE ON-THE-JOB TRAINING LITERATURE. Washington, D.C.: American Institutes for Research, 1974. 171 p.

>On-the-job training literature is reviewed, and references are organized under the following categories: literature reviews and bibliographies, handbooks and manuals, cost effectiveness, technique comparison studies, systems analysis of training, approaches to program evaluation, and military documents.

W61 Stimpson, D.V. "T-Group Training to Improve Counseling Skills." JOURNAL OF PSYCHOLOGY 89 (1975): 89-94.

>An interpersonal response tendency questionnaire measured the relative frequency of probing, advising, and accepting responses to a request for help.

W62 Stone, L.A., and Brosseau, J.D. "Cross-Validation of a System for Predicting Training Success of Medex Trainees." PSYCHOLOGICAL REPORTS 33 (1973): 917-18.

>A previously developed multiple-regression model for predicting

Staff Development & Administration

success of Medex trainees in a training program was cross-validated, using a new group of Medex trainees.

W63   Thies, J.B. "Hospital Personnel and Computer-Based Systems: A Study of Attitudes and Perceptions." HOSPITAL ADMINISTRATION 20 (Winter 1975): 17-26.

     This article reports summary findings from a multi-hospital systems evaluation project in which personnel attitudes toward computer-based systems were studied.

W64   Tobin, H.M., et al. THE PROCESS OF STAFF DEVELOPMENT: COMPONENTS FOR CHANGE. St. Louis: C.V. Mosby Co., 1974. xi, 174 p. Append. Illus.

     Topics covered in this text include continuing education and staff development; organization and administration; philosophy, purpose and goals; identifying learning needs; designing and implementing learning offerings; selecting teaching methods and aids; and evaluation.

W65   U.S. Department of Health, Education, and Welfare, National Institute of Mental Health, Alcohol, Drug Abuse, and Mental Health Administration. CONTINUING EDUCATION IN MENTAL HEALTH. DHEW Publication no. HSM 73-9126. Rockville, Md.: 1974. vii, 207 p.

     Reports of over one hundred projects describe training techniques for personnel working in mental health and related service delivery systems. Training designs, methodologies, and evaluation methods are described for continuing education programs for physicians, pyschiatrists, behavioral scientists, nurses, social workers, and related care-giving personnel.

W66   VanderKolk, C.J. "Comparison of Two Mental Health Counselor Training Programs." COMMUNITY MENTAL HEALTH JOURNAL 9 (1973): 260-69.

     Two groups of psychiatric attendants were trained using an integrated, didactic-experiential approach (n = 14) and a traditional training approach (n = 14). Both training groups demonstrated significantly greater change in interpersonal skill than the control group (n = 14) and the integrated group showed greater positive change than the traditional group.

W67   Wallace, C.J., et al. "Modeling and Staff Behavior." JOURNAL OF COUNSELING PSYCHOLOGY 41 (1973): 422-25.

     This study investigated the effects of modeling and instructions on the frequency of interaction between nursing staff and patients in a neuropsychiatric facility.

Staff Development & Administration

W68 Waring, M.L. "The Impact of Specialized Training in Alcoholism on Management-Level Professionals." JOURNAL OF STUDIES ON ALCOHOL 36 (1975): 406-15.

> In the year after attending an eight-week alcoholism training program, administrative and management-level social workers and nurses significantly increased their alcohol-related work activities.

W69 Wehmer, G., et al. "Evaluation of the Effects of Training of Paraprofessionals in the Treatment of Alcoholism: A Pilot Study." BRITISH JOURNAL OF ADDICTION 69 (1974): 25-32.

> A program for the training of paraprofessionals in alcoholism is outlined. Characteristics of persons attracted to the program are described; and the effects of the program on a pilot group of thirteen are evaluated.

W70 Wells, R.A. "Training in Facilitative Skills." SOCIAL WORK 20 (1975): 242-43.

> Systematic training in the facilitative qualities--empathy, respect, and genuineness--was devised to develop and enhance their use by trainees in the helping professions.

W71 Wolle, J.M. "Multidisciplinary Teams Develop Programming for Patient Education." HEALTH SERVICES REPORTS 89 (1974): 8-12.

> To promote more effective patient education programming in state health facilities, a state department of health convened a workshop to serve as a focal point for bringing together representatives from a variety of health facilities who were already interested in patient education but who needed help in further development plans. This paper reports on the workshop and the following aspects: purpose, financial arrangements, preworkshop questionnaire, participants, special materials, design, patient education plans, and follow-up plans.

X. CONSULTATION AND SUPERVISION

X1 Allen, L.A. "M for Management: Theory Y Updated." PERSONNEL JOURNAL 52 (1973): 1061-67.

> The author believes that managers can make good use of his Theory M by helping people realize that working toward team goals leads to greater personal rewards than singleminded pursuit of personal objectives. Professional managers may find, also, that they can help overcome resistance to change by maintaining good communications and by giving people more responsibility and the authority to make the most of their own decisions.

Staff Development & Administration

X2 Berkanovic, E., and Vander Haegen, E. "Power Strategies in Professional Organizations: The Case of a Mental Hospital." HOSPITAL ADMINISTRATION 19 (Spring 1974): 52-62.

> Formal changes in the authority structure of a state mental hospital were observed. From a structure that approximated the model of a professional organization, the hospital changed to a dual authority structure in which authority resided as much with the heads of the functional units of the hospital as it did with professional division heads.

X3 Bragg, J.E., and Andrews, I.R. "Participative Decision Making: An Experimental Study in a Hospital." JOURNAL OF APPLIED BEHAVIORAL SCIENCE 9 (1973): 727-35.

> Participative decision making was introduced into a hospital subsystem over an eighteen-month period. Attitudes improved; absence rates declined; and productivity increased.

X4 Evans, M.G. "Extensions of a Path-Goal Theory of Motivation." JOURNAL OF APPLIED PSYCHOLOGY 59 (1974): 172-78.

> This article extends and replicates the author's hypothesis concerning the way the behavior of superiors affects subordinates' perceptions of expectancies and instrumentalities in the path-goal theory of motivation. The role of motivation as an intervening variable between supervisory behavior and subordinate behavior and as a moderator in the behavior satisfaction relationship is explored.

X5 Freeborn, D.K., and Darsky, B.J. "A Study of the Power Structure of the Medical Community." MEDICAL CARE 12 (1974): 1-12.

> This study, based upon responses to a questionnaire given to the total population of physicians in private practice in Windsor, Ontario, analyzed the structure and bases of power within a local medical community. The purpose of the study was to determine the extent to which power is unitary or differentiated.

X6 Gaoni, B., and Neumann, M. "Supervision from the Point of View of the Supervisee." AMERICAN JOURNAL OF PSYCHOTHERAPY 28 (1974): 108-14.

> The authors describe the supervisory process from the point of view of the supervisee, four main stages of the learning process, and the modifications necessary with the development of the supervisee's individual and professional personality.

X7 Goin, M.K., and Kline, F.M. "Supervision Observed." JOURNAL OF NERVOUS AND MENTAL DISEASE 158 (1974): 208-13.

Staff Development & Administration

The meetings of five supervisors with their second-year psychiatric residents were videotaped, and the tapes were studied to determine what factors distinguish the outstanding supervisor from the moderately good one. According to the authors, this is the first report in the literature of the actual observation of supervision.

X8 Good, K.C., and Good, L.R. "Attitude Similarity and Liking for a Supervisor." JOURNAL OF PSYCHOLOGY 88 (1974): 313-16.

This study tested the hypothesis that a job supervisor who is attitudinally similar to the evaluator will be evaluated more positively than an attitudinally dissimilar one on the variables: evaluating employees, understanding of people, open-mindedness, judgment in a work situation, personal attractiveness, and desirability as a supervisor.

X9 Gruenfeld, L., and Kassum, S. "Supervisory Style and Organizational Effectiveness in a Pediatric Hospital." PERSONNEL PSYCHOLOGY 26 (1973): 531-44.

This study, conducted in a pediatric hospital, examines previous findings of the interactive effects of two dimensions of leadership style, initiation structure and consideration, and finds them generalizable to the supervision of female nurses.

X10 Heilbrun, A.B., Jr. "Interviewer Style, Client Satisfaction, and Premature Termination Following the Initial Counseling Contact." JOURNAL OF COUNSELING PSYCHOLOGY 21 (1974): 346-50.

Two alternative explanations for early defection of female counseling clients were tested. Evaluations of interviewer directiveness and interview satisfaction were obtained from clients immediately following the initial interview.

X11 Ilgen, D.R., and O'Brien, G. "Leader-Member Relations in Small Groups." ORGANIZATIONAL BEHAVIOR AND HUMAN PERFORMANCE 12 (1974): 335-50.

Influence of task organization and group composition factors on leader-member relations was investigated, with leader-member relations defined according to F.E. Fiedler's Contingency Model of Leadership.

X12 Ivancevich, J.M. "A Study of a Cognitive Training Program: Trainer Styles and Group Development." ACADEMY OF MANAGEMENT JOURNAL 17 (1974): 428-39.

Two groups of managers in a cognitive-oriented training program were studied. The structured or directive style was found to be more effective in achieving group cohesiveness, minimizing conflict, increasing communication, achieving group productivity, and encouraging a more favorable attitude toward the trainer than a less structured trainer style.

Staff Development & Administration

X13 Jones, A.P., et al. "Perceived Leadership Behavior and Employee Confidence in the Leader as Moderated by Job Involvement." JOURNAL OF APPLIED PSYCHOLOGY 60 (1975): 146-49.

> This study examines the effect of job involvement upon the relationship between perceived leader behaviors and confidence and trust in the leader. The high-job-involvement sample tended to have significantly lower correlations between confidence and trust and between leadership variables.

X14 Kadushin, A. "Supervisor-Supervisee: A Survey." SOCIAL WORK 19 (1974): 288-97.

> This article analyzes a nationwide survey of 750 supervisors and 750 supervisees. It is divided into the sections: project overview, supervisory power, satisfaction in supervision, functions and objectives, the ideal and the actual, classical dilemmas, and respondents' comments.

X15 Lindahl, R.L., et al. "A Survey of the Attitudes of Dentists Toward Expanding Auxiliaries' Duties." HEALTH SERVICES REPORTS 88 (1973): 423-26.

> This survey of dentists in North Carolina reports attitudes of dentists toward expanding the roles of auxiliaries.

X16 Lyon, H.L., and Ivancevich, J.M. "An Exploratory Investigation of Organizational Climate and Job Satisfaction in a Hospital." ACADEMY OF MANAGEMENT JOURNAL 17 (1974): 635-48.

> The organizational climate of a hospital and its impact on job satisfaction of nurses and administrators are investigated.

X17 McDonnell, J.F. "The Human Element in Decision Making." PERSONNEL JOURNAL 54 (1974): 188-90.

> The decision-making process in reaching organizational objectives is discussed in relation to bias, subjectiveness, perceived pressures, and obligations of the decision maker, as human elements that may impede goals.

X18 Marwell, G., and Schmitt, D.R. COOPERATION: AN EXPERIMENTAL ANALYSIS. New York: Academic Press, 1975. 218 p.

> This book reports results of six years of programmatic research designed to uncover factors that inhibit, maintain, or promote cooperation. Using the results of some thirty interrelated experiments, it considers the relationship between the concept of cooperation and the various operational definitions used in the literature.

X19 Mitchell, T.R. "Motivation and Participation: An Integration." ACADEMY OF MANAGEMENT JOURNAL 16 (1973): 670-79.

An integration of the rational model of expectancy theory and the "caring for people" philosophy of the participative approach to employee motivation is presented.

X20 Nash, A. "Hospital Values, Conflicts and Supervisory Practices." PERSONNEL JOURNAL 52 (1973): 1056-67.

The author's belief that hospitals differ from businesses and industrial organizations in being relatively free from the labor-management disputes found in profit-making enterprises is based on the hospital's value system. This system, centered around patient care, motivates employees to play a constructive role and comply more willing with hospital rules and regulations.

X21 Nash, K.B., and Mittlefehldt, V.A. "Supervision and the Emerging Professional." AMERICAN JOURNAL OF ORTHOPSYCHIATRY 5 (1975): 93-101.

A model for training supervisors of new professionals is presented. Concepts, principles, and methods of supervision are outlined with emphasis on such issues as entry, race, class, sex, and upward mobility of the emerging professional in the established mental health system.

X22 Patti, R.J. "Organizational Resistance and Change: The View from Below." SOCIAL SERVICE REVIEW 48 (1974): 367-83.

An analytic framework is outlined to assess elements of resistance to organizational change proposals coming from low-power practitioners in social agencies, where changes must be approved by an administrative superior. Implications for goal selection and organizational change strategies are presented.

X23 Pfeffer, J., and Salancik, G.R. "Determinants of Supervisory Behavior: A Role Set Analysis." HUMAN RELATIONS 28 (1975): 139-54.

This study examines some determinants of supervisory actions. It is argued that the behavior of the supervisor is constrained by the demands of others in his or her role set. Results are discussed in terms of Weick's (1969) concept that individuals in an organization will interlock behaviors to obtain a stable, mutually satisfying interaction.

X24 Schneier, C.E. "Behavior Modification: A Review and Critique." ACADEMY OF MANAGEMENT JOURNAL 17 (1974): 528-48.

The use of operant principles and behavior modification techniques in the management literature is reviewed and critiqued. Included are studies explaining operant principles, organiza-

Staff Development & Administration

tional research testing these principles, and work regarding applications of behavior modification in organizations.

X25 Shelton, J.E. "Counselor Characteristics and Effectiveness in Serving Economically Disadvantaged and Advantaged Males." COUNSELOR EDUCATION AND SUPERVISION 13 (1973): 129-36.

Personality characteristics, as determined by the Sixteen Personality Factor Questionnaire, were compared for junior and senior high school counselors serving economically disadvantaged and advantaged males.

X26 Shirley, R.C. "A Model for Analysis of Organizational Change." MSU (MICHIGAN STATE UNIVERSITY) BUSINESS TOPICS 22, no. 2 (1974): 60-67.

A generic framework is given for the analysis and management of organizational change.

X27 Szilagyi, A.D., and Sims, H.P., Jr. "An Exploration of the Path-Goal Theory of Leadership in a Health Care Environment." ACADEMY OF MANAGEMENT JOURNAL 17 (1974): 622-34.

Relationships between leader behavior and subordinate satisfaction and performance were studied at multiple occupational skill levels in a hospital.

X28 Van de Ven, A.H., and Delbecq, A.L. "The Effectiveness of Nominal, Delphi, and Interacting Group Decision Making Processes." ACADEMY OF MANAGEMENT JOURNAL 17 (1974): 605-21.

The conventional interacting group is compared with nominal and delphi groups in terms of the quantity of ideas generated and the perceived satisfaction of participants.

X29 Watson, J.P. THE INTERACTION BETWEEN LEADERSHIP, CLIMATE, AND SATISFACTION IN A PROFESSIONAL ORGANIZATION. Monterey, Calif.: Naval Postgraduate School, 1974. 77 p.

This study investigated the correlation between leadership style, organizational climate, and employee satisfaction. Measurements of leadership style, organizational climate, and employee satisfaction were compiled from a modification of a questionnaire developed by Taylor and Bowers.

X30 Wolfe, J., and Moe, B.L. "An Experimental Evaluation of a Hospital Supervisory Training Program." HOSPITAL ADMINISTRATION 18 (1973): 65-77.

This study describes an experimental evaluation for a supervisor's training program in a large, urban medical center. The program attempted to implement the commitment approach in that the supervisors of those being trained were also involved in the training process.

Y. PROFESSIONAL PREPARATION AND QUALITY CONTROL

Y1 Abou-Rass, M. "Effects of Varying Sequence and Amount of Training on Learning and Performance in Pre-Clinical Endodontics: Part 1: Design, Experimental Procedures, and Sequencing." JOURNAL OF DENTAL EDUCATION 38 (1974): 32-41.

>An educational research experiment was conducted at a school of dentistry to investigate the effects of sequencing and amounts of teaching on learning.

Y2 Adamson, T.E. "The Teaching and Use of the Problem-Oriented Medical Record in Medical Schools." JOURNAL OF MEDICAL EDUCATION 49 (1974): 905-7.

>Responses to 117 survey questionnaires, completed by deans of medical schools in the United States and Canada, reported on use and problems with instruction of the problem-oriented medical record.

Y3 Allan, F.N. "Communication of Drug Information to the Physician." FOOD DRUG COSMETIC LAW JOURNAL 29 (1974): 146-53.

>Where physicians obtain information about drugs, how well the physician uses the sources of information, how much influence the transmitters exert on his or her prescribing, and the reliability of the information are some of the issues raised in this article.

Y4 Anderson, J.D. "Human Relations Training and Group Work." SOCIAL WORK 20 (1975): 195-99.

>Research on new group approaches indicates that alienation can be combatted through the teaching of empathy. The author suggests a framework for adapting the new technique to group work.

Y5 Auter, S.B., and Zide, E.D. "Master's Level Professional Training in Clinical Psychology and Community Mental Health." PROFESSIONAL PSYCHOLOGY 5 (1974): 115-21.

>This article describes a master's-level training program for psychologists at Massachusetts General Hospital. The program consists of intensive and extensive clinical-community field experience and related case work.

Staff Development & Administration

Y6   Baker, E.J. "The Mental Health Associate: One Year Later." COMMUNITY MENTAL HEALTH JOURNAL 9 (1973): 203-14.

> Changes occurring over a one-year interval among a group of mental health associates, some of the first graduates of two-year programs, are examined. A group of graduates entering the field one year later is included. Comparisons are made of personal data, career objectives, obstacles encountered, utilization by agencies, and perceived characteristics of work situations that provided or failed to provide them with satisfactory working conditions.

Y7   Barr, D.M., and Gaus, C.R. "A Population-Based Approach to Quality Assessment in Health Maintenance Organizations." MEDICAL CARE 11 (1973): 523-28.

> This article presents an approach to quality assessment, which focuses on the accessibility, efficiency, and effectiveness of providing care. The approach is population-based because accessibility is included as a dimension of quality.

Y8   Barry, G.M. "Evaluation Technicians: Who Are They? Who Needs Them?" EDUCATIONAL TECHNOLOGY 14 (February 1974): 49-51.

> Due to the lack of skilled evaluators, semiprofessional evaluation technicians trained to evaluate innovative educational research are advocated as a means to translate research into educational programs and to provide ongoing evaluation.

Y9   Bellin, L.E. PSRO--Quality Control? Or Gimmickry?" MEDICAL CARE 12 (1974): 1012-18.

> This article discusses efforts to implement quality control and the barriers faced by health care practitioners. Barriers include lack of enthusiasm of major third-party payers, functional inadequacies of Medicare utilization review committees, and conflicts of interests of medical societies. Recommendations are made for making Professional Standards Review Organizations more effective.

Y10   Bradley, F.O. "A Modified Interpersonal Process Recall Technique as a Training Model." COUNSELOR EDUCATION AND SUPERVISION 14 (1974): 34-39.

> The use of a modified form of the Interpersonal Process Recall technique was studied as an influence on the counseling dimensions of level of regard, empathic understanding, unconditionality of regard, and congruence. Analysis of covariance revealed no significant differences between groups on each of the four variables.

Staff Development & Administration

Y11 Braun, S.H. "Ethical Issues in Behavior Modification." BEHAVIOR THERAPY 6 (1975): 51-62.

> Behavior modifiers are currently confronted with ethical issues traditionally faced by those with the power of behavior control. Arguments for and against the development of a code of ethics to regulate behavior modification and guidelines for the development of such a code are presented.

Y12 Bretz, R. THE UNIVERSITY OF TEXAS DENTAL BRANCH CLINICAL ENCOUNTER SYSTEM. Santa Monica, Calif.: The Rand Corp., 1975. 15 p.

> A university's abolishment of conventional courses from its dental curriculum and its institution of a self-pacing, or self-directed, program are described.

Y13 Carmody, J. ETHICAL ISSUES IN HEALTH SERVICES: A REPORT AND ANNOTATED BIBLIOGRAPHY. SUPPLEMENT 1, 1970-1973. BHSR-75-36, DHEW Publication no. HRA-74-3123. Rockville, Md.: Bureau of Health Services Research, 1974. 58 p.

> This bibliography covers the subjects of medical ethics, abortion, determination of death, euthanasia, genetic engineering, human experimentation, malpractice, and the right to health care.

Y14 Dacey, M.L., and Wintrob, R.M. "Human Behavior: The Teaching of Social and Behavioral Sciences in Medical Schools." SOCIAL SCIENCE AND MEDICINE 7 (1973): 943-57.

> This paper reviews the status of social and behavioral sciences in American medical schools.

Y15 Dashef, S.S., et al. "Time-Limited Sensitivity Groups for Medical Students." AMERICAN JOURNAL OF PSYCHIATRY 131 (1974): 287-92.

> This paper describes three phases of the group process, the role of the group leader, and the reactions of participating medical students to a voluntary unstructured sensitivity group.

Y16 Dyer, E.D., et al. "What Are the Relationships of Quality Patient Care to Nurses' Performance, Biographical and Personality Variables?" PSYCHOLOGICAL REPORTS 36 (1975): 255-66.

> Relationships among measures of quality patient care, nurse performance, and biographical and personality data were studied for 387 staff nurses in seven VA hospitals. Patient-care ratings were obtained by outside observers, and nurses' performance ratings were obtained from three supervisory levels, peers, and subordinates.

Staff Development & Administration

Y17   Emling, R.C., and Gellin, M.E. "An Evaluation of Programmed Text, Slide-Tape, and Lecture at Six Dental Schools." JOURNAL OF DENTAL EDUCATION 39 (1975): 72-77.

> Effectiveness of programmed instruction is compared to the lecture method.

Y18   Epstein, D.W., et al. "Issues in Training New Professionals." SOCIAL CASEWORK 55 (1974): 36-42.

> This paper describes an associate program for social workers, a skill-training program devised to prepare a new level of casework practitioners. Graduates will have responsibility for direct casework implementation, working under the supervision of a trained social worker with an M.S.W. degree. The curriculum, fieldwork training, program background, and evaluation are discussed.

Y19   Etzioni, A. "Alternative Conceptions of Accountability, Part 1." HOSPITAL PROGRESS 55 (June 1974): 34-39.

> This article discusses alternative conceptions of accountability in health administration and the implications of this analysis for the education of health administrators.

Y20   _____. "Alternative Conceptions of Accountability, Part 2." HOSPITAL PROGRESS 55 (July 1974): 56-59.

> This article continues a discussion of the alternative conceptions of accountability in health administration and also examines the consequences of an analysis of accountability for the education of health administrators.

Y21   Gadd, A.S. "Educational Aspects of Integrating Social Sciences in the Medical Curriculum." SOCIAL SCIENCE AND MEDICINE 7 (1973): 975-84.

> The application of theories and methods of educational psychology in integrating social medicine into medical schools is described in the context of a continuing study of a medical curriculum.

Y22   Golladay, F.L., et al. "Allied Health Manpower Strategies: Estimates of the Potential Gains from Efficient Task Delegation." MEDICAL CARE 11 (1973): 457-69.

> This study analyzes the potential impact of physician extenders on the productivity of primary care practice and considers the consequent implications for future health manpower requirements. It constructs and operates a simulation model of the representative practice, permitting one to synthesize the ex-

periences and insights of earlier demonstration projects. The model reveals that physician extenders could increase the productivity of a representative primary care practice by up to 74 percent. The article concludes with observations on the implications of physician extenders for future health manpower requirements.

Y23   Goulston, K. "Problems of Postgraduate Medical Education." AUSTRALIAN AND NEW ZEALAND JOURNAL OF MEDICINE 3 (1973): 230-38.

A questionnaire of forty-two general practitioners and fifty-four specialists in Australia surveyed time ideally devoted to postgraduate education, responsibility, sources, and motivation.

Y24   Graham, J.F., and Joiner, S. SURVEY OF HEALTH PROFESSIONS EDUCATION AS THEY RELATE TO LONGITUDINAL RESEARCH METHODOLOGY. Albuquerque: New Mexico University, School of Medicine, 1974. 169 p.

In this report the Ad Hoc National Coordinating Committee for Longitudinal Research in Health Professions Education identifies and describes various research groups in health professions education as they relate to longitudinal research.

Y25   Hess, S.W. "Communicating with Physicians." JOURNAL OF ADVERTISING RESEARCH 14 (1974): 13-18.

A purpose of this study was to obtain objectives measures of the actual time physicians spend with each kind of promotional media to better understand the communication process, improve the diffusion of new medical practice and technique, and aid pharmaceutical companies in sales promotion.

Y26   Hipple, J.L. "Comparison of Applicants and Nonapplicants to Human Relations Training Laboratories." JOURNAL OF YOUTH AND ADOLESCENCE 3 (1974): 161-68.

Students who applied to participate in a human relations training laboratory were compared with a group of students who did not apply by means of the administration of several relationship assessment instruments. The two groups were not significantly different in the manner by which they evaluated their relationship styles.

Y27   Hiss, R.G., and Peirce, J.C. "A Strategy for Developing Educational Objectives in Medicine: Problem-Solving Skills." JOURNAL OF MEDICAL EDUCATION 49 (1974): 660-65.

A formula is described for setting objectives in medical edu-

Staff Development & Administration

cation for problem-solving skills, therapeutic proficiency, procedural and interpretive skills, and enabling objectives. The format enables the creation of objectives in behavioral terms and permits the definition of the conditions under which the behavior will be observed. It is applicable to the objective-setting process for medical students, residents, and trainees in subspecialty programs.

Y28 Holder, L. "A Perspective on Non-Traditional Academic Programs in the Health Sciences." HEALTH EDUCATION MONOGRAPHS 3 (1975): 262-66.

In the 1970s nontraditional academic programs began to gain impetus, and schools of public health implemented nontraditional, off-campus programs relevant to personnel in health service agencies. This article introduces three programs that represent alternative approaches to off-campus public health degree programs.

Y29 Hollingshead, A.B. "Medical Sociology: A Brief Review." MILBANK MEMORIAL FUND QUARTERLY 51 (1973): 531-42.

A brief history of the development of medical sociology is presented. Five textbooks on medical sociology are reviewed, and some suggestions are made about issues in need of study for the future development of the field.

Y30 Hunt, S.M. "The Relationship between Psychology and Medicine." SOCIAL SCIENCE AND MEDICINE 8 (1974): 105-9.

This paper presents possible barriers to understanding that exist between academic psychology and medicine, and which may account for the relatively slow acceptance of psychology into medical curricula. Some implications for the teaching of psychology to medical students are discussed.

Y31 Kleinbach, G. "Social Structure and the Education of Health Personnel." INTERNATIONAL JOURNAL OF HEALTH SERVICES 4 (1974): 297-317.

This article analyzes the effect of a hierarchical structure of the health team on patient and community health outcomes. It assesses the role of the total school system in non-Socialist underdeveloped and developed countries in reproducing the hierarchical structure. The role of the health intellectual in the critique and formulation of educational and health policy also is discussed.

Y32 Klopfenstein, T.D. "The Core is Not the Part You Throw Away: Allied Health Education." NEW DIRECTIONS FOR COMMUNITY

Staff Development & Administration

COLLEGES No. 4 (1973): 35-43.

> Establishing a standard core of material as a basis for all later specialization is advocated as a means to permit students to acquire necessary skills without prematurely declaring a specialty. The core curriculum approach is described in relation to allied health education.

Y33 Leung, P. "Comparative Effects of Training in External and Internal Concentration on Two Counseling Behaviors." JOURNAL OF COUNSELING PSYCHOLOGY 20 (1973): 227-34.

> The author describes a training procedure that appears to facilitate training in the techniques of empathic understanding of the client and the ability to respond selectively to client statements during a counseling interview. The procedure is built around the methods used for training Zen Buddhist monks.

Y34 Levine, D.M., and Bonito, A.J. "Impact of Clinical Training on Attitudes of Medical Students: Self-Perpetuating Barrier to Change in the System?" BRITISH JOURNAL OF MEDICAL EDUCATION 8 (1974): 13-16.

> The attitudes of a sample of students and teachers in a medical school toward changes in the organization of medical practice were analyzed. Findings from this study suggest that attitudes of teachers and students vary by current or future primary activity and specialty.

Y35 Levine, D.M., et al. "Trends in Medical Education Research: Past, Present, and Future." MEDICAL EDUCATION 49 (1974): 129-36.

> A conference of medical educators and social scientists, convened to identify areas for future medical education research, specified five areas: selection of physicians--the applicants' and the medical schools' admission procedures; medical school socialization process; house officer training; medical schools as social institutions; and physician performance.

Y36 Lewis, C.E. "The State of the Art of Quality Assessment--1973." MEDICAL CARE 12 (1974): 799-806.

> Activities to assess the quality of care in the United States are reviewed. Recommendations are made for studying outcomes of care that are dependent upon patients' behaviors (change in habits, use of medication, etc.).

Y37 Linden, V. "Medicine and the Behavioral Sciences." SCANDANAVIAN JOURNAL OF SOCIAL MEDICINE 1, no. 3 (1973): 77-79.

> That medicine and the behavioral sciences are concerned with

Staff Development & Administration

the understanding and treatment of man and that communication and cooperation between these fields are essential for both the clients and the sciences is the proposition explored by the author. Ideological, practical, educational, and semantic factors are explored.

Y38   Lippard, V.W. A HALF-CENTURY OF AMERICAN MEDICAL EDUCATION: 1920-1970. New York: Josiah Macy Jr. Foundation, 1974. 134 p. Index.

Contents include: "Initial Impact of the Flexner Report," "Curriculum," "Medical Students and Their Environment," "From Graduation to Professorship," "Faculty Organization and Governance," "Rise of Academic Medical Centers," "Veterans Hospitals," "Impact of Research," "Licensure," "Specialization and Specialty Boards," "National Organizations and Foundations," and "New Schools."

Y39   Liston, E.H. "Psychiatric Aspects of Life-Threatening Illness: A Course for Medical Students." INTERNATIONAL JOURNAL OF PSYCHIATRY IN MEDICINE 5, no. 1 (1974): 51-56.

This article reports on a course for medical students on the psychiatric aspects of life-threatening illness. The course content includes seminar discussions and patient interviews. Student response to the course supported the view that instruction in this area should be a required component of medical student education.

Y40   Lyons, T.E., and Payne, B.C. "The Relationship of Physicians' Medical Recording Performance to Their Medical Care Performance." MEDICAL CARE 12 (1974): 714-20.

Within eight diagnostic categories, measures of good medical recording performance and good medical care performance are related. The overall relationship is not at all perfect, and the reliabilities of the measures may attenuate the degrees of obtained relationships.

Y41   McWhirter, J.J. "Counselor Preparation Through Small Group Interaction." SMALL GROUP BEHAVIOR 5 (1974): 23-29.

This investigation was designed to isolate and study the relationship between two small group training approaches and beginning counselors' offerings of the facilitative conditions of empathy, warmth, and genuineness. Findings suggest that small group sensitivity training is effective in helping counselors to increase their levels on a measure of empathy.

Y42   Monteiro, L.A. "Nursing's Acceptance of the Function of Family Plan-

Staff Development & Administration

ning Counselor." FAMILY COORDINATOR 23 (1974): 67-72.

> The pattern variables of the medical role, as specified by Parsons, are used as a frame of reference to suggest how the family planning counseling function was absorbed into the professional domain of nurses.

Y43 Morse, E.V., et al. "Hospital Costs and Quality of Care: An Organizational Perspective." MILBANK MEMORIAL FUND QUARTERLY 52 (1974): 315-46.

> Based on survey data gathered from 388 government, nonfederal, and voluntary general service hospitals, this study examines the impact of several dimensions of organizational structure on indicators of hospital efficiency and level of adoption of new medical technology.

Y44 Nathanson, C.A., and Becker, M.H. "Doctors, Nurses, and Clinical Records." MEDICAL CARE 11 (1973): 214-23.

> Detailed chart audits were carried out in five comprehensive child care clinics. Measures of clinic structure and staff attitudes were also obtained, making it possible to examine the relationships between chart quality and hospital goals, levels of clinic bureaucracy, or physicians' training and qualifications.

Y45 _____. "Job Satisfaction and Performance: An Empirical Test of Some Theoretical Propositions." ORGANIZATIONAL BEHAVIOR AND HUMAN PERFORMANCE 9 (1973): 267-79.

> A theoretical analysis of the relationship between job satisfaction and job performance is examined. The data used is from a study of 103 physicians providing routine pediatric care in ambulatory clinics.

Y46 _____. "Work Satisfaction and Performance of Physicians in Pediatric Outpatient Clinics." HEALTH SERVICES RESEARCH 8 (1973): 17-26.

> Sources and consequences of variations in work satisfaction are investigated in a study of approximately one hundred physicians in six pediatric outpatient clinics, half of them associated with teaching hospitals and half with community hospitals. Implications for potential conflict between outpatient care and academic aims in teaching hospitals are discussed, and avenues of further research are suggested.

Y47 Ott, J.E., et al. "Patient Management by Telephone by Child Health Associates and Pediatric House Officers." JOURNAL OF MEDICAL EDUCATION 49 (1974): 596-600.

Staff Development & Administration

This study used simulated telephone conversations to assess the ability of pediatric house officers and child health associate interns to determine the nature and severity of the caller's problem and to provide the caller with information in response to the question asked.

Y48 Parkhouse, J. GRADUATE MEDICAL EDUCATION IN THE EUROPEAN REGION, EURO 6301 (1). Copenhagen: World Health Organization, Regional Office for Europe, 1974, 87 p. Append. Tables.

Contents include definition of specialities in the European region, requirements for specialization, policy concerning specialist certification, internship requirements, and comments and recommendations.

Y49 Pattishall, E.G. "Basic Assumptions for the Teaching of Behavioral Science in Medical Schools." SOCIAL SCIENCE AND MEDICINE 7 (1973): 923-26.

Some basic assumptions are postulated on the evolution and teaching of the social and behavioral sciences in medical education. These assumptions are discussed in terms of the scientific and educational imbalance of human biology, medical education, and patient care.

Y50 THE ROLE OF THE CONSUMER IN ASSURING QUALITY HEALTH CARE. Albuquerque: New Mexico Regional Medical Program, 1973.

A task force outlined appropriate roles for American consumers who want to improve the quality of health care they receive. Possible improvements in access to health care, the health care process, compliance with health care instructions, and health care continuity were considered, as well as consumer roles in overcoming barriers to quality health care.

Y51 Sadoff, R.L. "Comprehensive Training in Forensic Psychiatry." AMERICAN JOURNAL OF PSYCHIATRY 131 (1974): 223-25.

A university's program of training in legal psychiatry is presented as a model of an integrated program of training, research, and service. The author stresses the need for formal training in this subspecialty instead of relying, as in the past, on self-training of preceptorships.

Y52 Shaprio, I.S. "The Teaching Role of Health Professionals in a Formal Organization." HEALTH EDUCATION MONOGRAPHS 1, no. 36 (1973): 40-48.

Experiences in the Health Insurance Plan of Greater New York (HIP), since its inception in 1947, are discussed in relation

Staff Development & Administration

to general principles of role determination and change among health professionals. The framework for the discussion is Dr. Stanley E. Seashore's comments on "characteristics of organizations that bear upon their power to change people." Examples are given of the influence and uses of the normative element, the economic element, communications, consensual validation, group pressure, identification, personal influence, role prescriptions, rewards and penalties, and participation within HIP to encourage development of educational programs and attitudes among professional staff. Specific efforts of HIP in cancer education are summarized.

Y53 Sheldrake, P. "Behavioural Science in the Medical Curriculum." SOCIAL SCIENCE AND MEDICINE 7 (1973): 967-73.

The views of students and staff on a course in behavioral sciences taught in the third preclinical year of a medical school curriculum are reported.

Y54 Shemberg, K.M., and Keeley, S.M. "Training Practices and Satisfaction with Preinternship Preparation." PROFESSIONAL PSYCHOLOGY 5 (1974): 98-105.

This paper describes a survey conducted to broaden the data base on training attitudes and practices of internship facilities for clinical psychologists. A second purpose was to learn how these facilities evaluate the adequacy of preinternship clinical training in the academic setting.

Y55 Spitzer, W.O., et al. "The Burlington Randomized Trial of the Nurse Practitioner." NEW ENGLAND JOURNAL OF MEDICINE 290 (1974): 251-56.

A one-year, randomized, controlled trial was conducted in a large suburban practice of two family physicians to assess the effects of substituting nurse practitioners for physicians in primary-care practice. An assessment was made of quality of care and patient satisfaction. Cost-effectiveness of this method of primary care and restrictions on reimbursement for the nurse-practitioner services are discussed.

Y56 Suinn, R.M. "Traits for Selection of Paraprofessionals for Behavior-Modification Consultation Training." COMMUNITY MENTAL HEALTH JOURNAL 10 (1974): 441-49.

This paper specifies characteristics associated with success in behavior-modification consultation. Consultants were undergraduate paraprofessionals who completed a program in consultation. Training procedures, the psychometric and other selection measures, and statistical results when comparing the selection tools with different criteria are discussed.

Staff Development & Administration

Y57 Talbert, L.M., and Bishop, E.H. "Educational Committee Panel Presentations: Medical Education by Self-Instructional Methods." AMERICAN JOURNAL OF OBSTETRICS AND GYNECOLOGY 117 (1973): 727-30.

> Benefits of a self-instructional package used in the Department of Obstetrics and Gynecology at the Medical School of the University of North Carolina are described.

Y58 Toigo, R. "Child Care Manpower Development: A Literature Review." CHILD CARE QUARTERLY 4 (1975): 6-17.

> This review of the literature on child care manpower development focuses on identification of social background characteristics of child care workers and the child care role from the standpoints of initial training and practice norms.

Y59 White, K.L. "Caveats for PSRO's." WESTERN JOURNAL OF MEDICINE 120 (1974): 338-43.

> Emphasis on "process" standards directed at physicians' practices in relation to Professional Standards Review Organizations is questioned by the author. He points out a need for more work to identify efficacious procedures; population-based, problem-oriented information systems; caution against using predetermined "process" standards based on unvalidated data collected from unrepresentative samples; a need to focus on hospitals and systems before focusing on individual physicians and their individual patients; and an educational rather than a punitive approach.

Y60 Wolf, S. "The Place of the Person in Medical Education." COMPREHENSIVE PSYCHIATRY 15 (1974): 119-22.

> The author explores his concept that the key persons in the process of medical education, the patient, the teacher, and the student, have become depersonalized during the educational experience.

Y61 Wolle, J.M. "Patient Education." JOURNAL OF THE AMERICAN COLLEGE HEALTH ASSOCIATION 22 (1974): 231-33.

> The author suggests areas in health education in which a college health nurse can work to educate the consumers of college health services. Six steps for developing a plan for patient education are presented: knowledge of population served, educational needs determination, location of available resources, selection of methods and materials, the program implementation, and program evaluation.

Y62 World Health Organization, Regional Office for Europe. MODERN

Staff Development & Administration

MEDICAL TEACHING METHODS: REPORT ON A SEMINAR CONVENED BY THE REGIONAL OFFICE FOR EUROPE OF THE WORLD HEALTH ORGANIZATION, EURO 6202. Copenhagen: 1974. 37 p.

This publication covers the topics trends and issues in medical education, curriculum planning, self-instruction and programmed instruction in medical education, technical aids and methods of communication, computer scoring and analysis of objective-type tests, and group teaching.

Chapter 10
RESEARCH AND EVALUATION METHODS

The scientific methodologies of research and evaluation apply to patient education as they do to any application of medical or social sciences in which human subjects are studied. There are but a few exceptions to the usual applications of randomized control procedures and methods of design and measurement as applied in patient education. These exceptions are discussed in the following recent works:

Faden, R.R., and Faden, A.I. "Ethical Issues in Public Health Policy: Health Education and Life-Style Interventions." HEALTH EDUCATION MONOGRAPHS 6 (1978): whole issue.

Figa-Talamanca, I. "Problems in the Evaluation of Training of Health Personnel." HEALTH EDUCATION MONOGRAPHS 3 (1975): 232-50.

German, P.S., and Chwalow, A.J. "Ethical Problems of Patient Education Strategies for Hypertension Control." INTERNATIONAL JOURNAL OF HEALTH EDUCATION 19 (1976): 195-201.

Green, L.W. "Determining the Impact and Effectiveness of Health Education for Federal Policy." HEALTH EDUCATION MONOGRAPHS 6 Suppl. 1 (1978): 28-66.

_____. "Evaluation and Measurement: Some Dilemmas for Health Education." AMERICAN JOURNAL OF PUBLIC HEALTH 67 (1977): 155-61.

_____. "Evaluation of Patient Education Programs: Criteria and Measurement Techniques." In RX: EDUCATION FOR THE PATIENT, edited by E. Slepcevitch, pp. 89-98. Carbondale: Southern Illinois University, 1975.

_____. "Research Methods Translatable to the Practice Setting: From Rigor to Reality and Back." In NEW DIRECTIONS IN PATIENT COMPLIANCE, edited by S. Cohen, pp. 141-51. Lexington, Mass.: Lexington Books, D.C. Heath Co., 1979.

Mahoney, M.J. "Experimental Methods and Outcome Evaluation." JOURNAL OF CONSULTING AND CLINICAL PSYCHOLOGY 46 (1978): 660-72.

Recent reviews of the literature related to evaluations of patient education applying criteria of clinical trials methodology and some of the additional considerations identified in the foregoing works reveal the major gaps in research methods available and appropriately used:

Bertram, D.A., and Brooks-Bertram, P.A. "The Evaluation of Continuing Medical Education: A Literature Review." HEALTH EDUCATION MONOGRAPHS 5 (1977): 330-62.

Best, J.A., and Block, M. "Compliance in the Control of Cigarette Smoking." In COMPLIANCE IN HEALTH CARE, edited by R.B. Haynes et al., pp. 202-22. Baltimore: Johns Hopkins University Press, 1979.

Blanchard, E.B., and Young, L.D. "Clinical Applications of Biofeedback Training: A Review of Evidence." ARCHIVES OF GENERAL PSYCHIATRY 30 (1974): 573-89.

Craighead, W.E., et al. BEHAVIOR MODIFICATION: PRINCIPLES, ISSUES, AND APPLICATIONS. Boston: Houghton-Mifflin, 1976.

Frankle, B.L., et al. "Treatment of Hypertension with Biofeedback and Relaxation Techniques." PSYCHOSOMATIC MEDICINE 40 (1978): 276-93.

Frumkin, K., et al. "Nonpharmacologic Control of Essential Hypertension in Man: A Critical Review of the Experimental Literature." PSYCHOSOMATIC MEDICINE 40 (1978): 294-320.

Green, L.W. "Educational Strategies to Improve Compliance with Therapeutic and Preventive Regimens: The Recent Evidence." In COMPLIANCE IN HEALTH CARE, edited by R.B. Haynes et al., pp. 157-73. Baltimore: Johns Hopkins University Press, 1979.

_____. "Methods Available To Evaluate the Health Education Components of Preventive Health Programs." In PREVENTIVE MEDICINE U.S.A., pp. 162-71. New York: Prodist, 1976.

Green, L.W., and Faden, R.R. "The Potential Impact of Patient Package Inserts on Patients and Drug Consumers." DRUG INFORMATION JOURNAL 2 Suppl. (1977): 64-70.

Green, L.W., and Fedder, D. "Drug Information, the Pharmacist, and the Community." AMERICAN JOURNAL OF PHARMACEUTICAL EDUCATION 41 (1977): 444-48.

Green, L.W., et al. "How Cost-Effective Are Smoking Cessation Methods?" WORLD SMOKING AND HEALTH 3 (1978): 33-40.

Green, L.W., et al. "Patient Education: An Inquiry into the State of the Art." SECOND NATIONAL SYMPOSIUM ON PATIENT EDUCATION, PROCEEDINGS, edited by W. Squyres. New York: Springer Publishing Co., 1980.

Kazdin, A.E., and Wilson, G.T. EVALUATION OF BEHAVIOR THERAPY: ISSUES, EVIDENCE, AND RESEARCH STRATEGIES. Cambridge, Mass.: Ballinger, 1978.

Ley, P. The Use of Techniques and Findings from Social and Experimental Psychology to Improve Doctor-Patient Communication." In his HEALTH EDUCATION AND PRIMARY CARE: CONFERENCE REPORT. Leeds, England: Department of Community Medicine and Leeds Polytechnic, 1975.

McFall, R.M. "Smoking-Cessation Research." JOURNAL OF CONSULTING AND CLINICAL PSYCHOLOGY 46 (1978): 703-12.

Miller, P.M. BEHAVIORAL TREATMENT OF ALCOHOLISM. New York: Pergamon Press, 1976.

Nelson, R.O. "Methodological Issues in Assessment Via Self-Monitoring." BEHAVIORAL ASSESSMENT, edited by J.D. Cone and R.P. Hawkins, pp. 217-40. New York: Bruner/Mazel, 1977.

Stunkard, A.J. THE PAIN OF OBESITY. Palo Alto, Calif.: Bull, 1976.

Weiss, S.M., ed. PROCEEDINGS OF THE NATIONAL HEART AND LUNG INSTITUTE WORKING CONFERENCE ON HEALTH BEHAVIOR. DHEW Publication No. (NIH) 77-868. Bethesda, Md.: National Institutes of Health, 1977.

Williams, R.B., Jr., and Gentry, W.D. BEHAVIORAL APPROACHES TO MEDICAL TREATMENT. Cambridge, Mass.: Ballinger, 1977.

Wilson, G.T. "Cognitive Factors in Lifestyle Changes: A Social Learning Perspective." In BEHAVIORAL MEDICINE: CHANGING HEALTH LIFESTYLES, edited by P.O. Davidson and S.M. Davidson, pp. 3-37. New York: Bruner/Mazel, 1980.

Z. METHODS

Z1   Aitken, R.C.B. "Methodology of Research in Psychosomatic Medicine." PSYCHOTHERAPY AND PSYCHOSOMATICS 22 (1973): 80-88.

This paper reviews problems in the methodology of clinical psychosomatic research. Concentration is on symptomatology rather than on etiology of disease and on diagnosis and treatment of known psychopathology, particularly mood disturbance, rather than on underlying speculative mechanisms.

Z2   Alpander, G.G., and Gutmann, J.E. "A Model for Measuring the Impact of Change on an Organization." HOSPITAL AND COMMUNITY PSYCHIATRY 25 (1974): 719-23.

In a study at a psychiatric hospital, a model was developed for measuring organizational climate after the introduction of a geographical unit system brought a major change in the hospital's structure. Areas studied included gaps in role perception between workers and supervisors, relationships among members and leaders of ward teams, motivational factors, and staff perceptions of the impact of change.

Z3   AMBULATORY MEDICAL CARE RECORDS: UNIFORM MINIMUM BASIC DATA SET. A REPORT OF THE UNITED STATES NATIONAL COMMITTEE ON VITAL AND HEALTH STATISTICS. Vital and Health Statistics Series 4: Documents and Committee Reports, no. 16. DHEW Publication no. HRA 75-1453. Rockville, Md.: National Center for

Research and Evaluation Methods

Health Statistics, 1974. vii, 16 p.

> Contents include summary of recommendations, purposes of ambulatory medical care data, the minimum basic data set, the maintenance of confidentiality, uses and limitations of the minimum data set and its relationship to an ambulatory care encounter form, and implementation and future revision of the Minimum Basic Data Set.

Z4 Attkisson, C.C., et al. "A Working Model for Mental Health Program Evaluation." AMERICAN JOURNAL OF ORTHOPSYCHIATRY 44 (1974): 741-53.

> This article describes a conceptual model of human service program evaluation that integrates three components of the evaluation process: levels of evaluative activity, functional roles of the evaluator, and program information capability.

Z5 Baker, E.L., and Alkin, M.C. "ERIC/AVCR Annual Review Paper: Formative Evaluation of Instructional Development." AV COMMUNICATION REVIEW 21 (1973): 389-418.

> In this paper formative evaluation is defined as oriented primarily to developers in order to provide them with information to modify and improve programs. Topics covered include formative evaluation, research on formative evaluation, formative evaluation models, and procedures for formative evaluation.

Z6 Balaban, R.M. "The Contribution of Participant Observation to the Study of Process in Program Evaluation." INTERNATIONAL JOURNAL OF MENTAL HEALTH 2 (Summer 1973): 59-70.

> The various conceptual and practical limitations of experimental and quasi-experimental designs in the study of broadly aimed programs led to this examination of the nature of process study within the context of program evaluation. This presentation of process analysis serves as a background and orientation for participant observation, a key method in the study of process. This article also presents the contributions of participant observation and its data to the study of process.

Z7 Balinsky, W., and Berger, R. "A Review of the Research on General Health Status Indexes." MEDICAL CARE 13 (1975): 283-93.

> This article reviews the literature on general health status indexes. Common objectives and constraints are presented, and the expanding role of general health status indexes is discussed.

Z8 Banner, D.K. "The Politics of Evaluation Research." OMEGA--INTER-

Research and Evaluation Methods

NATIONAL JOURNAL OF MANAGEMENT SCIENCE 2 (1974): 763-74.

> This study examines the requirements for social planners and administrators to prove the efficacy of particular programs before they are refunded by federal monies. Evaluation research as a legitimizing function and the political environment surrounding evaluation research are described.

Z9 Barron, F.H. "Behavioral Decision Theory: A Topical Bibliography for Management Scientists." INTERFACES 5 (November 1974): 56-62.

> This topical bibliography provides an introduction to the literature of behavioral decision theory and presents a selected bibliography.

Z10 Berdie, D.R., and Anderson, J.F. QUESTIONNAIRES: DESIGN AND USE. Metuchen, N.J.: Scarecrow Press, 1974. 225 p. Illus.

> Topics in this volume include advantages and limitations of questionnaires, suggestions for designing a study using questionnaires, questionnaire format and question phrasing, considerations of response rate, and some recommendations on analyses of questionnaire data.

Z11 Berner, E.S., et al. "A New Approach to Evaluating Problem-Solving in Medical Students." JOURNAL OF MEDICAL EDUCATION 49 (1974): 666-72.

> This paper describes the development of a technique to assess clinical problem-solving skills in medical students. Students were required to read an extensive data base and construct problem lists, order diagnostic tests, and plan for the management of the patient. The technique was designed to be administered to large groups of students, to be scored easily and to overcome some of the difficulties inherent in previously developed means of assessment.

Z12 Bernstein, I.N., and Freeman, H.E. ACADEMIC AND ENTREPRENEURIAL RESEARCH: THE CONSEQUENCES OF DIVERSITY IN FEDERAL EVALUATION STUDIES. New York: Russell Sage Foundation, 1975. 208 p. Bibliog.

> This study of the expenditure of federal funds for evaluation research reviews federally supported evaluations of programs, including evaluations of social-change experiments and research-demonstration programs.

Z13 Berry, R.E., Jr. "Cost and Efficiency in the Production of Hospital Services." MILBANK MEMORIAL FUND QUARTERLY 52 (1974): 291-313.

Research and Evaluation Methods

The findings of a research effort designed to identify and measure factors affecting the cost and efficiency of the short-term general hospital system in the United States are summarized. The empirical analysis involves data on approximately 6,000 hospitals for the years 1965, 1966, and 1967 and a model that expresses hospital cost as a function of the level of output, quality of services provided, scope of services provided, factor prices, and relative efficiency.

Z14   Blum, H.L. "Evaluating Health Care." MEDICAL CARE 12 (1974): 999-1011.

The author suggests that in-depth analysis at two points in health care delivery, the entry to care and the outcomes of care, is most likely to reveal the basis for failures. Interventions most likely to provide good results are discussed.

Z15   Burck, H.D., and Peterson, G.W. "Needed: More Evaluation, Not Research." PERSONNEL AND GUIDANCE JOURNAL 53 (1975): 563-69.

The authors advocate the need for more evaluation. Differences between evaluation and research are illustrated, five basic steps for good evaluation are provided, and examples are given.

Z16   Caro, F.G. "Evaluative Researchers and Practitioners: Conflicts and Accommodations." JOURNAL OF RESEARCH AND DEVELOPMENT IN EDUCATION 8 (1975): 55-62.

Conflicts in research-practitioner relationships are explored, and potential accommodations to these conflicts are suggested.

Z17   Cattell, R.B. "How Good Is the Modern Questionnaire? General Principles for Evaluation." JOURNAL OF PERSONALITY ASSESSMENT 38 (1974): 115-29.

Eight basic requirements for construction of factor-true scales are detailed, and it is shown that one or more is missing in 80 percent of recently published studies.

Z18   Cattell, R.B., and Child, D. MOTIVATION AND DYNAMIC STRUCTURE. New York: Holt, Rinehart & Winston, 1975. 290 p. Bibliog. Indexes.

The authors demonstrate the application of the dynamic structure theory to such fields as conflict, clinical psychology, learning, occupational psychology, and group dynamics.

Z19   Cooper, B.S., and Worthington, N.L. COMPARISONS OF COST

Research and Evaluation Methods

AND BENEFIT INCIDENCE OF GOVERNMENT MEDICAL CARE PROGRAMS, FISCAL YEARS 1966 AND 1969. Staff Paper no. 18. DHEW Publication no. SSA 75-11852. U.S. Department of Health, Education, and Welfare, Social Security Administration, Office of Research and Statistics, 1974. 91 p. Bibliog. Tables.

> This publication includes reports on health insurance for the aged (Medicare), temporary disability insurance, workmen's compensation, public assistance, general hospital and medical care, maternal and child health, Veteran's Administration, and medical vocational rehabilitation.

Z20 Costello, R.M., et al. "Attitudinal Ambivalence with Alcoholic Respondents." JOURNAL OF CONSULTING AND CLINICAL PSYCHOLOGY 42 (1974): 303-4.

> This investigation examines attitude ambivalence measurement in the context of defining outcome criteria upon which to measure the effects of aversive conditioning in alcoholism. Subjects were thirty chronic alcoholics, the total resident membership of an alcoholic rehabilitation center for one month in 1973. Each subject was asked to rate two slides, one with alcohol content and one without, on a four-point rating scale. Other findings presented are applicable to attitude scaling, theory about alcoholism and maintenance of disruptive drinking habits, and aversive conditioning to alcoholism.

Z21 Dansereau, D.F., et al. LEARNING STRATEGIES: A REVIEW AND SYNTHESIS OF THE CURRENT LITERATURE. Fort Worth: Texas Christian University, Institute for the Study of Cognitive Systems, 1974. 90 p.

> This report reviews and synthesizes psychological and educational research on learning strategies. It contains an overview of strategy modification, a review of factors influencing strategy selection and use, a review of learning strategies, and recommendations involving future research.

Z22 Davidson, P.O., and Neufeld, R.W.J. "Response to Pain and Stress: A Multivariate Analysis." JOURNAL OF PSYCHOSOMATIC RESEARCH 18 (1974): 25-32.

> This study was designed to compare the effects of a pain stimulus and a psychological threat stimulus on a variety of measures chosen from across response domains, psychological, physiological, and behavioral. It is a preliminary, empirical demonstration that it is possible, using multiple measures and appropriate statistical analyses, to discriminate patterns of response to pain from other types of stressors.

Research and Evaluation Methods

Z23   Derogatis, L.R., et al. "The Hopkins Symptom Checklist (HSCL): A Self-Report Symptom Inventory." BEHAVIORAL SCIENCE 19 (1974): 1-15.

> This report describes the historical evolution, development, rationale, and validation of the Hopkins Symptom Checklist, a self-report symptom inventory. Standard indexes of scale reliability are presented, and a broad range of criterion-related validity studies, in particular an important series reflecting sensitivity to treatment with psychotherapeutic drugs, are reviewed and discussed.

Z24   Dolfman, M.L. "Toward Operational Definitions of Health." JOURNAL OF SCHOOL HEALTH 44 (1974): 206-9.

> An attempt is made to develop an operational model of the word health. The model contains concepts fundamental to an understanding of the meaning of health.

Z25   Elinson, J. "Toward Sociomedical Health Indicators." SOCIAL INDICATORS RESEARCH 1 (May 1974): 59-71.

> This paper describes attempts to develop sociomedical health indicators, including measures of social disability; typologies of presenting symptoms; measures focusing on behavioral expressions of sickness; research based on operational definitions of positive mental health; happiness and perceived quality of life; and assessments of met and unmet needs for health care.

Z26   Figà-Talamanca, I. "Problems in the Evaluation of Training Health Personnel." HEALTH EDUCATION MONOGRAPHS 3 (1975): 232-50.

> Evaluation of training programs is presented as an ongoing educational process. The framework specifies sources and methods of data collection, improvement in trainee performance, relevance of the program to the community, involvement of appropriate groups in planning, the planning process, utilization and diffusion of evaluative feedback, and consequences of the training activity for the sponsoring organization and consumer.

Z27   Frey, D.H. "Being Systematic When You Have but One Subject: Ideographic Method, N = 1, and All That." MEASUREMENT AND EVALUATION IN GUIDANCE 6 (April 1973): 35-43.

> This article provides patient educators with parameters for making their single subject or ideographic investigations more systematic and more scientific.

Z28   Fry, L.N. "Participant Observation and Program Evaluation." JOUR-

NAL OF HEALTH AND SOCIAL BEHAVIOR 14 (1973): 274-78.

> The contributions that participant observation can make to program evaluation are analyzed, based on research in a therapeutic drug community. Participant observation includes (1) gaining access to data, (2) evoking behavior, (3) identifying psychologically with the people being studied, (4) connecting concepts with indicators, and (5) formulating hypotheses. The implication of the study is that participant observation should be integrated into a network of research techniques.

Z29 Glouberman, D. "Person Perception and Scientific Objectivity." EUROPEAN JOURNAL OF SOCIAL PSYCHOLOGY 3 (1973): 241-53.

> The participant observer model is described as optimal for the study of human beings, rather than the "predict and control" model for psychology and the "medical model" of psychiatry that involve the perception of people as objects.

Z30 Goldberg, G.A. "Implementing University Hospital Ambulatory Care Evaluation." JOURNAL OF MEDICAL EDUCATION 50 (1975): 435-42.

> This article reports on an observation of the clinics of a single university hospital center to determine a rationale for and impediments to implementing a medical care evaluation program. The importance of quality assurance mechanisms as well as barriers to implementing quality assurance programs are detailed.

Z31 Gormally, J., and Hill, C.E. "Guidelines for Research on Carkhuff's Training Model." JOURNAL OF COUNSELING PSYCHOLOGY 21 (1974): 539-47.

> Several methodological issues related to Carkhuff's human relations training model are discussed with an emphasis on developing research guidelines. Each discussion of an issue is accompanied by research guidelines for evaluating the present literature and conducting further research. Several research suggestions are offered that might resolve logical gaps as well as extend the model.

Z32 Gough, H.G. "Estimation of Locus-of-Control Scores from the California Psychological Inventory." PSYCHOLOGICAL RESEARCH 35 (1974): 343-48.

> Rotter's Locus of Control measure was correlated with eighteen standard and four new scales of the California Psychological Inventory in samples of 141 males and 220 females.

Research and Evaluation Methods

Z33   Green, L.W. "Toward Cost-Benefit Evaluations of Health Education: Some Concepts, Methods, and Examples." HEALTH EDUCATION MONOGRAPHS 2 Suppl. 1 (1974): 34-64.

> This paper analyzes some conceptual and technical problems of evaluation that are particularly relevant to health education. It presents methods and examples of cost-benefit analysis applicable to the educational components of a variety of health programs.

Z34   Green, L.W., and Figà-Talamanca, I. "Suggested Designs for Evaluation of Patient Education Programs." HEALTH EDUCATION MONOGRAPHS 2 (1974): 54-71.

> Problems related to the relationship between evaluative research and planning of an educational program and the problem of setting goals and deciding on the criteria of success are discussed briefly. Problems associated with the methodology of evaluative research are examined in detail. Six evaluative designs are presented on the basis of their appropriateness for patient education programs.

Z35   Greenbaum, H.H. "The Audit of Organizational Communication." ACADEMY OF MANAGEMENT JOURNAL 17 (1974): 739-54.

> A conceptual and methodological structure for examining communication processes in organizations is presented. Components of the communication system are identified; a taxonomical table of activities is suggested; and an audit program is outlined.

Z36   Greenberg, J.S. "Behavior Modification and Values Clarification and Their Research Implications." JOURNAL OF SCHOOL HEALTH 45 (1975): 91-95.

> Behavior modification and values clarification are defined. Optimum conditions for their use are described and contrasts and similarities between the two methods are drawn.

Z37   Greif, E.B., and Hogan, R. "The Theory and Measurement of Empathy." JOURNAL OF COUNSELING PSYCHOLOGY 20 (1973): 280-84.

> This paper reviews several studies that support the idea that empathy is an important aspect of interpersonal behavior and moral conduct and studies that provide further evidence for the validity of an empathy scale. Minres factor analyses were performed to determine the underlying structure of the empathy scale and its relationship to the California Psychological Inventory from which it can be scored. Characteristics of empathic persons and uses of the empathy scale in future research are discussed.

Research and Evaluation Methods

Z38 Greist, J.H., et al. "Suicide Risk Prediction: A New Approach." LIFE-THREATENING BEHAVIOR 4 (1974): 212-23.

> This article reports on a pilot test of a computer program written to collect suicide risk factors. A computer interview was conducted with persons having thoughts of suicide, and the data collected was used to provide risk predictions. More than half of the patients interviewed preferred the computer to a doctor as an interviewer. A retrospective study compared risk predictions made by the computer with predictions by experienced clinicians.

Z39 Guba, E.G. "Problems in Utilizing the Results of Evaluation." JOURNAL OF RESEARCH AND DEVELOPMENT IN EDUCATION 8 (1975): 42-54.

> Seven factors that impede the use of results of evaluations of educational and social action programs are discussed.

Z40 Gullen, W.H., and Garrison, G.E. "Factors Influencing Physicians' Responses to Mailed Questionnaires." HEALTH SERVICES REPORTS 88 (1973): 510-14.

> This experiment was conducted to determine which of several kinds of questionnaire mailings would elicit the best rate of response from physicians. The results support the assumption usually made in designing mail surveys that the more attractive and finished the format, the more personal the address, and the higher the class of postage, the better the response rate will be.

Z41 Guttentag, M., et al. THE EVALUATION OF TRAINING IN MENTAL HEALTH. New York: Behavioral Publications, 1975. 131 p.

> Contents include conceptualization of the evaluation problem, organization of evaluation research, a paradigm for evaluation of training programs for community mental health, goal attainment scoring and quantification of values, review of the history of the evaluation of training in psychology, and issues and recommendations.

Z42 Hambleton, R.K. "Testing and Decision-Making Procedures for Selected Individualized Instructional Programs." REVIEW OF EDUCATIONAL RESEARCH 44 (1974): 371-400.

> The testing models of three instructional programs, IPI, Project PLAN, and Mastery Learning, are described and compared. On the basis of a review of the models, the author outlines several research methods that could contribute to the quality of testing and decision making within the context of these and other individualized instructional programs.

Research and Evaluation Methods

Z43 Hardy, M.E. "Theories: Components, Development, Evaluation." NURSING RESEARCH 23 (1974): 100-107.

> The roles of concepts, statements of relationship, and models in theory development are examined. Criteria for evaluating theories are outlined, and the tentative nature of theories is discussed.

Z44 Harris, A.H., and Brady, J.V. "Instrumental (Operant) Conditioning of Visceral and Autonomic Functions." SEMINARS IN PSYCHIATRY 5 (1973): 369-76.

> This article discusses the different types of studies that have been done in the investigation and application of "visceral learning" phenomena.

Z45 Hartnagel, T.F. "Measuring the Significance of Others. A Methodological Note." AMERICAN JOURNAL OF SOCIOLOGY 80 (1974): 397-401.

> Data on self-concept and significant others using the semantic-differential technique are examined for the presence of multicolinearity. It is concluded that the presence of a number of fairly high first-order correlations among the presumed independent variables invalidates the use of partial correlations as measures of the concept, significant others.

Z46 Harvey, T.R. "An Heretical Approach to Evaluation." JOURNAL OF HIGHER EDUCATION 45 (1974): 628-34.

> This paper suggests that, in situations of fiscal crisis, inadequate prior evaluation, and time constraints, faculty evaluation cannot merely follow the traditional models of systematic evaluation. An alternate evaluation approach is suggested, and its usefulness for limiting the extent and scope of analysis is described.

Z47 Hickey, T., and Spinetta, J.J. "Bridging Research and Application." GERONTOLOGIST 14 (1974): 526-30.

> This paper takes a cognitive approach to the problems involved in "building bridges" between the researcher and the practitioner.

Z48 Horowitz, M. "Stress Response Syndromes: A Character Style and Dynamic Psychotherapy." ARCHIVES OF GENERAL PSYCHIATRY 31 (1974): 768-81.

> This report models an approach to organizing clinical knowledge about psychopathology and treatment. The approach limits disorder to stress response syndromes, personality to

Research and Evaluation Methods

obsessional and hysterical neurotic styles, and treatment to focal dynamic psychotherapy. An information processing approach to working through conflicted ideas and feelings is developed.

Z49 Hulka, B.S., and Cassel, J.C. "The AAFP-UNC Study of the Organization, Utilization, and Assessment of Primary Medical Care." AMERICAN JOURNAL OF PUBLIC HEALTH 63 (1973): 494-501.

This study was designed to determine the relative effectiveness of various systems for the delivery of primary medical care. Two methodologic approaches are used: a household survey to determine the factors that influence entry into the health care system and the indicator case model to determine the impact of medical care on the patient once he or she has entered the system. The indicator case model is applied to four indicator conditions: pregnancy, infancy, diabetes mellitus, and congestive heart failure. Within the model, the elements of utilization, cost-convenience, physician performance, communication from physician to patient, patient compliance, physician awareness of patient concerns, patient attitudes towards physicians and outcomes--functional, symptomatic, and medical status--are identified as measurements of medical care components.

Z50 Jaccard, J., et al. "A Multitrait-Multimethod Analysis of Four Attitude Assessment Procedures." JOURNAL OF EXPERIMENTAL AND SOCIAL PSYCHOLOGY 11 (1975): 149-54.

This article describes a multitrait-multimethod analysis of four attitude assessment procedures: the Likert method of summated ratings, Thurstone's method of equal appearing intervals, Guilford's self-rating procedure, and the semantic differential technique. Implications of the findings are discussed.

Z51 Jeffrey, D.B. "Some Methodological Issues in Research on Obesity." PSYCHOLOGICAL REPORTS 35 (1974): 623-26.

The author advocates moving beyond studies that demonstrate that obesity can be controlled to studying problems of patient dropouts, standardized improvement measures, long-term maintenance, cost-effectiveness analysis, research strategies, the development of behavioral predictors, and an analysis of both successes and failures.

Z52 Jernstedt, G.C., and Newcomer, J.P. "Blood Pressure and Pulse Wave Velocity Measurement for Operant Conditioning of Autonomic Responding." BEHAVIOR RESEARCH METHODS AND INSTRUMENTATION 6 (1974): 393-97.

Research and Evaluation Methods

A procedure for measuring a correlate of the blood pressure information obtained with conventional sphygmomanometric systems is described. It involves the measurement of speed of propagation through the arterial system of the pressure pulse from the heart's contraction.

Z53 Klemmack, D.L., et al. "An Empirical and Critical Assessment." JOURNAL OF HEALTH AND SOCIAL BEHAVIOR 15 (1974): 267-70.

Three measures of well-being, the Life Satisfaction scale, the Social Isolation scale, and the Willingness to Live scale, were evaluated for empirical similarity. Implications for health-related and gerontological studies are discussed.

Z54 Kosecoff, J., and Fitz-Gibbon, C. "Many a Slip." JOURNAL OF EDUCATIONAL EVALUATION 4 (December 1973): 1-6.

Objectives of evaluation efforts are explored. The recommendation is made that evaluators and the sponsors of evaluation studies attend not only to the technical excellence of their instruments and analyses but also to the organization of evaluations that are compatible with the decision context of the program being evaluated.

Z55 Krieger, S.R., et al. "Personal Constructs, Threat and Attitudes Toward Death." OMEGA--JOURNAL OF DEATH AND DYING 5 (1974): 299-310.

This study used a personal construct approach in the assessment of threat of death. Two experiments are reported in which the authors investigated the relationship of the measure, the Threat Index, to a number of self-report variables, the Lester Fear of Death scale, and the Templer Death Anxiety scale.

Z56 Larkin, E.J. "Three Models of Evaluation." CANADIAN PSYCHOLOGIST 15 (1974): 89-94.

Three types of evaluative research, evaluation by objectives, a systems approach, and a model proposed by Ackoff et al. (1962) are described.

Z57 Larsons, M.S. "Some Problems in Dissonance Theory Research." CENTRAL STATES SPEECH JOURNAL 24 (1973): 183-88.

This article briefly surveys the types of research based on the dissonance model of self-persuasion, reports two experiments that contradict many self-persuasion studies, and reviews several trends in self-persuasion research that may have led to the contradictory findings.

Research and Evaluation Methods

Z58 Lynch, J.J. "Biofeedback: Some Reflections on Modern Behavioral Science." SEMINARS IN PSYCHIATRY 5 (1973): 551-62.

> This paper reviews the current status of biofeedback research within the context of the evolution of behavioral science research in general. Methodological and conceptual problems inherent in current biofeedback research are discussed in terms of earlier conditioning theory and research. Future implications of biofeedback research for the practice of clinical psychology and psychiatry are also discussed.

Z59 Mackler, B. "Two Kinds of Research on Evaluation." PSYCHOLOGICAL REPORTS 34 (1974): 289-90.

> Formative and summative evaluation are described in detail and are compared.

Z60 McLaughlin, C.P., and Sheldon, A. THE FUTURE AND MEDICAL CARE: A HEALTH MANAGER'S GUIDE TO FORECASTING. Cambridge, Mass.: Ballinger Publishing Co., 1974. 125 p. Figs. Tables.

> This book covers the following topics: society, medical care, and the future of technology; forecasting in health; methods of forecasting and models; and a study of health care forecasting.

Z61 McWhirter, J.J. "Two Measures of the Facilitative Conditions: A Correlation Study." JOURNAL OF COUNSELING PSYCHOLOGY 20 (1973): 317-20.

> Counselor trainees (n = forty-five) were rated by coached patients using the Barrett-Lennard Relationship Inventory and by trained judges using the Truax rating scales on measures of empathy, warmth, and genuineness.

Z62 Merskey, H. "The Perception and Measurement of Pain." JOURNAL OF PSYCHOSOMATIC RESEARCH 17 (1973): 251-55.

> This article considers ways in which pain might be measured and how measures of the relative contributions of stimuli and of emotion might be derived.

Z63 Metsch, J.M., and Veney, J.E. "Measuring the Outcome of Consumer Participation." JOURNAL OF HEALTH AND SOCIAL BEHAVIOR 14 (1973): 368-74.

> This paper reports on the development of a methodology to measure the outcome of consumer participation through content analysis of consumer advisory board minutes.

Z64 Meyers, L.S., and Grossen, N.E. BEHAVIORAL RESEARCH: THEORY, PROCEDURE, AND DESIGN. 2d ed. Psychology Series. San Francisco: Freeman, 1978. 374 p. Illus. Tables.

> Intended for use in a first course, this text is designed to provide students with a basic understanding of the experimental design and research procedures used in the modern behavioral sciences. The four sections describe the methodological theory guiding scientists in the investigation of behavior, outline the ways in which they collect and analyze data, discuss research methodologies employed by behavioral scientists, and consider the means by which behavioral scientists communicate their knowledge to one another.

Z65 Monteiro, L.A., and Wessen, A.F. MONITORING HEALTH STATUS AND MEDICAL CARE. Cambridge, Mass.: Ballinger Publishing Co., 1976. 144 p. Tables.

> This book demonstrates the use of the survey method to study the affiliation between patients and physicians.

Z66 Moos, R.H. EVALUATING TREATMENT ENVIRONMENTS: A SOCIAL ECOLOGICAL APPROACH. New York: Wiley-Interscience, 1974. 416 p.

> This book presents methods for measuring and changing treatment milieus and for relating the characteristics of those milieus to human adaptation and functioning. Emphasis is placed on both subjective (satisfaction, morale, helping behavior) and objective (dropout, release) effects of treatment programs. It also discusses the clinical utility of research data about program milieus as an aid to teaching, planning innovative approaches to treatment, identifying trouble spots, and helping patients and staff change their social environments.

Z67 Moxley, R.A., Jr. "Formative and Non-Formative Evaluation." INSTRUCTIONAL SCIENCE 3 (1974): 243-83.

> Formative and nonformative evaluation are examined from the viewpoints of adaptation to improve individual performance and maintenance to preserve an existing system. Distinctions are made in terms of the experimental methods, criterion-referenced measures, accessible events, and individual behavior.

Z68 Pankratz, L., and Pankratz, D. "Nursing Autonomy and Patients' Rights: Development of a Nursing Attitude Scale." JOURNAL OF HEALTH AND SOCIAL BEHAVIOR 15 (1974): 211-16.

> A sixty-nine item attitude scale was administered to 702 nurses in five different settings to study nursing autonomy

Research and Evaluation Methods

and advocacy, patients' rights, and rejection of traditional role limitations.

Z69 Payne, J.W. "Alternative Approaches to Decision Making Under Risk: Moments Versus Risk Dimensions." PSYCHOLOGICAL BULLETIN 80 (1973): 439-53.

> The relative merits of explanations of approaches of individual decision making under risk are discussed.

Z70 Porter, A.C., and Chibucos, T.R. "Common Problems of Design and Analysis in Evaluative Research." SOCIOLOGICAL METHODS AND RESEARCH 3 (1975): 235-57.

> Strengths and weaknesses of several competing analysis strategies for quasi-experiments are considered. A multiple analysis approach that attempts to capitalize on the unique strengths and control the unique weaknesses of several analysis strategies is recommended.

Z71 Roberts, K.H., and O'Reilly, C.A., III. "Measuring Organizational Communication." JOURNAL OF APPLIED PSYCHOLOGY 59 (1974): 321-26.

> This research reports on initial attempts to develop instrumentation that can be used to compare dimensions of communication within and across organizations. Seven samples, with a total of more than 12,000 respondents, were used to develop a thirty-five-item questionnaire measuring sixteen facets of communication.

Z72 Rubin, D.B. "Estimating Causal Effects of Treatments in Randomized and Non-Randomized Studies." JOURNAL OF EDUCATIONAL PSYCHOLOGY 66 (1974): 688-701.

> A discussion of matching, randomization, random sampling, and other methods of controlling extraneous variation is presented with the objective of specifying the benefits of randomization in estimating causal effects of treatments.

Z73 Seiler, L.H. "The 22-Item Scale Used in Field Studies of Mental Illness: A Question of Method, a Question of Substance, and a Question of Theory." JOURNAL OF HEALTH AND SOCIAL BEHAVIOR 14 (1973): 252-64.

> The Midtown Manhattan Study twenty-two-item scale is assessed for its usefulness in measuring mental illness. Use of the measurement for examining the relationship between stressful life experiences and mental illness is discouraged.

Z74 Sinning, W.E. EXPERIMENTS AND DEMONSTRATIONS IN EXERCISE

PHYSIOLOGY. Philadelphia: W.B. Saunders Co., 1975. 162 p. Illus.

> To supplement texts used in courses on the physiology of exercise, this laboratory manual covers the topics of muscle contraction and its neural control, oxygen consumption and energy expenditure, body composition, physical fitness testing, kidney function, lung ventilation and related acid-base balance factors, and cardiovascular function.

Z75   Stahl, S.M., et al. "A Model for the Social Sciences and Medicine: The Case for Hypertension." SOCIAL SCIENCE AND MEDICINE 9 (1975): 31-38.

> Findings in the area of hypertension are synthesized from the disciplines of sociology and psychology to formulate a single model for guiding future research. The intervening role of perception for studying social and psychological factors in hypertension is explored.

Z76   Starfield, B. "Measurement of Outcome: A Proposed Scheme." MILBANK MEMORIAL FUND QUARTERLY 52 (1974): 39-50.

> This article proposes a scheme for describing health status that is based on the development of a profile rather than a single index.

Z77   Teevan, J.J., Jr. "On Measuring Anomia: Suggested Modification of the Srole Scale." PACIFIC SOCIOLOGICAL REVIEW 18 (1975): 159-70.

> This paper discusses briefly the concept of anomia and examines empirically a measurement problem of the Srole anomia scale.

Z78   Templer, D.I., and Lester, D. "An MMPI Scale for Assessing Death Anxiety." PSYCHOLOGICAL REPORTS 34 (1974): 238.

> A Minnesota Multiphasic Personality Inventory scale was constructed for measuring death anxiety. The authors feel this scale is useful in research and clinical situations in which the nature of the assessment should be disguised.

Z79   Thoresen, C.E., and Anton, J.L. "Intensive Experimental Research in Counseling." JOURNAL OF COUNSELING PSYCHOLOGY 21 (1974): 553-59.

> Intensive counseling research designs that examine treatment processes and effects with individuals over time are analyzed, and alternative designs and analysis methods are presented.

Research and Evaluation Methods

Z80 Uyeno, D.H. "Health Manpower Systems: An Application of Simulation to the Design of Primary Health Care Teams." MANAGEMENT SCIENCE 20 (1974): 981-89.

> This article describes a method developed to evaluate alternative primary health care team compositions and to examine skill levels for new categories of personnel. A simulation model of a general primary health care delivery unit was developed as part of the procedure, and an application of the simulation model and the evaluation procedure was made to the area of pediatrics.

Z81 Wallace, M.J., Jr., et al. "Measurement Modifications for Assessing Organizational Climate in Hospitals." ACADEMY OF MANAGEMENT JOURNAL 18 (1975): 82-97.

> The degree to which the Halpin and Croft Organizational Climate Description Questionnaire generalizes from business and educational settings to hospital settings is investigated. Two criteria established for the instrument were reliability and factor structure replication from previously investigated settings to hospital settings.

Z82 Ware, J.E., Jr., et al. THE MEASUREMENT OF HEALTH AS A VALUE: PRELIMINARY FINDINGS REGARDING SCALE RELIABILITY, VALIDITY AND ADMINISTRATION PROCEDURES. Measuring Health Concepts Research Project, MHC-74-11. Carbondale: Southern Illinois University School of Medicine, 1974. 64 p.

> After pretesting various methods of value measurement, health was added to the Rokeach Value Survey, and the interrelationships among value rankings were studied. The sample population consisted of residents and university students (n = 433). Four value dimensions were identified and defined for measurement: self-preservation, psychological orientation, social harmony, and enjoyment.

Z83 Ware, J.E., Jr., et al. MEASURES OF PERCEPTIONS REGARDING HEALTH STATUS: PRELIMINARY FINDINGS. Carbondale: Southern Illinois University, School of Medicine, 1974. 96 p.

> A scale to measure perceptions regarding health status and response to health state (lack of health anxiety and acceptance of the patient role) was administered to a sample of adults.

Z84 Weiss, C.H. "Between the Cup and the Lip." JOURNAL OF EDUCATIONAL EVALUATION 4 (December 1973): 1-6.

> The usability of much evaluation research is questioned and recommendations are made concerning the objectives of evaluation research.

Research and Evaluation Methods

Z85   Williams, R.L. "Explaining a Health Care Paradox." POLICY SCIENCES 6 (1975): 91-101.

> The link between physician density and measures of health based on outcome is analyzed through a single case-type. The apparently paradoxical findings in previous studies may be a result of failing to correct not only for differences in case severity but also for variations in sample errors.

Z86   Worden, J.W., et al. "Survival Quotient as a Method for Investigating Psychosocial Aspects of Cancer Survival." PSYCHOLOGICAL REPORTS 35 (1974): 719-26.

> This article describes the development of the Survival Quotient, an index of relative longevity applicable across different tumor sites with widely differing life expectancies. One site (lung) is used with case samples to illustrate the procedure.

Z87   Zyzanski, S.J., et al. "Scale for the Measurement of 'Satisfaction' with Medical Care: Modifications in Content, Format and Scoring." MEDICAL CARE 12 (1974): 611-20.

> Modifications and additional evaluation of a scale to measure attitudes toward physicians and primary medical care are reported. The complete scale is published with scale values and direction of affect, for use by other investigators.

Appendix 1
RESOURCES FOR PATIENT EDUCATION

Patient education programs are supported best by local experience and locally produced materials tailored to the educational needs of specific groups of patients as determined by the diagnostic steps outlined in this book and elsewhere (Green, L.W., et al. HEALTH EDUCATION PLANNING: A DIAGNOSTIC APPROACH, Palo Alto, Mayfield Publishing Co., 1979). Before embarking on the local production of programs and materials, however, it is only logical and prudent that one should review what has been produced and accomplished by others in relation to similar health problems or goals. There is likely to be a need to adapt what has been done elsewhere to make it appropriate to the local circumstances, but there is no need to start from scratch or to reinvent the wheel.

The following lists of sources identify agencies, institutions, and companies who produce educational materials or provide other forms of assistance in relation to health education. Most offer free or inexpensive materials or services. Many have local chapters, branches, affiliates, or representatives who can be contacted most conveniently by telephone. These, of course, are in addition to the local health department, regional health services agency, or state health department which are dependable sources of health education materials and often other forms of assistance as well.

ACCIDENT PREVENTION, INJURY CONTROL

Aetna Life and Casualty
Public Relations and Advertising Department
151 Farmington Avenue
Hartford, Conn. 06115

    Free loan films (safety).

American Insurance Association
Engineering and Safety Service
85 John Street
New York, N.Y. 10038

    Leaflets, pamphlets, film list.

Resources for Patient Education

American National Red Cross
Seventeenth and D Streets, N.W.
Washington, D.C. 20006
(Or you may contact local Red Cross chapters).

    Films, pamphlets, textbooks.

Center for Disease Control, CEMA
1600 Clifton Road, N.E.
Atlanta, Ga. 30333

    Pamphlets.

CNA Financial Corp.
Communications Department
310 South Michigan Avenue
Chicago, Ill. 60604

    Pamphlet.

Connecticut General Life Insurance Co.
Advertising and Public Relations--319
Hartford, Conn. 06115

    Pamphlets.

Institute for Safer Living
American Mutual Liability Insurance Co.
Wakefield, Mass. 01880

    Pamphlets.

Insurance Institute for Highway Safety
711 Watergate Office Building
2600 Virginia Avenue, N.W.
Washington, D.C. 20037

    Highway loss reduction information.

Kemper Insurance
Advertising and Public Relations Department
110 Tenth Avenue
Fulton, Ill. 61252

    Pamphlets.

Liberty Mutual Insurance Co.
Public Relations Dept.
175 Berkeley Street
Boston, Mass. 02117

    Pamphlets.

Resources for Patient Education

National Easter Seal Society for Crippled Children and Adults
2023 West Ogden Avenue
Chicago, Ill. 60612

    Leaflets on safety and home accidents.

National Safety Council
Director of Public Information
425 North Michigan Avenue
Chicago, Ill. 60611

    Films, pamphlets, posters.

Travelers Insurance Companies
Marketing Services
One Tower Square
Hartford, Conn. 06115

    Booklets of street and highway accident data.

Union Central Life Insurance Co.
P.O. Box 179
Cincinnati, Ohio 45201

    Booklet.

U.S. Department of Agriculture
Independence Avenue
Washington, D.C. 20205

    Leaflets on farm and home safety.

AGING

Administration on Aging
Office of Human Development
U.S. Department of Health, Education and Welfare
Washington, D.C. 20201

    Pamphlets, leaflets.

National Council on Aging
1828 L Street, N.W.
Washington, D.C. 20036

    Books, bibliographies, pamphlets.

Resources for Patient Education

ALCOHOLISM

Aetna Life & Casualty
151 Farmington Avenue
Hartford, Conn. 06115

    Free loan film.

Al-Anon Family Group Headquarters
P.O. Box 182
Madison Square Station
New York, N.Y. 10010

    Books, pamphlets, braille, audiotapes, newsletter.

Alcoholics Anonymous World Services
Box 459
Grand Central Station
New York, N.Y. 10017

    Books, pamphlets, catalog.

Association for the Advancement of Health Education
1201 Sixteenth Street, N.W.
Washington, D.C. 20036

    Books, pamphlets, charts, audiovisuals, catalog of additional teaching materials.

Kemper Insurance Companies
Advertising and Public Affairs Department
Long Grove, Ill. 60049

    Pamphlets on employees with a drinking problem.

National Council on Alcoholism
2 Park Avenue
New York, N.Y. 10016

    Pamphlets, books.

National Institute on Alcohol Abuse and Alcoholism
National Institute of Mental Health
5600 Fishers Lane
Rockville, Md. 20852

    Pamphlets, reprints, reports.

Resources for Patient Education

ALLERGIES

Allergy Foundation of America
801 Second Avenue
New York, N.Y. 10017

National Institute of Allergy and Infectious Diseases
Office of Information
Bethesda, Md. 20014

    Pamphlets, reprints, reports.

ARTHRITIS AND RHEUMATISM

Arthritis Foundation
GPO Box 2525
New York, N.Y. 10036

    Pamphlets, reprints.

BIRTH CONTROL

American Academy of Pediatrics
1801 Hinman Avenue
Evanston, Ill. 60294

    Catalog.

American Medical Association
535 North Dearborn Street
Chicago, Ill. 60610

    Pamphlets.

Association for Voluntary Sterilization
14 West Fortieth Street
New York, N.Y. 10018

    Pamphlets.

Bureau of Community Health Services
Parklawn Building, Rm. 7A-20
5600 Fishers Lane
Rockville, Md. 20852

    Pamphlets.

Resources for Patient Education

CORE Communications in Health
1290 Avenue of the Americas
New York, N.Y. 10019

    Pamphlets, films or slides, other materials.

Eaton Laboratories
Division of Morton-Norwich Products
Norwich, N.Y. 13815

    Pamphlets.

Medfact
P.O. Box 458
Massillon, Ohio 44646

    Miscellaneous materials.

Milex Central
(Doctor Discusses Series)
1873 Grove Street
Glenview, Ill. 60025

    Books.

Milner-Fenwick
3800 Liberty Heights Avenue
Baltimore, Md. 21215

    Miscellaneous materials.

Montreal Health Press
P.O. Box 1000, Station G
Montreal, Quebec, Canada H2W 2N1

    Pamphlets.

OMNI Education
190 West Main Street
P.O. Box 220
Sommerville, N.J. 08876

    Pamphlets, films or slides, other materials.

Ortho Pharmaceutical Corp.
Route 202
Raritan, N.J. 08869

    Pamphlets.

Resources for Patient Education

Planned Parenthood--World Population
810--Seventh Avenue
New York, N.Y. 10019

    Pamphlets.

Planned Parenthood Association, Chicago
55 East Jackson Boulevard
Chicago, Ill. 60604

    Pamphlets.

Planned Parenthood Federation of America
515 Madison Avenue, 5th Floor
New York, N.Y. 10022

    Miscellaneous materials.

Public Affairs Pamphlets
381 Park Avenue S.
New York, N.Y. 10016

    Pamphlets.

Rocky Mountain Planned Parenthood
2030 East Twentieth Avenue
Denver, Colo. 80205

    Pamphlets.

Searle Laboratories
Division of G.D. Searle and Co.
Box 5110
Chicago, Ill. 60680

    Pamphlets.

Sex Information Education Council of United States
Behavioral Publications
72 Fifth Avenue
New York, N.Y. 10011

    Pamphlets.

BIRTH DEFECTS

National Foundation--March of Dimes
800 Second Avenue
New York, N.Y. 10017

Resources for Patient Education

CANCER

American Cancer Society
Vice President for Public Education
219 East Forty-second Street
New York, N.Y. 10017

Office of Cancer Communications
National Cancer Institute
National Institutes of Health
Public Health Services
U.S. Department of Health, Education and Welfare
Rockville, Md. 20852

CEREBRAL PALSY

United Cerebral Palsy Association
66 East 34th Street
New York, N.Y. 10016

CHILD CARE AND DEVELOPMENT

American Academy of Pediatrics
1801 Hinman Avenue
Evanston, Ill. 60294

    Pamphlets.

American Association of Poison Control Centers
Committee on Educational Activities
c/o Academy of Medicine of Cleveland
Poison Information Center
10525 Carnegie Avenue
Cleveland, Ohio 44106

    Pamphlets.

American Dental Association
211 East Chicago Avenue
Chicago, Illinois 60611

    Pamphlets.

American National Red Cross
Office of Public Relations
Washington, D.C. 20006

    Pamphlets.

Resources for Patient Education

Association for the Aid of Crippled Children
Division of Publications
345 East Forty-sixth Street
New York, N.Y. 10017

    Books, pamphlets, reprints.

Carnation Co.
Medical Marketing Department
5045 Wilshire Boulevard
Los Angeles, Calif. 90036

    Miscellaneous materials.

Celanese Fibers Marketing Co.
Manager, Consumer and Retail Information
522 Fifth Avenue
New York, N.Y. 10036

    Pamphlets.

Channing L. Bete Co.
45 Federal Street
Greenfield, Mass. 01301

    Pamphlets.

Childsafe
Wisconsin Hospital Association
P.O. Box 4387
Madison, Wis. 53711

    Pamphlets, films or slides.

Child Study Association of America, Wel-Met
50 Madison Avenue
New York, N.Y. 10010

    Pamphlets, book lists.

Community Services Administration
Social and Rehabilitation Service
U.S. Dept. of Health, Education and Welfare
Washington, D.C. 20201

    Pamphlets, leaflets.

Consortium on Early Childbearing and Childrearing
1145 Nineteenth Street, N.W., Suite 618
Washington, D.C. 20036

    Pamphlets.

Resources for Patient Education

Consumer Product Safety Commission
Washington, D.C. 20207

> Pamphlets, catalogs.

CORE Communications in Health
Robert J. Brady Co.
Bowie, Md. 20715

> Pamphlets, films or slides, other materials.

Council on Family Health
633 Third Avenue
New York, N.Y. 10017

> Pamphlets.

Department of Pediatrics and Child Health
Howard University College of Medicine
Washington, D.C. 20001

> Pamphlets.

Gerber Products Co.
Professional Communication Department
445 State Street
Freemont, Mich. 49412

> Pamphlets.

J.B. Lippincott
East Washington Square
Philadelphia, Pa. 19105

> Pamphlets, films or slides.

John Hancock Mutual Life Insurance Co.
Manager, Community Relations
200 Berkeley Street
Boston, Mass. 02117

> Booklet in English and Spanish.

Johnson and Johnson
Department P
501 George Street
New Brunswick, N.J. 08903

> Pamphlets, folder, and chart on child development from birth to two years.

Resources for Patient Education

Juvenile Products Manufacturers Association
53 East Main Street
Moorestown, N.Y. 08057

    Pamphlets.

Mead Johnson Laboratories
Division of Mead Johnson and Co.
2404 West Pennsylvania Street
Evansville, Ind. 47721

    Pamphlets.

MEDCOM
12601 Industry Street
Garden Grove, Calif. 92641

    Films or slides.

Mendota Mental Health Institute
Child-Adolescent Center
Home and Community Treatment Project
Child Management Techniques
301 Troy Drive
Madison, Wis. 53704

    Pamphlets.

Metropolitan Life
Health and Welfare Division
One Madison Avenue
New York, N.Y. 10010

    Pamphlets, films or slides.

Milex Central (Doctor Discusses Series)
1873 Grove Street
Glenview, Ill. 60026

    Books.

National Association for Children with Learning Disabilities
5225 Grace Street
Pittsburgh, Pa. 15236

    Pamphlets.

National Center for Prevention and Treatment of Child Abuse and Neglect
University of Colorado Medical Center
Denver, Colo. 80210

    Pamphlets.

Resources for Patient Education

National Congress of Parents and Teachers
700 North Rush Street
Chicago, Ill. 60611

    Pamphlets on children and school problems.

National Easter Seal Society for Crippled Children and Adults
2023 West Ogden Avenue
Chicago, Ill. 60612

    Reprints, leaflets, pamphlets, bibliographies.

National Foundation--March of Dimes
1275 Mamaroneck Avenue
P.O. Box 2000
White Plains, N.Y. 10036

    Pamphlets, catalogs.

National Safety Council
425 North Michigan Avenue
Chicago, Ill. 60611

    Pamphlets.

Office of Child Development
Division of Public Education
P.O. Box 1182
Washington, D.C. 20210

    Pamphlets.

Parent's Magazine Films
52 Vanderbilt Avenue
New York, N.Y. 10017

    Films or slides.

Practical Parenting Publications
Box 18
Columbia, Mo.

    Pamphlets, catalogs, other materials.

Professional Research
660 South Bonnie Brae
Los Angeles, Calif. 90057

    Films or slides, other materials.

Resources for Patient Education

Public Affairs Pamphlets
381 Park Avenue S.
New York, N.Y. 10016

    Pamphlets.

Ross Laboratories
Division of Abbott Laboratories
Columbus, Ohio 43216

    Pamphlets, Spanish-language materials.

Videodetics Corporation
2121 South Manchester Avenue
Anaheim, Calif. 92802

    Films or slides, other materials.

Wisconsin Department of Health and Social Services
Division of Health
Section of Child Behavior and Development
P.O. Box 309
Madison, Wis. 53701

    Pamphlets.

CYSTIC FIBROSIS

National Cystic Fibrosis Research Foundation
60 East Forty-second Street
New York, N.Y. 10017

    Leaflets, films.

DENTAL HEALTH

American Dental Association
Bureau of Dental Health Education
211 East Chicago Avenue
Chicago, Ill. 60611

    Catalog listing pamphlets, charts, posters, and audiovisual materials.

National Dairy Council
Nutrition Education Division
111 North Canal Street
Chicago, Ill. 60606
(Or you may contact nearest dairy council or national headquarters.)

    Material by grade level.

Resources for Patient Education

DIABETES

American Diabetes Association
1 West Forty-eighth Street
New York, N.Y. 10020

> Pamphlets, books, catalogs, other materials.

American Medical Association
535 North Dearborn Street
Chicago, Ill. 60610

> Pamphlets, catalogs.

Ames Co.
Division of Miles Laboratories
Elkart, Ind. 46514

> Pamphlets, films or slides, other materials, Spanish-language materials.

Becton Dickenson and Co.
Rutherford, N.J. 07070

> Pamphlets, films or slides, other materials.

Central Ohio Diabetes Association
1406 Presidential Drive
Columbus, Ohio 43212

> Pamphlets, other materials.

Children's Hospital
561 South Seventeenth Street
Columbus, Ohio 43205

> Pamphlets.

CORE Communications in Health
Robert J. Brady Co.
Bowie, Md. 20715

> Pamphlets, films or slides, catalogs, other materials.

Diabetes Education Center
4959 Excelsior Boulevard
Minneapolis, Minn. 55416.

> Pamphlets, films or slides, books, catalogs, other materials.

Resources for Patient Education

Diabetes in the News
3553 West Peterson Avenue
Chicago, Ill. 60659

    Pamphlets.

Education for Health
205 Deerwood Lane
Minneapolis, Minn. 55427

    Pamphlets.

Eli Lilly and Co.
307 East McCarthy
P.O. Box 618
Indianapolis, Ind. 46206

    Pamphlets, catalogs.

Geigy Pharmaceuticals
Division of CIBA--GEIGY Corp.
Ardsley, N.Y. 10502

    Pamphlets.

Medfact Films
P.O. Box 458
Massilon, Ohio 44646

    Films or slides, catalogs.

Medic Alert Foundation
P.O. Box 1009
Turlock, Calif. 95380

    Pamphlets.

Mercy Medical Center
Patient Education Division
Dubuque, Iowa 52001

    Films or slides.

North Carolina Diabetes Association
408 Tyron Street
Charlotte, N.C. 28202

    Pamphlets, catalogs.

Resources for Patient Education

Pfizer
Pfizer Labs Division
235 East Forty-second Street
New York, N.Y. 10017

    Pamphlets, catalogs.

Pritchett and Hull
2996 Grandview, N.E.
Atlanta, Ga. 30305

    Pamphlets, films or slides, catalogs, other materials.

Public Affairs Committee
Film Library
22 East Twenty-third Street
New York, N.Y. 10016

    Films or slides.

South Texas Diabetes Association
P.O. Box 1638
Pasadena, Tex. 77501

    Books.

Train-Aide
1015 Grandiew
Glendale, Calif. 91201

    Pamphlets, films or slides, catalogs, other materials.

Trainex Corp.
P.O. Box 116
Garden Grove, Calif. 92642

    Films or slides, catalogs.

DRUG DEPENDENCE AND ABUSE

American Social Health Association
1740 Broadway
New York, N.Y. 10019

    Pamphlets.

Connecticut General Life Insurance Co.
Advertising and Public Relations--319
Hartford, Conn. 06115

    Pamphlets.

Resources for Patient Education

Kemper Insurance Companies
Advertising and Public Affairs Department
Long Grove, Ill. 60049

    Film.

National Institute of Drug Abuse, DHEW
11400 Rockville Pike
Rockville, Md. 20852

    Publications and pamphlets on drug abuse.

Pharmaceutical Manufacturers Association
Public Relations Division
1155 Fifteenth Street, N.W.
Washington, D.C. 20005

    Publications on prescription drug industry, drug abuse, and other health-related subjects.

ENVIRONMENTAL POLLUTION

Aetna Life & Casualty
Public Relations and Advertising Department
151 Farmington Avenue
Hartford, Conn. 06115

    Pamphlet, film.

American Lung Association
1740 Broadway
New York, N.Y. 10019
(Or you may contact local Lung Association chapters).

CNA Financial Corp.
Communications Department
310 South Michigan Avenue
Chicago, Ill. 60604

    Reprint.

Public Health Service
Environmental Health Service
U.S. Department of Health, Education, and Welfare
Rockville, Md. 20852

    Reprints, pamphlets.

Resources for Patient Education

EPILEPSY

Epilepsy Foundation of America
1828 L Street, N.W., Suite 406
Washington, D.C. 20036

    Pamphlets, films.

EYESIGHT

American Foundation for the Blind
15 West Sixteenth Street
New York, N.Y. 10011

    Books, pamphlets, journals, newsletters, films.

American Optometric Association
Division of Public Information
7000 Chippewa Street
St. Louis, Mo. 63119

    Pamphlets, posters, films, transcriptions, scripts for broadcast use. Catalogs.

Better Vision Institute
230 Park Avenue
New York, N.J. 10017

    Packet of materials.

National Society for the Prevention of Blindness
79 Madison Avenue
New York, N.Y. 10016

    Films, pamphlets, exhibits, radio and television spots.

FAMILY LIFE EDUCATION

American Medical Association
Department of Health Education
535 North Dearborn Street
Chicago, Ill. 60610

    Pamphlet series.

American Social Health Association
1740 Broadway
New York, N.Y. 10019

    Pamphlets, posters.

Resources for Patient Education

Kimberley-Clark Corp.
Life Cycle Center
Neenah, Wis. 54956

    Materials to supplement health and family life education courses.

Mental Health Materials Center
419 Park Avenue S.
New York, N.Y. 10016

    Selective reference guides, plays and other publications for mental health and family life education programs.

Planned Parenthood Federation of America
Publication Section
810 Seventh Avenue
New York, N.Y. 10019

    Pamphlets, reprints, films. Some Spanish-language material available.

Public Affairs Committee
381 Park Avenue S.
New York, N.Y. 10016

    Pamphlets on child guidance, family well-being, marriage and special family concerns.

FIRST AID

American National Red Cross
Seventeenth and D Streets, N.W.
Washington, D.C. 20006
(Or you may contact local Red Cross chapters.)

    Films, pamphlets, textbooks, exhibits, radio scripts.

Johnson and Johnson
Health Care Division
501 George Street
New Brunswick, N.J. 08901

    Folder and chart.

FOOT HEALTH

American Podiatry Association
20 Chevy Chase Circle, N.W.
Washington, D.C. 20010

    Leaflets, pamphlets, films, exhibits.

Resources for Patient Education

GENETIC DISEASE

Foundation for Research and Education in Sickle Cell Diseases
423 West 120th Street
New York, N.Y. 10027

>Pamphlets.

National Genetics Foundation
250 West Fifty-seventh Street
New York, N.Y. 10019

>Leaflets on genetic counseling and the NGF network.

National Tay-Sachs and Allied Diseases
Public and Professional Information
122 East Forty-second Street
New York, N.Y. 10017

>Pamphlet, reprint.

HEALTH (GENERAL)

Aetna Life and Casualty
Public Relations and Advertising Department
151 Farmington Avenue
Hartford, Conn. 06115

>Leaflets, posters, films.

American Medical Association
Department of Health Education
535 North Dearborn Street
Chicago, Ill. 60610

>Films, pamphlets, posters.

American National Red Cross
Seventeenth and D Streets, N.W.
Washington, D.C. 20006
(Or you may contact local Red Cross chapters.)

>Films, pamphlets, textbooks.

American Osteopathic Association
Editorial Department
212 East Ohio Street
Chicago, Ill. 60611

>Pamphlets, reprints, publications.

Resources for Patient Education

American Public Health Association
1015 Eighteenth Street, N.W.
Washington, D.C. 20036

> Educational qualifications of health workers, reprints, health guides, control handbooks, housing manuals, standard laboratory procedures.

CNA Financial Corp.
Communications Department
310 South Michigan Avenue
Chicago, Ill. 60604

> Booklet.

Commercial Union Assurance Co.
Public Relations Department
One Beacon Street
Boston, Mass. 02108

> Books, pamphlets, posters.

Consumers Union of United States
256 Washington Street
Mount Vernon, N.Y. 10550

> Reprints, books, booklets, and consumer education materials for classroom use.

Kimberley-Clark Corp.
Life Cycle Center
Neenah, Wis. 54956

> Material to supplement health and family life courses.

Liberty Mutual Insurance Co.
Public Relations Department
175 Berkeley Street
Boston, Mass. 02117

> Pamphlets.

Medical Services Administration
Social and Rehabilitation Service
U.S. Department of Health, Education and Welfare
Washington, D.C. 20201

Mental Health Materials Center
419 Park Avenue S.
New York, N.Y. 10016

> Selected reference guides, plays, and other publications for education programs. Leaflets.

Resources for Patient Education

Metropolitan Life Insurance Co.
Health and Welfare Division
1 Madison Avenue
New York, N.Y. 10010

>Catalogs and pamphlets, films and filmstrips related to personal, family, school, and community health.

Pharmaceutical Manufacturers Association
Public Relations Division
1155 Fifteenth Street, N.W.
Washington, D.C. 20005

>Variety of publications of prescription drug industry, drug abuse, and other health-related subjects.

Prudential Insurance Company of America
Public Relations Department
Prudential Plaza
Newark, N.J. 07101

>Leaflets.

Public Affairs Committee
381 Park Avenue S.
New York, N.Y. 10016

>Pamphlets, films (list available).

Social and Rehabilitation Service
U.S. Department of Health, Education and Welfare
Washington, D.C. 20201

>General information, films, exhibits, radio and television spots.

Tampax Incorp.
Department HI
Educational Director
5 Dakota Drive
Lake Success, N.Y. 11040

>Educational materials on menstruation and the menstral cycle.

U.S. Government Printing Office
Washington, D.C. 20402

>List of government publications on health, smoking and health, and baby and child care.

Resources for Patient Education

HEALTH CAREERS

American Association for Rehabilitation Therapy
Box No. 93
North Little Rock, Ark. 72115

    Pamphlets.

American Dental Association
Division of Career Guidance
211 East Chicago Avenue
Chicago, Ill. 60611

    Career information.

American Dietetic Association
620 North Michigan Avenue
Chicago, Ill. 60611

    Careers in dietetics.

American Hospital Association
Division of Health Careers
840 North Lake Shore Drive
Chicago, Ill. 60611

    Pamphlets, posters.

American Medical Association
Department of Health Education
535 North Dearborn Street
Chicago, Ill. 60610

    Health career information.

American Medical Women's Association
1740 Broadway
New York, N.Y. 10019

    Pamphlets, films.

American Osteopathic Association
Office of Education
212 East Ohio Street
Chicago, Ill. 60611

    Pamphlets, reprints.

Resources for Patient Education

American Physical Therapy Association
1156 Fifteenth Street, N.W.
Washington, D.C. 20005

    Career information.

American Podiatry Association
Council on Education
20 Chevy Chase Circle, N.W.
Washington, D.C. 20015

    Leaflets, pamphlets, films, exhibits.

American Public Health Association
1015 Eighteenth Street, N.W.
Washington, D.C. 20036

    Education qualifications of health workers, reprints.

Association for the Advancement of Health Education
1201 Sixteenth Street, N.W.
Washington, D.C. 20036

    Careers in health education and school nursing.

CNA Financial Corp.
Communications Department
310 South Michigan Avenue
Chicago, Ill. 60604

    Reprints.

National Association for Retarded Citizens
P.O. Box 6109
2709 Avenue E East
Arlington, Tex. 76011

    Pamphlet on mental retardation.

National Easter Seal Society for Crippled Children and Adults
2023 West Ogden Avenue
Chicago, Ill. 60612

    Pamphlets.

National Health Council
1740 Broadway
New York, N.Y. 10019

    Guidance to many sources of information on health careers and training programs.

Resources for Patient Education

National League for Nursing
10 Columbus Circle
New York, N.Y. 10019

> Lists of accredited nursing programs and scholarship sources.

Social and Rehabilitation Service
U.S. Department of Health, Education and Welfare
Washington, D.C. 20201

HEARING

National Association of Hearing and Speech Agencies
919 Eighteenth Street, N.W.
Washington, D.C. 20006

> Information on speech and hearing problems. Pamphlets, posters, television spots, films. (Postcard requests preferred).

National Easter Seal Society for Crippled Children and Adults
2023 West Ogden Avenue
Chicago, Ill. 60612

> Pamphlet on "Do's and Don'ts" for deaf children.

HEART DISEASE

American Heart Association
7320 Greenville Avenue
Dallas, Tex. 75231
(Or you may contact local Heart Association chapters.)

> Pamphlets, posters.

National Heart, Lungs and Blood Institute
Public Health Service
U.S. Department of Health, Education and Welfare
Bethesda, Md. 20014

HOSPITAL SERVICES

American Hospital Association
Hospital Research and Educational Trust
840 North Lake Shore Drive
Chicago, Ill 60611

> Pamphlets, posters, films, training programs, resource catalog.

Resources for Patient Education

American Osteopathic Association
Office of Hospital Affairs
212 East Ohio Street
Chicago, Ill. 60611

    Pamphlets.

HYPERTENSION

Abbott Film Service
Scientificom Distribution Center
708 North Dearborn Street
Chicago, Ill. 60610

    Films or slides, catalogs.

American Heart Association
7320 Greenville Avenue
Dallas, Tex. 75231

    Pamphlets, films or slides, catalogs, other materials.

Auburn University
Educational Television Department
Auburn, Ala. 36830

    Pamphlets, films or slides, other materials.

Boehringer Ingleheim
33 West Tarrytown Road
Elmsford, N.Y. 10523

    Pamphlets, catalogs, other materials.

Bristol Laboratories
Division of Bristol-Myers Co.
P.O. Box 657
Syracuse, N.Y. 13201

    Pamphlets.

CIBA Pharmaceutical Co.
Division of CIBA-GEIGY Corp.
556 Morris Avenue
Summit, N.J. 07901

    Pamphlets, films or slides, catalogs.

Resources for Patient Education

CORE Communications in Health
Robert J. Brady Co.
Bowie, Md. 20715

    Pamphlets, films or slides, catalogs, other materials, Spanish-language materials.

Education for Health
205 Deerwood Lane
Minneapolis, Minn. 55427

    Pamphlets, films or slides, other materials, catalogs.

Lawren Productions
P.O. Box 1452
Burlingame, Calif. 94010

    Films or slides, catalogs.

Lee Creative Communications
P.O. Box 1367
Rochester, N.Y. 14603

    Films or slides, catalogs.

Ellin Lieberman, M.D.
Childrens Hospital of Los Angeles
P.O. Box 54700
Los Angeles, Calif. 90054

    Films or slides, Spanish-language materials.

Mallinckrodt
Pharmaceutical Products Division
St. Louis, Mo. 63147

    Pamphlets.

Medfact
P.O. Box 458
420 Lake Avenue N.E.
Massilon, Ohio 44646

    Pamphlets, films or slides, catalogs.

Media Medica
555 Fifth Avenue
New York, N.Y. 10017

    Pamphlets.

Resources for Patient Education

Milex Central
1873 Grove Street
Glenview, Ill. 60025

    Books.

National High Blood Pressure Information Center
120/80 National Institutes of Health
Bethesda, Md. 20014

    Pamphlets, films or slides, other materials, catalogs, Spanish-language materials.

National Kidney Foundation
315 Park Avenue S.
New York, N.Y. 10010

    Pamphlets.

Professional Research
660 South Bonnie Brae Street
Los Angeles, Calif. 90057

    Pamphlets, films or slides, other materials, catalogs, Spanish-language materials.

Public Affairs Pamphlets
381 Park Avenue S.
New York, N.Y. 10016

    Pamphlets.

G.D. Searle and Co.
P.O. Box 5110
Chicago, Ill. 60680

    Books.

Smith, Kline and French Labs
Division of Smith Kline Corporation
1500 Spring Garden Street
Philadelphia, Pa. 19101

    Pamphlets, Spanish-language materials.

Trainex Corp.
P.O. Box 116
Garden Grove, Calif. 92642

    Pamphlets, films or slides, catalogs.

Resources for Patient Education

Video Communication
Suite 904, Watergate Office Building
2600 Virginia Avenue, N.W.
Washington, D.C. 20037

    Films or slides, catalogs.

MENTAL HEALTH

Connecticut Mutual Life Insurance Co.
Human Relations Program
140 Garden Street
Hartford, Conn. 06115

    Cartoon booklets available to teachers. Hartford lectures by professional staff of Institute of Living.

Mental Health Materials Center
419 Park Avenue S.
New York, N.Y. 10016

    Selective reference guides, plays, and other publications for mental health and family life education programs.

Metropolitan Life Insurance Co.
Health and Welfare Division
1 Madison Avenue
New York, N.Y. 10010

    Booklets and films.

National Association for Mental Health
(and its affiliated state and local associations)
43 West Sixty-first Street
New York, N.Y. 10019

    Pamphlets, books, films, plays.

National Association for Retarded Citizens
P.O. Box 6109
2709 Avenue E East
Arlington, Tex 76011

    Pamphlets.

National Institute of Mental Health
5600 Fisher Lane
Rockville, Md. 20852

    Pamphlets, bibliographies, bulletins.

Resources for Patient Education

MULTIPLE SCLEROSIS

National Multiple Sclerosis Society
Public Relations Department
257 Park Avenue S.
New York, N.Y. 10010
(Or you may contact local National Multiple Sclerosis Society chapters.)

    Pamphlets, medical papers, audiovisuals, loan films, posters.

MUSCULAR DYSTROPHY

Muscular Dystrophy Associations of America
Public Information Department
810 Seventh Avenue
New York, N.Y. 10019

NUTRITION AND WEIGHT CONTROL

Abbot Laboratories
North Chicago, Ill. 60064
Pharmaceutical Products Division

    Pamphlets.

American Association for Health, Recreation, and Physical Education
1201 Sixteenth Street, N.W.
Washington, D.C. 20036

American Bakers Association
Public Relations Department
Suite 560
1700 Pennsylvania Avenue, N.W.
Washington, D.C. 20056

    Posters, wall plaques, decals, films, pamphlets.

American Cancer Society
37 South Wabash
Chicago, Ill. 60603

    Pamphlets.

American Dental Association
211 East Chicago Avenue
Chicago, Ill. 60611

    Pamphlets.

Resources for Patient Education

American Diabetic Association
18 East Forty-eighth Street
New York, N.Y. 10017

    Miscellaneous materials.

American Dietetic Association
620 North Michigan Avenue
Chicago, Ill. 60611

    Pamphlets, booklets, materials for teachers.

American Dry Milk Institute
130 North Franklin Street
Chicago, Ill. 60606

    Pamphlets.

American Egg Board
205 Touhy Avenue
Park Ridge, Ill. 60068

    Pamphlets.

American Heart Association
44 East Twenty-third Street
New York, N.Y. 10010

    Pamphlets, films or slides, Spanish-language materials.

American Home Economic Association
2010 Massachusetts Avenue, N.W.
Washington, D.C. 20036

    Pamphlets, catalogs, reprints.

American Institute of Baking
Consumer Service Department
400 East Ontario Street
Chicago, Ill. 60611

    Pamphlets.

American Meat Institute
Department of Public Relations
59 East Van Buren Street
Chicago, Ill. 60605

    Pamphlets, catalogs.

Resources for Patient Education

American Medical Association
Section of Nutrition Information
535 North Dearborn Street
Chicago, Ill. 60610

    Pamphlets, catalogs.

Appleton-Century-Crofts Educational Division
Meredith Corp.
440 Park Avenue S.
New York, N.Y. 10016

    Pamphlets.

Armour Food Co.
Consumer Services Department
Greyhound Towers
Phoenix, Ariz. 85077

    Pamphlets.

Arthritis and Rheumatism Foundation
1212 Avenue of the Americas
New York, N.Y. 10036

    Pamphlets, films or slides, catalogs, other materials.

Bantrae--The No Meat "Meat"
Textured Vegetable Protein Products
8888 Rogers Street
Gloucester, Mass. 01930

    Pamphlets.

Best Foods--Consumer Service Department
Division of CPC International
International Plaza
Englewood Cliffs, N.J. 07632

    Pamphlets, films or slides, catalogs, other materials.

Blue Cross Association
840 North Lake Shore Drive
Chicago, Ill. 60611

    Pamphlets.

Borden Company
Marketing Services
50 West Broad Street
Columbus, Ohio 43215

    Pamphlets.

Resources for Patient Education

Bureau of Nutrition
Department of Health, City of New York
93 Worth Street, Room 714
New York, N.Y. 10013

    Pamphlets, other materials.

California Literacy
243 East Main Street
Aloama, Calif. 91801

    Pamphlets, catalogs.

California Prune Advisory Board
World Trade Center
San Francisco, Calif. 94111

    Pamphlets.

California Tree Fruit Agreement
701 Fulton Avenue
Sacramento, Calif. 95072

    Pamphlets.

Campbell Soup Co.
Food Service Products Division
375 Memorial Avenue
Camden, N.J. 08101

    Pamphlets, catalogs, other materials.

Carnation Co.
Medical Marketing Department
5045 Wilshire Boulevard
Los Angeles, Calif. 90036

    Pamphlets, other materials.

Cereal Institute
135 South LaSalle Street
Chicago, Ill. 60603

    Pamphlets, filmstrips, catalogs, other materials.

Chicago Dietetic Supply
405 East Shawnut
La Grange, Ill. 60525

    Pamphlets.

Resources for Patient Education

CORE Communications in Health
Robert Brady Co.
Bowie, Md. 20715

    Pamphlets.

Del Monte Kitchens
Del Monte Corp.
215 Premont Street
San Francisco, Calif. 94119

    Pamphlets.

Diabetes Education Center
4959 Excelsior Boulevard
Minneapolis, Minn. 55416

    Pamphlets, films or slides, catalogs.

Dietary Department
University Hospital
University of Wisconsin
Madison, Wis. 53706

    Pamphlets.

Equitable Life Assurance Society of the United States
1285 Avenue of the Americas
New York, N.Y. 10019

    Pamphlets.

Fleishmann's Corn Oil Margarines
1000 Donnelley Drive
Elm City, N.C. 27822

    Pamphlets.

Fleishmann's Margarine
P.O. Box 1407
Elm City, N.C. 27822

    Pamphlets.

Food and Agriculture Organizations of the United States
North American Regional Office
1325 C Street, S.W.
Washington, D.C. 20025

    Pamphlets.

Resources for Patient Education

Food and Nutrition Board
National Academy of Science
2101 Constitution Avenue, N.W.
Washington, D.C. 20418

    Pamphlets, catalogs.

Food and Nutrition Information and Educational Materials Center
National Agriculture Library
10301 Baltimore Boulevard, Room 304
Beltsville, Md. 20705

    Pamphlets.

Food Council of America
1750 Pennsylvania Avenue N.W.
Washington, D.C. 20005

    Pamphlets.

Geigy Pharmaceuticals
Division of CIBA-GEIGY Corp.
Ordsley, N.Y. 10502

    Pamphlets.

General Foods Cooperation
Consumer Service Department
250 North Street
White Plains, N.Y. 10602

    Pamphlets.

Gerber Products Co.
445 State Street
Fremont, Mich. 49412

    Pamphlets, other materials.

Good Food
P.O. Box 3838
Hollywood, Calif. 90028

    Pamphlets.

Good Housekeeping
Bulletin Service
959 Eighth Avenue
New York, N.Y. 10010

    Pamphlets.

Resources for Patient Education

Green Giant Co.
Home Services Department
5601 Green Valley Drive
Minneapolis, Minn. 55437

    Pamphlets.

H.J. Heinz
Consumer Relations
P.O. Box 57
Pittsburgh, Pa. 15230

    Pamphlets, other materials.

Harshe--Rothman and Druck
California Avocado Advisory Board
3345 Wilshire Boulevard
Los Angeles, Calif. 90010

    Pamphlets.

Information Division
Food and Nutrition Service
U.S. Drug Administration
Washington, D.C. 20705

    Pamphlets.

Institute of Nutrition
Allied Health Sciences Building
University of North Carolina--Chapel Hill
Chapel Hill, N.C. 27514

    Pamphlets, other materials.

International Apple Institute
Public Relations
2430 Pennsylvania Avenue, N.W.
Washington, D.C. 20037

    Pamphlets, films or slides, catalogs.

Kellogg Co.
Department of Home Economic Services
Battle Creek, Mich 49016

    Pamphlets.

Knox Gelatin
Subsidiary of Thomas J. Lipton
Englewood Cliffs, N.J. 07632

    Pamphlets.

Resources for Patient Education

Kraft Foods
500 Peshtigo Court
Chicago, Ill. 60690

    Pamphlets.

Lamb Educational Center
200 Clayton Street
Denver, Colo. 80206

    Pamphlets, other materials.

Libby, McNeil and Libby
200 South Michigan Avenue
Chicago, Ill. 60604

    Pamphlets.

Long Term Care Education Section
Division of Long Term Care
Health Services Administration
Parklawn Building, Rm. 11A-25
5600 Fishers Lane
Rockville, Md. 20852

    Pamphlets.

Mead Johnson Nutritionals
2402 Pennsylvania Avenue
Evansville, Ind. 47721

    Pamphlets.

Merck, Sharp and Dome
Educational Services
West Point, Pa. 19486

    Pamphlets, catalogs.

Merrell-National Laboratories
Division of Richardson-Merrill
Cincinnati, Ohio 45215

    Pamphlets.

Milex Central
1873 Grove Street
Glenview, Ill. 60025

    Books.

Resources for Patient Education

Nabisco
425 Park Avenue
New York, N.Y. 10022

    Miscellaneous materials.

National Academy of Sciences
National Research Council
Food and Nutrition Board
2101 Constitution Avenue
Washington, D.C. 20418

    Pamphlets, catalogs, other materials.

National Canners Association
Consumer Services
1133 Twentieth Street, N.W.
Washington, D.C. 20036

    Pamphlets, films or slides.

National Council on the Aging
1828 L Street, N.W.
Washington, D.C. 20036

    Pamphlets.

National Dairy Council
Nutrition Education Division
111 North Canal Street
Chicago, Ill. 60606

    Pamphlets, films, filmstrips, displays by grade level.

National Dairy Council
6300 North River Road
Rosemont, Ill. 60018

    Pamphlets, films or slides, catalogs, other materials.

National Foundation--March of Dimes
Public Education Department
P.O. Box 2000
White Plains, N.Y. 10602

    Pamphlets, films or slides, catalogs, other materials.

National Live Stock and Meat Board
36 South Wabash
Chicago, Ill. 60603

    Pamphlets, catalogs.

Resources for Patient Education

Nutra-Mate Textured Vegetable Protein
A.E. Staley Manufacturing Co.
Food Service Division
2011 Swift Drive
Oak Brook, Ill. 60521

    Pamphlets.

Nutrition Foundation
99 Park Avenue
New York, N.Y. 10016

    Pamphlets, other materials.

Nutrition Foundation
888 Seventeenth Street, N.W.
Washington, D.C. 20006

    Pamphlets, films or slides, catalogs, other materials.

Nutrition Section
Office of Clinical Services
Bureau of Community Health Services
Health Services Administration
Parklawn Building, Rm. 12-30
5600 Fishers Lane
Rockville, Md. 20852

    Pamphlets.

Nutrition Section
Division of Health
P.O. Box 309
Madison, Wis. 53701

    Pamphlets, films or slides, catalogs.

Overeaters Anonymous
3730 Motor Avenue
Los Angeles, Calif. 90034

    Pamphlets.

Pacifica Vegetable Oil Corp.
Saffola Products Division
World Trade Center
San Francisco, Calif. 94111

    Pamphlets.

Resources for Patient Education

Penwalt Corp.
Pharmaceutical Division
Bill Route 53/4
Woodridge, Ill. 60515

    Pamphlets.

Penwalt Pharmaceuticals Corp.
P.O. Box 1212
Rochester, N.Y. 14603

    Pamphlets.

Pet Incorp.
Office of Consumer Affairs
Pet Plaza
400 South Fourth Street
St. Louis, Mo. 63166

    Pamphlets.

Prudential Insurance Company of America
Prudential Plaza
Chicago, Ill. 60601

    Pamphlets, other materials.

Quaker Oats Co.
Consumer Services
Merchandise Mart Plaza
Chicago, Ill. 60654

    Pamphlets.

Ralston Purina Co.
Nutrition Service
Checkerboard Square
St. Louis, Md. 63102

    Pamphlets.

Rice Council
P.O. Box 22302
9317 Richmond Avenue
Houston, Tex. 77027

    Pamphlets.

Roche Labs
Division of Hoffmann-LaRoche
Nutley, N.J. 07110

    Pamphlets.

Resources for Patient Education

Sandoz Pharmaceuticals
East Hanover, N.J. 07936

    Pamphlets.

Standard Brands
DMS
23 East Twenty-second Street
New York, N.Y. 10010

    Pamphlets.

Stokely, Van Camp
Home Economics Department
941 Meridian
Indianapolis, Ind. 46206

    Pamphlets, other materials.

Sugar Information
254 West Thirty-first Street
New York, N.Y. 10001

    Pamphlets.

Sunkist Growers
Consumer Service
P.O. Box 7888
Valley Annex
Van Nuys, Calif. 91409

    Pamphlets, other materials.

Swift and Co.
Public Relations Department
115 West Jackson Boulevard
Chicago, Ill. 60604

    Pamphlets.

Syntex Laboratories
1344 Elmwood Avenue
Wilmette, Ill. 60091

    Pamphlets.

United Fresh Fruit and Vegetable Association
777 Fourteenth Street, N.W.
Washington, D.C. 20005

    Pamphlets, catalogs.

Resources for Patient Education

U.S. Department of Agriculture
Agricultural Research Service
Human Nutrition Research Branch and Home Economics Research Branch
Washington, D.C. 20201

    Pamphlets, publications list.

U.S. Department of Agriculture
Bureau of Human Nutrition and Home Economics
Home and Garden
Washington, D.C. 20201

    Pamphlets.

U.S. Department of Health Education and Welfare
Children's Bureau
Office of Child Development
U.S. Government Printing Office
Washington, D.C. 20402

    Pamphlets, films or slides, other materials.

U.S. Government Printing Office
Superintendent of Documents
Washington, D.C. 20402

    Pamphlets, catalogs.

Upjohn Co.
7171 Portage Road
Kalamazoo, Mich. 49001

    Pamphlets, catalogs.

Vitamin Information Bureau
383 Madison Avenue
New York, N.Y. 10017

    Pamphlets, films or slides, catalogs, other materials.

Wheat Flour Institute
Home Economics Department
14 East Jackson Boulevard
Chicago, Ill. 60604

    Pamphlets, films or slides.

Resources for Patient Education

OBESITY (See also NUTRITION)

Connecticut General Life Insurance Co.
Advertising and Public Relations--319
Hartford, Conn. 06115

    Pamphlet.

Liberty Mutual Insurance Co.
Public Relations Dept.
175 Berkeley Street
Boston, Mass. 02117

    Pamphlets.

Metropolitan Life Insurance Co.
Health and Welfare Division
1 Madison Avenue
New York, N.Y. 10010

    Booklets, films, weight tables.

OCCUPATIONAL THERAPY

American Occupational Therapy Association
6000 Executive Boulevard
Rockville, Md. 20852

    Pamphlets, films, poster, list of publications.

PARKINSON'S DISEASE

National Parkinson Foundation, Supporting the National Parkinson Institute for Treatment and Rehabilitation. Also supports the Bob Hope Parkinson Research Institute
1501 Northwest Ninth Avenue
Miami, Fla. 33136

    Leaflets.

PHYSICAL FITNESS

American Medical Association
Committee on Exercise and Physical Fitness
535 North Dearborn Street
Chicago, Ill. 60610

Resources for Patient Education

Association for the Advancement of Health Education
1201 Sixteenth Street, N.W.
Washington, D.C. 20036

    Films, pamphlets, award system.

Commercial Union Assurance Co.
Public Relations Department
One Beacon Street
Boston, Mass. 02108

    Booklets, posters.

Liberty Mutual Insurance Co.
Public Relations Department
175 Berkeley Street
Boston, Mass. 02117

President's Council on Physical Fitness and Sports
400 Sixth Street, N.W.--Room 3030
Washington, D.C. 20201

REHABILITATION

American Association for Rehabilitation Therapy
Box No. 93
North Little Rock, Ark. 72115

    Pamphlets.

CNA Financial Corp.
Communications Department
310 South Michigan Avenue
Chicago, Ill. 60604

    Reprints.

Rehabilitation Services Administration
Social and Rehabilitation Service
U.S. Department of Health, Education and Welfare
Washington, D.C. 20201

    Leaflets, pamphlets, reports, films.

Resources for Patient Education

RESPIRATORY DISEASES

American Lung Association
1740 Broadway
New York, N.Y. 10019
(Or you may contact local Lung Association chapters.)

    Pamphlets, audiovisuals, Spanish-language materials.

SCHOOL HEALTH

American Medical Association
Department of Health Education
535 North Dearborn Street
Chicago, Ill. 50610

Association for the Advancement of Health Education
1201 Sixteenth Street, N.W.
Washington, D.C. 20036

    Books, pamphlets, charts, audiovisuals, catalog of additional teaching materials.

Licensed Beverage Industries
Division of Educational Studies
485 Lexington Avenue
New York, N.Y. 10017

    Alcohol education reprints.

Office of Education, DHEW
400 Maryland Avenue, S.W.
Washington, D.C. 20016

    Pamphlets, leaflets.

Public Health Service
Public Inquiries Office of Information
U.S. Department of Health, Education and Welfare
Rockville, Md. 20852

    Pamphlets, leaflets, Directory of National Organizations with Interest in School Health.

Resources for Patient Education

SMOKING AND HEALTH

American Heart Association
7320 Greenville Avenue
Dallas, Tex. 75231
(Or you may contact local Heart Association chapters.)

    Brochures.

American Lung Association
1740 Broadway
New York, N.Y. 10019
(Or you may contact local Lung Association chapters.)

    Pamphlets, audiovisuals, Spanish-language materials.

National Clearinghouse for Smoking and Health
U.S. Public Health Service
5600 Fishers Lane
Rockville, Md. 20852

    Brochures, pamphlets, and films.

TUBERCULOSIS

American Lung Association
1740 Broadway
New York, N.Y. 10019
(Also contact local Lung Association chapters.)

    Pamphlets, display material, loan films, Spanish-language materials.

URINARY TRACT INFECTIONS

Ames Co.
Division of Miles Laboratories
Elkhart, Ind. 46514

    Pamphlets.

Burroughs Wellcome Co.
Research Triangle Park
Durham, N.C. 27709

    Miscellaneous materials.

Resources for Patient Education

CORE Communications in Health
Robert J. Brady Co.
Bowie, Md. 20715

    Pamphlets, films or slides, other materials.

Eaton Laboratories
Division of Morton-Norwich Products
Norwich, Conn. 13815

    Pamphlets.

Merrell-National Laboratories
Division of Richardson-Merrell
Cincinnati, Ohio 45215

    Miscellaneous materials.

National Kidney Foundation
315 Park Avenue S.
New York, N.Y. 10010

    Pamphlets.

Riker Laboratories
19901 Nordhoff Street
Northridge, Calif. 91324

    Pamphlets.

Warner, Chilcott
Division of Warner-Lambert
Morris Plains, N.J. 07950

    Pamphlets.

VENEREAL DISEASE

American Medical Association
Department of Health Education
535 North Dearborn Street
Chicago, Ill. 60610

    Pamphlets, posters, exhibits, books, catalog, consultants, workshops, conferences.

American Social Health Association
1740 Broadway
New York, N.Y. 10019

    Pamphlets, some titles in Spanish.

Resources for Patient Education

Connecticut General Life Insurance Co.
Advertising and Public Relations--319
Hartford, Conn. 06115

    Pamphlet.

Institute for Sex Education
22 East Madison Street, Suite 805
Chicago, Ill. 60602

    Pamphlets, books, films, reviews of materials, training programs for teachers and health professionals.

National Center for Disease Control, DHEW
1600 Clifton Road, N.E.
Atlanta, Ga. 30333

    Pamphlets, display material, films, consultants.

AUDIOVISUAL COMMERCIAL SUPPLIERS

Alfred Higgens Products
9100 Sunset Boulevard
Los Angeles, Calif. 90069

Ames Co.
Division of Miles Laboratory
1127 Myrtle Street
Elkhart, Ind. 46514

American Learning Systems
P.O. Box 2173
Columbus, Ga. 31902

American Video Network
660 South Bonnie Brae Street
Los Angeles, Calif. 90057

Audience Planners
208 South LaSalle Street
Chicago, Ill. 60604

AV Scientific Aids
639 North Fairfax Avenue
Los Angeles, Calif. 90036

Beckton Dickinson and Co.
Rutherford, N.J. 07070

Bluestone Video Makers
4018 Twenty-second Street
San Francisco, Calif. 94114

BNA Communications
8371 Bernice Drive
Strongsville, Ohio 44136

Brookhaven Memorial Hospital
101 Brookhaven Hospital Road
Patchogue, N.Y. 11772

Channing L. Bete
45 Federal Street
Greenfield, Mass. 01301

Churchill Films
662 North Robertson Boulevard
Los Angeles, Calif. 90069

CORE Communications in Health
Robert J. Brady Co.
Rt. 197
Bowie, Md. 20715

Resources for Patient Education

Diabetes Education Center
4959 Escelsior Boulevard
Minneapolis, Minn. 55416

Fairview General Hospital
c/o Greater Cleveland Hospital Assn.
1021 Euclid Avenue
Cleveland, Ohio 44115

Family Communications
4802 Fifth Avenue
Pittsburgh, Pa. 15213

Film and Video Service
ACOG
P.O. Box 299
Wheaton, Ill. 60187

Health Films Library
P.O. Box 309
One West Wilson Street
Madison, Wis. 53701

Hospital Audio Visual Education
606 Halstead Avenue
Mamaroneck, N.Y. 10543

Illinois, University of. Medical Center
Public Information Office
P.O. Box 6998
1737 West Polk Street
Chicago, Ill. 60680

Indiana University School of Medicine
Medical Educational Resources Program
1100 West Michigan Street
Indianapolis, Ind. 46202

Kansas, University of
College of Health Sciences and Hospital
Thirty-ninth and Rainbow Streets
Kansas City, Kans. 66103

Lee Creative Communications
P.O. Box 1367
Five South St. Regis Drive
Rochester, N.Y. 14618

Martland Hospital
Health Education Project
College of Medicine and Dentistry
  of New Jersey
65 Bergen Street
Newark, N.J. 07107

MEDCOM
1633 Broadway
New York, N.Y. 10019

MEDFACT
420 Lake Avenue, N.E.
Massillon, Ohio 04646

Milner-Fenwick
3800 Liberty Heights Avenue
Baltimore, Md. 21215

Mississippi Methodist Hospital and
  Rehabilitation Center
1675 Lakeland Drive, Suite 501
Jackson, Miss. 39216

Motion
4437 Klingle Street, N.E.
Washington, D.C. 20016

Oakwoods Media
2243 South Eleventh Street
Kalamazoo, Mich. 49009

Ormont Drug and Chemical Co.
520 South Dean Street
Englewood, N.J. 07631

Parent's Magazine Films
80 New Bridge Road
Bergerfield, N.J. 07621

Patient Information Library
Physicians' Art Service
343-B Serramonte Plaza Office Center
Daly City, Calif. 94015

Resources for Patient Education

Pelican Films
3010 Santa Monica Boulevard, #440
Santa Monica, Calif. 90404

Pritchett and Hull Associates
2996 Grandview Avenue, N.E.
Atlanta, Ga. 30305

Professional Research
660 South Bonnie Brae Street
Los Angeles, Calif. 90057

Professional Research
Davis Videocassette Co.
1470 South Woodhaven Drive
Baton Rouge, La. 70815

Public Television Library
475 L'Enfant Plaza, S.W.
Washington, D.C. 20024

Pyramid Films
Box 1048
2801 Colorado Boulevard
Santa Monica, Calif. 90406

Research Media
96 Mount Auburn Street
Cambridge, Mass. 02138

Teach'Em
625 North Michigan Avenue
Chicago, Ill. 60611

Toronto, University of
Division of Instructional Media Services
Eight Taddlecreek Road
Toronto, Ontario, Canada M5S 1A8

Train-Aide
1015 Grandview Avenue
Glendale, Calif. 91201

Trainex Patient Video Corp.
10 Perimeter Place, N.W.
Atlanta, Ga. 30339

VIDCOM
10 Perimeter Place, Suite 150
Atlanta, Ga. 30339

Walnat Co.
3430-32 North Illinois Street
Indianapolis, Ind. 46208

Wells National Services Corp.
200 Park Avenue
New York, N.Y. 10017

Appendix 2
HEALTH NEWSLETTERS

The following list is a compilation by the Health Policy Center of Georgetown University of selected newsletters which cover national health activity. The list includes private, federal, and commercial publications. Arranged in alphabetical order by title, the entries include pertinent subscription information. Additionally, each entry includes a brief summary which indicates subject focus and depth of coverage as well as any unique features the newsletter offers a reader.

"The Blue Sheet." DRUG RESEARCH REPORTS: 1152 National Press Building, Washington, D.C. 20045. Weekly.

> Government agencies and activities in drug, medical, and allied research fields; government and private foundation grants. Capsule one-liners of weekly news; occasional supplements.

CITY HEALTH OFFICERS NEWS. U.S. Conference of City Health Officers, 1620 Eye Street, N.W., Washington, D.C. 20006. Monthly.

> Federal health regulations, legislation, and impact on local health care delivery. Local noted health department programs; annual supplement of vital statistics summary for urban centers of 100,000.

COMMUNITY MENTAL HEALTH CENTER NEWS. Morris Associates, 1346 Connecticut Avenue, N.W., Washington, D.C. 20036. Monthly.

> Developments at federal level affecting community mental health centers. Subject focus.

COTH REPORT. Association of American Medical Colleges, One Dupont Circle, Washington, D.C. 20036. Monthly.

> News, commentary, and analysis for health care executives and medical educators; includes information on current federal and state legislation. Review of pertinent surveys, studies and reports; hospital orientation.

Health Newsletters

DEVELOPMENTAL DISABILITIES. Morris Associates, Governmental Affairs Consultants, 1346 Connecticut Avenue, N.W., Washington, D.C. 20036. Periodically.

> News of congressional and administration actions affecting the DD program. Subject focus.

DRUG RESEARCH REPORTS. See entry under "The Blue Sheet."

ENVIRONMENTAL HEALTH LETTER. Gershon Fishbein, ed., National Press Building, Washington, D.C. 20045. Semimonthly.

> Broad coverage of developing environmental issues. Special supplements as news warrants.

FOCAL POINTS. Bureau of Health Education, Center for Disease Control, Public Health Service, U.S. Department of Health, Education and Welfare, Atlanta, Ga. 30333. Monthly.

> Descriptive information of health education projects being carried out at local, state, and national levels. Current activities and concerns of the Bureau of Health Education. Compiles unpublished notes from the field, providing notice of health education projects and programs in progress.

HEALTH DAILY. Capitol Publications, 2430 Pennsylvania Avenue, N.W., Suite G-12, Washington, D.C. 20037. Daily.

> Coverage of major federal health news developments. Daily publication; computerized quarterly; and cumulative annual index.

HEALTH EDUCATION REPORTS. Plus Publications, 2626 Pennsylvania Avenue, Washington, D.C. 20037. Semimonthly.

> Recent health education legislation, activities of federal agencies related to health education.

HEALTH MANPOWER REPORT. Education News Service, Division of Capitol Publications, Suite G-12, 2430 Pennsylvania Avenue, N.W., Washington, D.C. 20037. Biweekly.

> Federal health-general, current developments.

HEALTH PERSPECTIVES. Consumer Commission of the Accreditation of Health Services, 381 Park Avenue S., New York, N.Y. 10016. Bimonthly.

> Consumer health issues; in-depth analysis of single topic per issue. Consumer orientation.

HEALTH PLANNING & MANPOWER REPORTS. Plus Publications, 2626 Penn-

Health Newsletters

sylvania Avenue, N.W., Washington, D.C. 20037. Biweekly.

> Federal developments and activity relating to health planning and manpower legislation. Subject focus.

HEALTH PLANNING LETTER. McGraw-Hill, 457 National Press Building, Washington, D.C. 20045. Semimonthly.

> Health planning and resource development. Special reports on major regulatory development. Health planning across the nation.

HEALTH RESOURCES NEWS. DHEW-Public Health Service, Health Resources Administration, 5600 Fishers Lane, Rockville, Md. 20852. Monthly.

> Federal health news; resource development, statistics, research, manpower. Selected HRA reading list in most issues.

HEALTH SERVICES INFORMATION. Healthcare Publication, 976 National Press Building, Washington, D.C. 20045. Weekly.

> Combines news reporting with in-depth analysis of federal regulations and legislation. Covers state and local news; scoops on innovative health programs.

HEALTH SYSTEMS. Morris Associates, 1346 Connecticut Avenue, N.W., Washington, D.C. 20036. Semimonthly.

> Developments at federal level affecting health services delivery and funding. Reports on major health actions; a view of federal health policy trends.

HOME HEALTH HIGHLIGHTS. National Association of Home Health Agencies, 605 Bannock Street, Denver, Colo. 80204. Biweekly.

> Federal legislation and regulatory actions affecting Home Health agencies. Subject focus.

HMO & HEALTH SERVICES REPT. Girard Associates, 399 Howard Boulevard, Mt. Arlington, N.J. 07856. Monthly.

> Offers history of legislation, discussion of major issues.

LEGISLATIVE ROUNDUP. American Medical Association, 535 North Dearborn Street, Chicago, Ill. 60610. Weekly.

> National medical legislation; coverage of hearings with emphasis on AMA participation. Special issues devoted to selected topic of particular interest to the medical profession.

MEDICARE AND MEDICAID BULLETIN. American Hospital Association, 840 North Lake Shore Drive, Chicago, Ill. 60611. Monthly.

Health Newsletters

    Actions pertinent to Medicare and Medicaid; specifics of AHA action.

MEDICARE/MEDICAID MONTHLY. Healthcare Publications, 976 National Press Building, Washington, D.C. 20045. Monthly.

    Wrap-up of the month's events surrounding the two largest and most controversial federal programs. Subject focus.

MH-MR REPORT. Morris Associates, Governmental Affairs Consultants, 1346 Connecticut Avenue, N.W., Washington, D.C. 20036. Semimonthly.

    Mental health and mental retardation activities in Congress and federal agencies. Subject focus.

NATIONAL HEALTH INSURANCE REPORTS. Plus Publications, 2626 Pennsylvania Avenue, N.W., Washington, D.C. 20037. Biweekly.

    NHI and health policy developments.

NEWS OF THE COOPERATIVE HEALTH STATISTICS SYSTEM. DHEW-Public Health Service, Health Resources Administration, 5600 Fishers Lane, Rockville, Md. 20852. Bimonthly.

    Review of health data. Attention to state information systems.

OCCUPATIONAL HEALTH & SAFETY LETTER. Gershon Fishbein, ed., National Press Building, Washington, D.C. 20045. Semimonthly.

    As title indicates, OSHA focus.

"The Pink Sheet." F-D-C REPORTS. 1152 National Press Building, Washington, D.C. 20045. Weekly.

    Actions of Congress and regulatory agencies affecting drug and cosmetic industry, industry trends, and financial analysis. Capsulated news summaries.

PSRO LETTER. McGraw-Hill, 457 National Press Building, Washington, D.C. 20045. Semimonthly.

    Utilization review and PSRO and general health care provider regulations. "PSRO Around the Nation."

SOPHE NEWS AND VIEWS. Society for Public Health Education. 693 Sutter Street, San Francisco, Calif. 94103. Quarterly.

    Organizational announcements, federal developments, trends, and administrative implications for health educators.

Health Newsletters

WASHINGTON ACTIONS ON HEALTH. 1740 N Street, N.W., Washington, D.C. 20036. Weekly.

> Significant federal health documents. Access to document retrieval; handy extractions from CONGRESSIONAL RECORD and FEDERAL REGISTER.

WASHINGTON DEVELOPMENTS. American Hospital Association, 1 Farragut Square S., Washington, D.C. 20006. Semimonthly.

> Federal news of health issues in Congress; activities of agencies noted. News wrap-up page; occasional legislative status report.

WASHINGTON DRUG AND DEVICE LETTER. Capitol Publications, 2430 Pennsylvania Avenue, N.W., G-12, Washington, D.C. 20037. Weekly.

> Federal developments on regulation and legislation. Only publication to combine drug and device reporting.

WASHINGTON INFORMATION NATIONAL HEALTH INSURANCE. Girard Associates, 399 Howard Boulevard, Mt. Arlington, N.J. 07856. Monthly.

> NHI policy developments.

WASHINGTON NEWSLETTER. American Public Health Association, 1015 Eighteenth Street, N.W., Washington, D.C. 20036. Monthly.

> Summary and status of federal legislation; general health news.

WASHINGTON REPORT ON HEALTH LEGISLATION. McGraw-Hill, 437 National Press Building, Washington, D.C. 20045. Weekly.

> Update and review of federal health legislation. Tables; e.g., listing of new health bills, comparisons of NHI proposals.

WASHINGTON REPORT ON LONG TERM CARE. McGraw-Hill, 457 National Press Building, Washington, D.C. 20045. Weekly.

> Federal activity regarding nursing homes, facilities for mental health. Special reports on major regulatory developments.

WASHINGTON REPORT ON MEDICINE AND HEALTH. McGraw-Hill, 457 National Press Building, Washington, D.C. 20045. Weekly.

> Federal health activity, including legislation. Special reports on budgets, appropriations, and NHI.

Appendix 3
CONSULTATION IN PATIENT EDUCATION

The following accredited schools of public health have health education faculty members who can provide consultation on patient education.

University of California
School of Public Health
19 Earl Warren Hall
Berkeley, Calif. 94720

University of California at Los Angeles
School of Public Health
Los Angeles, Calif. 90024

Columbia University
School of Public Health
600 West 168th Street
New York, N.Y. 10032

Harvard University
School of Public Health
677 Huntington Avenue
Boston, Mass. 02115

University of Hawaii
School of Public Health
1960 East-West Road
Honolulu, Hawaii 96822

University of Illinois at the Medical Center
School of Public Health
P.O. Box 6998
Chicago, Ill. 60680

Johns Hopkins University
School of Hygiene and Public Health
615 North Wolfe Street
Baltimore, Md. 21205

Loma Linda University
School of Health
Loma Linda, Calif. 92354

University of Massachusetts
Division of Public Health
School of Health Sciences
Amherst, Mass. 01002

University of Michigan
School of Public Health
Ann Arbor, Mich. 48104

University of Minnesota
School of Public Health
1360 Mayo Memorial Building
412 Union Street, S.E.
Minneapolis, Minn. 55455

University of North Carolina
School of Public Health
Chapel Hill, N.C. 27514

University of Oklahoma
Division of Public Health
College of Health
Health Sciences Center
P.O. Box 26901
Oklahoma City, Okla. 73190

University of Pittsburgh
Graduate School of Public Health
111 Parran Hall
Pittsburgh, Pa. 15261

Consultation in Patient Education

University of Puerto Rico
School of Public Health
Medical Sciences Campus
GPO Box 5067
San Juan, P.R. 00936

University of Texas at Houston
School of Public Health
P.O. Box 20186
Astrodome Station
Houston, Tex. 77025

Tulane University
School of Public Health and Tropical
    Medicine
1430 Tulane Avenue
New Orleans, La. 70112

University of Washington
School of Public Health and
    Community Medicine
F356d Health Sciences Building
Mail Drop SC-30
Seattle, Wash. 98195

Yale University
Department of Epidemiology and
    Public Health
School of Medicine
60 College Street
New Haven, Conn. 06510

In addition, the faculty of the following accredited graduate programs in community health education can provide similar consultation:

California State University of
    Northridge
Department of Health Sciences
Northridge, Calif. 91324

Hunter College
Community Health Education Program
Institute for Health Sciences
105 East 106th Street
New York, N.Y. 10029

University of Missouri
Division of Community Health Education
202 Clark Hall
Columbia, Mo. 65201

New York University
School of Education, Health, Nursing
    and the Arts
Washington Square
New York, N.Y. 10003

San Jose State University
Department of Health Science
125 South Seventh Street
San Jose, Calif. 95114

University of Tennessee
College of Education
Health and Safety Department
Knoxville, Tenn. 37916

Consultation on patient education is available also from state and local health departments, university departments of health education other than those listed above, voluntary health associations, and professional health associations such as SOPHE, the Society for Public Health Education, the national membership association of professional health educators. This society represents the complete range of professional skills and experience in health education practice. The members are employed by federal, state, and local government agencies, international organizations, voluntary health agencies (national, state and local), medical care and nursing home facilities, and academic institutions. Expertise ranges from preparation of communication materials to health service planning, administration, and evaluation. The society membership is represented in most major communities of the United States. There are fifteen affiliated chapter organizations of SOPHE Fellows.

Consultation in Patient Education

For further information about the SOPHE Consultation Service, its availability, and/or appropriateness for specific programs, contact:

Mr. Arthur Jack Grimes, M.P.H.
Chairman, Consultation Service
c/o American Institute of Biological Sciences
1401 Wilson Boulevard
Arlington, Va. 22209
Phone: (703) 527-6776

or

Mr. James P. Lovegren, M.P.H.
Executive Director
Society for Public Health Education
693 Sutter Street
San Francisco, Calif. 94103
Phone: (415) 673-7266

INDEXES

AUTHOR INDEX

This index is alphabetized letter by letter and refers the reader to applicable entry numbers of annotated materials or to page numbers when citations are listed within the introductory text preceding each chapter.

A

Abelle, B.E. S1
Abernathy, W.J. U1
Ablon, J. C1
Abou-Rass, M. Y1
Abrams, R.D. W1
Abramson, R. P1
Adams, D.W. D19
Adamson, J.D. J1
Adamson, T.E. Y2
Aday, L.A. F1, K1
Addington, W.W. U33
Adelson, R. W2
Aderman, D. B14
Adkins, J.R. T85
Adler, N.E. D1
Adler, R. J2
Aiba, F.H. P2
Aitken, R.C.B. Z1
Ajzen, I. p. 68, I1
Alkhateeb, W. p. 120
Alkin, M.C. Z5
Allan, F.N. Y3
Allen, L.A. X1
Almond, R. p. 165
Alpander, G.G. Z2
Alterman, A.I. N1

Altshuler, A. p. 121
American Cancer Society A1
American Hospital Association p. 123
American Public Health Association
   p. 90, p. 123, p. 165
Andersen, M.D. B1
Andersen, R. F1
Anderson, C.L. p. 87
Anderson, H.E., Jr. I2
Anderson, J.D. Y4
Anderson, J.F. Z10
Anderson, J.G. F2
Anderson, N.H. Q15
Anderson, R. p. 87
Anderson, W.F. E1
Andreasen, N.J.C. B4
Andrews, I.R. X3
Angers, W.P. L1
Annas, G.J. P3
Anthony, W.A. W3
Anton, J.L. Z79
Antonovsky, A. F3, p. 88
App, H. E2
Arkowitz, H.S. Q17, T1
Arnold, W.E. Q1
Arthur D. Little Co. (see Little, Arthur D., Co.
Asken, M.J. B2
Astrachan, B.M. R1
Attkisson, C.C. Z4

Author Index

Autor, S.B. Y5
Azrin, N.H. T2

B

Back, K.W. W4
Baizerman, M. D2
Baker, E.J. Y6
Baker, E.L. Z5
Baker, F. U38
Balaban, R.M. Z6
Balch, P. J3, R2, U18
Balinsky, W. Z7
Bandura, A. T3
Banner, D.K. Z8
Barber, B. P30
Barber, W.H. W5
Bardo, H.R. T4
Barglow, P. D3
Barr, D.M. Y7
Barrish, I.J. T5
Barron, F.H. Z9
Barry, G.M. Y8
Barsky, A.J. G11, p. 123
Bartkus, D.E. F2
Bartle, S.H. A2
Bartow, J.C. U2
Bashshur, R.L. S2
Bassin, A. W6
Bean, P. B3
Becker, M.H. G1, G2, p. 68, 13, J26, p. 97, M1, P4, Y44, Y45, Y46
Beckhard, R. p. 163
Bednar, R.L. R3
Beebe, S.A. Q2
Bell, C.E. T6
Bell, C.H. p. 164
Bellack, A.S. G3
Bellin, L.E. Y9
Belsky, R. D4
Bender, E.I. p. 88
Bennis, W.G. p. 88, p. 164
Benton, A.L. T49
Berdie, D.R. Z10
Berger, B. p. 69
Berger, R. Z7
Berkanovic, E. F4, L2, X2
Berkman, B. A3
Berman, H.J. U3

Berner, E.S. Z11
Bernstein, I.N. Z12
Bernstein, L. P5
Berry, R.E., Jr. Z13
Bertcher, H. W7
Bertram, D.A. p. 216
Best, J.A. T7, T8, p. 216
Beutler, L.E. P6
Bickman, L. Q43
Bigelow, D.A. H14
Bindra, D. T9
Bishop, E.H. Y57
Bishop, L.F. A2
Bittker, T.E. S25
Black, L.F. p. 120
Blackwell, B. G4
Blanchard, E.B. T10, T11, T12, T13, T86, p. 216
Blane, H. p. 181
Blanton, G.W. W8
Bleecker, E.R. T14, T15
Bloch, A.D. S28
Blocher, D.H. W22
Block, B. P1
Block, M. p. 216
Bloom, J.R. p. 123
Bloxom, A.L. R22
Bluford, R.J. D5
Blum, H.L. V1, Z14
Blumenfeld, W.S. W9
Bodenheimer, T. L3
Boisvert, M.J. T16
Boller, J.D. W10
Bond, M.R. J4
Bonito, A.J. Y34
Bootzin, R. T58
Borgman, M.F. A4
Borman, L.D. p. 88
Boston Women's Health Book Collective D6
Bowden, C.L. P7
Braby, R. W11
Bracht, N.F. U4
Bracken, M.B. M2, P8, p. 121
Bradley, F.O. Y10
Brady, J.P. T17
Brady, J.V. Z44
Bragg, J.E. X3
Brand, F.N. G5
Branscomb, A.B. U5

Author Index

Branscomb, E.W. U5
Brashear, D.B. D7
Braun, S.H. Y11
Brennan, M.E. D16
Bretz, R. Y12
Bridges-Webb, C. F5
Britton, M. Q3
Brock, G.W. W19
Brodland, G.A. B4
Broll, L. Q4
Brooks-Bertram, P.A. p. 181, p. 216
Brosseau, J.D. W62
Brown, B.S. T18
Brown, J.S. F6
Brown, P.A. D14
Browning, P.L. C2, P9
Bryant, N.H. p. 121
Bryson, S. P10
Budzynski, T.H. T19, T20
Burck, H.D. Z15
Burgess, A.W. F7
Burgess, J.H. W47
Burgoon, M. Q26
Burke, W.M. F8
Burke, W.W. p. 164
Burkett, J.R. W60
Burleson, G. Q5
Burt, D.W. C3
Burt, M.R. Q25
Butler, A.K. T51
Butts, S.V. J5

C

Calia, V.F. W12
California Department of Health D21
Callahan, D. P25
Campbell, J.D. H1
Canada, R.M. W13
Canfield, E. W14
Canfield, R.E. P11
Canter, L. W15
Caplan, G. p. 88
Caplan, R. p. 45
Caplan, R.M. W16
Carey, R.G. B5
Carlson, K.W. Q6
Carlton, B. F9
Carmody, J. Y13
Caro, F.G. Z16

Caron, H.S. p. 120
Carroll, J.B. S3
Carson, R.A. B6
Carter, D.K. P12
Carter, R.E. W27
Casata, D.M. p. 181
Cassel, J.C. Z49
Cassell, J.L. O1
Cattell, R.B. Z17, Z18
Cauffman, J.G. V2, V3
Chafetz, M.E. p. 181
Chalmers, D.K. S4
Chambers, D.W. W17
Chance, J. O17
Chez, R.A. S33
Chibucos, T.R. Z70
Child, D. Z18
Chiles, J.A. O2
Chotlos, J. J5
Chwalow, A.J. p. 215
Cialdini, R.B. Q7
Clack, R.J. U24, W24
Clark, M. p. 70
Clark, R.D. III R4
Clark, T.L. W18
Cline, F.W. M3
Clipson, C.W. U6
Cobb, A.B. A5
Cobb, S. p. 97
Cobliner, W.G. H2
Coburn, D. E3
Cody, J. P10
Cohen, S. p. 45, p. 215
Coke, E.U. S5
Colby, M.A. I12
Collen, F.B. U7
Comazzi, A. C20
Cone, R.P. p. 217
Connell, R.W. P13
Connor, H.E., Jr. C25
Coombs, R. J25
Cooper, B.S. Z19
Corah, N.L. E4
Corbus, H.F. P13
Corning, M.E. U8
Costello, R.M. J6, Z20
Counte, M.A. U9
Cox, F.M. p. 165
Craig, K.D. J7
Craig, T.J. F16

299

Author Index

Craighead, W.E. p. 89, p. 216
Crane, D.P. W9
Crawford, S. U10
Creer, T.L. F10
Crisp, A.H. C4
Crowder, A.S. p. 4
Cull, J.G. C13
Cullen, J.W. p. 46, p. 166
Cunningham, M.R. Q41
Curtin, M.E. J11
Cuskey, W.R. Q8
Cutter, H.S. P50
Czaczkes, J.W. O9

D

Dacey, M.L. Y14
Daggett, D.R. V4
Dales, L.G. E5
Dalis, G.T. p. 69
D'Altroy, L. p. 121
Dalzell-Ward, A.J. E6
Dandurand, G.L. U10
Danish, S.J. W19
Dansereau, D.F. Z21
Darsky, B.J. X5
Dashef, S.S. Y15
D'Augelli, A.R. P14, W20
Davidson, P.O. J30, p. 217, Z22
Davies, R.K. A6
Davis, M.S. P33, T21
Davis, N.J. D8
Dayani, E.C. P46
DeAraujo, G. I4
Deeds, S.G. p. 5, S6
Delbecq, A.L. X28
Deliege, D. O3
Derogatis, L.R. Z23
Deykin, E. G6
Dickens, H.O. D9
Dillon, P.B. p. 98
Distefano, M.K., Jr. U11, W21
Dohrenwend, B.P. B7
Dohrenwend, B.S. B7
Dolfman, M.L. Z24
D'Onofrio, C.A. J8, p. 163
Dorken, H. U12
Dorroh, T. P15
Dowd, E.T. W22
Drouin, B. G7

Drumheller, S.J. Q9
Dudley, D.L. J9
Duncan, B.A.B. T64
Dyer, E.D. W23, Y16

E

Eagly, A. I6
Edgerton, J.W. W34
Edwards, G. J10
Eiben, R. W24
Einhorn, R.F. I5
Ekvall, S.W. C11
Elder, R.G. H3
Elder, S.T. T22
Elinson, J. Z25
Elms, A.C. H7
Elwood, T.W. E7
Emling, R.C. Y17
Emrick, C.D. G8
Enelow, A.J. p. 45, p. 166
Engel, B.T. A7, T14, T15, T23, T81
Engelhardt, H.T., Jr. D10
English, G.E. J11
Epstein, D.W. Y18
Erickson, R.C. N2
Etzioni, A. Y19, Y20
Etzwiler, D.D. G9, P16
Evans, M. C5
Evans, M.G. X4
Ewing, J.A. T24
Eyberg, S.M. M4

F

Faden, A.I. p. 69, p. 215
Faden, R.R. p. 69, p. 215, p. 216
Fan, M. F12
Farberow, N.L. B13
Farley, F.H. O12
Farmer, E.D. W25
Fedder, D. p. 216
Feldman, R.A. R5
Felton, B. J12
Feurzeig, W. S7
Fiester, A.R. G10
Figà-Talamanca, I. D11, p. 215, Z26, Z34
Fink, D.J. U13

Author Index

Fiori, F. U14
Fishbein, M. p. 68, 11
Fisher, D.F. S8
Fitz-Gibbon, C. Z54
Fitzhugh, Z.A. W26
Flannery, R.B., Jr. T25
Flegle, J.M. p. 121
Fletcher, C.M. Q10
Fletcher, S.W. P17
Flomenhaft, K. W27
Flowers, R.V. p. 121
Folman, R. P45
Fonoroff, A. p. 88
Forehand, R. M5
Foreyt, J.P. p. 89
Forrest, G.G. C6
Fort, J. C7
Foster, G.M. p. 70
Fox, B.H. p. 46
Frank, J.D. C8
Frankenberg, R. P18
Frankle, B.L. p. 216
Franks, C.M. T26
Frantz, R.A. p. 124
Frazier, P.J. K2
Freeborn, D.K. X5
Freedle, R.O. S3
Freeman, H.E. p. 68, p. 164, Z12
Freidson, E. K3, p. 164
Freire, P. p. 70
French, A.P. O4
French, W.L. p. 164
Frey, D.H. Z27
Fries, J.F. p. 167
Frumkin, K. p. 216
Fry, L.J. W28
Fry, L.N. Z28
Fuchs, V.E. L5
Fudge, R.P. p. 123
Fuller, D.S. P19
Fulton, M. P20
Furman, S. A8

G

Gadd, A.S. Y21
Gaines, W.G. W37
Galli, N. p. 4
Gallicchio, J. p. 46, p. 89, p. 122

Galvin, M.E. F12
Gaoni, B. X6
Gardner, J.E. U15
Gardner, M.E. p. 123
Garrett, L.D. W48
Garrison, G.E. Z40
Gartner, A. p. 88, p. 165
Gatchel, R.J. T27
Gaus, C.R. Y7
Gaylin, W. T28
Geiger, O.G. T29
Geist, R.A. C9
Gellin, M.E. Y17
Gentry, D.L. T30
Gentry, J.T. L6
Gentry, W.D. p. 217
German, P.S. A9, p. 215
Gilandas, A.J. W29
Gillum, R.F. G11, p. 123
Glaser, F.B. L7
Glass, D.C. J13
Gliva, G.E. M6
Glogow, E. P21
Glouberman, D. Z29
Gluckstern, N.B. W30
Glueck, B.C. T75
Goin, M.K. X7
Gold, R.A. U16
Goldberg, G.A. Z30
Goldberg, R.T. A10
Goldman, M. J17
Goldstein, G.S. T31
Golladay, F.L. Y22
Good, K.C. X8
Good, L.R. X8
Goodacre, D. R6
Goodstadt, M. S9
Gordon, D.W. U17
Gordon, S.B. T32
Gordon, W.A. A11
Gornally, J. Z31
Gotestam, K.G. C10, T55
Gottfried, A.W. T33
Gouge, A.L. C11
Gough, H.G. Z32
Goulston, K. Y23
Gove, W.R. J14
Graham, D. F14
Graham, J.F. Y24
Graning, H.M. L8

Author Index

Grant, I.   A12
Grant, R.H.   p. 165
Graves, V.   S10
Gray, B.H.   A14
Gray, J.E.   T66
Green, L.W.   p. 4, p. 5, D20, F13, p. 67, p. 89, p. 97, M1, p. 122, p. 123, p. 165, p. 181, p. 215, p. 216, Z33, Z34, p. 235
Green, P.F.   p. 97
Greenbaum, H.H.   Z35
Greenberg, J.S.   Z36
Greenberg, S.W.   L7
Greenblatt, D.J.   C12
Gregory, I.D.   Q11
Greif, E.B.   Z37
Greist, J.H.   Z38
Griffiths, W.   p. 163
Grimm, J.W.   P24
Gross, A.E.   Q4, Q12
Grossen, N.E.   Z64
Grosser, C.G.   p. 88
Grotjahn, M.   W31
Gruenfeld, L.   X9
Guba, E.G.   Z39
Gullen, W.H.   Z40
Gump, L.R.   P22
Gust, T.   W57
Gutmann, J.E.   Z2
Guttentag, M.   Z41
Gygi, C.   T34

H

Haan, N.   J15
Haefner, D.P.   E8, T44
Haffly, J.E.   L1
Hagberg, B.   G16, N3
Hagen, E.E.   p. 70
Hagen, R.L.   T35
Hall, L.A.   T32
Hall, R.G.   T36
Hall, S.M.   T36
Hallbauer, E.S.   T37
Hambleton, R.K.   Z42
Hamilton, D.L.   W17
Hammerman, S.R.   Q13
Hammond, A.H.   p. 120

Hardy, M.E.   Z43
Hardy, R.E.   C13
Hardy, W.E., Jr.   L9
Harper, R.   U18
Harr, E.   p. 23
Harris, A.H.   Z44
Harris, C.L.   p. 124, p. 166
Harris, M.B.   T37
Harris, R.M.   B10
Hart, L.K.   p. 124
Hartley, J.   S11
Hartman, H.   F3
Hartnagel, T.F.   Z45
Harvey, T.R.   Z46
Hassell, J.   p. 120
Hawker, A.   J32
Hawkins, R.P.   p. 217
Hawley, B.P.   O5
Haynes, R.B.   p. 46, G12, G19, p. 97, p. 122, p. 123, p. 216
Healey, J.M., Jr.   P3
Health Education and Welfare, U.S. Department of   A35, A36, F32, O18, S39, W65
Health Insurance Institute   L10
Health Policy Studies, Center for   L4
Health Resources Associates   p. 88
Health Statistics, National Center for   Z3
Hearnshaw, T.   S12
Heestand, D.E.   W8
Heidecker, B.   p. 120
Heilburn, A.B., Jr.   X10
Held, J.P.   R7
Hendershot, G.E.   P24
Henderson, J.B.   p. 46, p. 166
Hendrick, C.   Q20
Hepner, D.M.   p. 163
Hepner, J.O.   p. 163
Herbert-Jackson, E.W.   M10
Herman, C.P.   J16
Herzlich, C.   F14
Hess, K.   p. 166
Hess, S.W.   Y25
Hewitt, J.   J17
Hiatt, H.H.   L11
Hickey, T.   Z47
Hiemstra, R.   W32
Hill, C.E.   Z31

Author Index

Hilton, B. P25
Himmelfarb, S. I6
Hipple, J.L. Y26
Hirsch, E.O. W33
Hirschman, R. F15
Hiss, R.G. Y27
Hoekelman, R.A. H9
Hoffman, A.L. A13
Hogan, R. Z37
Hoge, A.F. V5
Holcer, P. T48
Holder, L. Y28
Hollingshead, A.B. Y29
Hollister, W.G. W34
Holmstrom, L.L. F7
Holtz, E. p. 88
Hopper, A.E. J31
Horan, P.M. A14
Horenstein, D. D12
Hornstein, H.A. p. 164
Hornstra, R.K. O6
Horowitz, M. Z48
Hospital Medical Education, Association for W53
Houston, B.K. D12
Howard, J. p. 88
Howe, L.W. p. 69
Howe, R. U19
Howell, M.C. p. 166
Hoyt, J.L. Q44
Huber, E.G. G13
Huber, S.C. D13
Huberty, D.J. M7
Huffine, C.L. F16
Hughson, B. C14
Hulka, B.S. O7, Z49
Humphrey, G.B. V5
Hunt, S.M. Y30
Hurst, J.C. S13
Hurvitz, N. p. 166
Hyerstay, B.J. N2

I

Idelson, R.K. A15
Ikard, F.F. O16
Ilgen, D.R. X11
Illich, I. p. 88
Insko, C.A. Q14
Institute of Medicine p. 181

Inui, T.S. W35
Ireton, H.R. p. 181
Isom, R. p. 46
Ivancevich, J.M. X12, X16
Ivey, A.E. S14, S15, W30

J

Jaccard, J. I7, Z50
Jackson, M. V6
Jaco, E.G. p. 68
Jacobs, A. R8
Jacobs, S.H. T38
Jacobson, G.R. C15
Jeffrey, D.B. p. 89, T39, Z51
Jencks, S.F. p. 166, p. 181
Jenkins, C.D. A16
Jenny, J. I8, O8
Jernstedt, G.C. Z52
Jesudasan, K. p. 122
Johnson, E.M. W36
Johnson, J.E. J18
Johnson, L.A. T29
Johnson, S.M. M4
Johnston, B.L. p. 122
Joiner, S. Y24
Jones, A.P. X13
Jones, E.W. L12
Jones, F.H. C16
Jones, R.J. p. 122
Jongsma, E.A. W37
Judkins, B.A. p. 124
Julian, J. p. 163

K

Kadushin, A. X14
Kahana, E. J12
Kahn, R.L. F38
Kakalik, J.S. L13
Kalmer, H. p. 3
Kaluzny, A.D. U20
Kane, F.J. J19
Kane, R.L. L14
Kapche, R. T40
Kaplan, M.F. Q15
Kaplan De-Nour, A. O9
Kapp, R.A. T59
Karpen, M.L. R9
Kasl, S.V. G14

Author Index

Kassum, S.   X9
Katkin, S.   U21
Katz, A.   p. 88
Katz, R.C.   p. 89
Kay, R.L.   p. 120
Kaye, J.D.   W38
Kayser, J.S.   P26
Kazdin, A.E.   T41, p. 216
Keegan, D.L.   A17
Keeley, S.M.   Y54
Kegeles, S.S.   F17
Kelly, A.   L15
Kenigsberg, D.   J20
Kessler, S.   T42
Khan, A.U.   A18, A19, T43
Kiell, N.   J21
Killilea, M.   p. 88
Kilmann, P.R.   R10
Kilmann, R.H.   J22
Kilty, K.M.   P27
Kimball, C.P.   J23
Kimberly, J.R.   U9
King, M.   G15
Kinsella, N.A.   Q16
Kinsman, R.A.   B8
Kirscht, J.P.   F18, H4, p. 122, T44
Kirshenbaum, H.   p. 69
Klausmeier, H.J.   p. 68
Kleinbach, G.   Y31
Klement, J.J.   S16
Klemmack, D.L.   Z53
Kline, F.M.   X7
Kline, J.   G15
Klopfenstein, T.D.   Y32
Kluckhohn, F.R.   p. 70
Knopf, A.   H5
Knowles, M.   p. 68
Knox, W.J.   P28
Knutson, A.L.   p. 69
Koch, J.H.   S17
Koch, M.F.   A20
Kohle, K.   W39
Kopel, S.A.   Q17
Kosa, J.   p. 69, p. 88
Kosecoff, J.   Z54
Kraegel, J.M.   U22
Krantz, D.S.   J24
Kreisman, J.J.   C17
Kress, J.R.   U23

Krieger, S.R.   Z55
Kriss, M.   Q18
Kroll, H.W.   T45
Kundu, M.R.   S18
Kupfer, D.J.   C18
Kupst, M.J.   P29
Kurtz, R.R.   W40

L

Labov, W.   Q19
LaDou, J.   E9
Lally, J.J.   P30
Lamb, D.   U24
Lambert, G.   A21
Langlie, J.K.   p. 97
Larkin, E.J.   Z56
Larsons, M.S.   Z57
Latos, D.L.   p. 122
Laugharne, E.   p. 120
Lawson, T.E.   S19
Lawson, V.K.   p. 120
Lazarus, R.S.   T46
Lazes, P.M.   p. 121
LeBow, M.D.   T47
Lee, E.A.   U25
Leibowitz, J.M.   T48
Leitch, C.J.   A31
Leonard, R.C.   p. 69
Lesser, A.L.   M6
Lester, B.J.   J14
Lester, D.   P31, Z78
Leung, P.   Y33
Levin, E.M.   W40
Levin, H.S.   T49
Levin, L.S.   p. 88, p. 166
Levine, B.A.   T50
Levine, D.M.   p. 5, p. 122, Y34, Y35
Levine, E.S.   A22
Levine, S.   p. 68, p. 163
Levitz, L.S.   O10
Lewin, K.   p. 69
Lewis, A.B., Jr.   U26
Lewis, C.E.   Y36
Lewis, J.A.   p. 166
Lewis, M.A.   p. 166
Ley, P.   p. 216
Leyhe, D.L.   F19
Lillie, D.C.   P32

Author Index

Lind, E. A32
Lindahl, R.L. X15
Lindberg, F.H. N4
Linden, V. Y37
Lindsay, C.A. W41
Linn, L.S. P33
Lipke, L.A. R9
Lippard, V.W. Y38
Lippey, G. S20
Liska, A.E. I9, I10, I11
Liston, E.H. Y39
Litman, G.K. C19
Little, Arthur D., Co. p. 88
Logan, M.H. H6
Lohr, W. p. 166
Lomazzi, F. J2
Long, R. W32
Long, R.M. I2
Luban-Plozza, B. C20
Lubin, A.W. R11
Lubin, B. O6, R11
Lucas, R.A. A23
Ludwig, A.M. C21
Luecke, J.R. S31
Lukes, S.J. T83
Lurie, H.J. W5
Lurie, O.R. F20
Luscutoff, S.A. H7
Lynch, J.J. A24, Z58
Lyon, H.L. X16
Lyons, R. C14
Lyons, T.E. Y40

M

McCall, R.J. R12
McCarthy, B.W. D14
McCleaf, J.E. I12
MacDonald, M.L. A25, T51
McDonnell, J.F. X17
McFall, R.M. p. 217
McFarland, R.A. J25
McGarry, J. Q20
McGinnies, E. Q21, Q49, Q50
McGrane, H.F. S21
McInnis, T.L. W49
McIntosh, J. P34
Mackie, M. H8
McKinlay, J.B. P35
McKinney, J.P. 113
Mackler, B. Z59

McLachlan, J.F.C. Q22
McLaughlin, C.P. Z60
McLaughlin, F.E. W42
McLeish, J. R13
McWhirter, J.J. Y41, Z61
McWilliams, J. C22
Maddock, J. C23
Magnuson, W.G. L16
Mahoney, M.J. p. 90, O11, T52, T53, T70, p. 215,
Maiman, L.A. I3, J26, p. 97, p. 122
Mallenby, T.W. C24
Malmquist, A. G16, N3
Malo-Juvera, D. S22
Mangiaracina, J. W54
Mann, G.V. A26, A27
Mannino, F.V. M8
Marchant, H. S38
Maris, R. C25
Markle, S. p. 69
Marks, S.E. Q23
Marmor, J. C26
Martin, J. p. 166
Martin, R.D. W43
Marwell, G. X18
Mash, E.J. T54
Maslow, A.H. p. 69
Matsuno, A.S. J27
Mattmiller, E.D. D17
Mayo, J.A. F21
Mazzullo, J.M. P36
Mead, M. p. 70
Means, R.L. p. 70
Mechanic, D. p. 69, p. 88, L23, L24, p. 97, U27, U28
Medved, E. p. 120
Meichenbaum, D. S23
Melin, G.L. C10, T55
Mellett, P. J28
Meloan, J.B. V7
Mersky, H. Z62
Messner, S. S24
Metcalf, W.K. W51
Metsch, J.M. V8, V9, Z63
Mettlin, C. Q24
Metzner, R.J. S25
Meyer, S. O15
Meyers, L.S. Z64
Michal-Smith, H. C27

Author Index

Michener, H.A.   Q25
Mico, P.   p. 4, p. 164
Milgram, G.G.   S26
Milio, N.   p. 166
Miller, D.   p. 46
Miller, G.R.   Q26
Miller, J.P.   W28
Miller, L.D.   Q29
Miller, M.H.   K4
Miller, P.M.   F23, J29, T56, T57, p. 217
Miller, W.B.   D15, E10
Mims, F.   W44
Minkler, M.   p. 98
Minnigerode, F.A.   P26
Missett, M.A.   Q27
Mitchell, J.A.   P37
Mitchell, J.H.   G17
Mitchell, M.M.   p. 120
Mitchell, T.R.   X19
Mittlefehldt, V.A.   X21
Moe, B.L.   X30
Mogielnicki, R.P.   B9
Molnar, G.D.   A20
Monteiro, L.A.   K5, Y42, Z65
Montgomery, L.J.   W45
Moore, M.   W46
Moos, R.H.   Z66
Morgenstern, M.   C27
Moriwaki, S.Y.   C28
Morris, D.   p. 166
Morris, D.A.   p. 182
Morris, L.D.   p. 182
Morse, E.V.   Y43
Moxley, R.A., Jr.   Z67
Moynihan, D.P.   p. 88
Mullen, P.D.   p. 98, M9, p. 164
Munan, L.   L15
Mushkin, S.J.   p. 98

N

Nagi, S.Z.   V10
Najman, J.   B11
Nash, A.   X20
Nash, H.   Q28
Nash, K.B.   X21
Nathanson, C.A.   Y44, Y45, Y46
Navarro, V.   p. 88
Neeman, M.   E11

Neeman, R.L.   E11
Neidermayer, H.   J7
Nellis, D.H.   W47
Nelson, R.O.   p. 217
Neufeld, R.W.J.   J30, Z22
Neumann, M.   X6
Neuringer, C.   B10
Newcomer, J.P.   Z52
Newhouse, J.P.   L17
New Mexico Regional Medical Program   Y50
Nicassio, P.   T58
Nimmer, W.H.   T59
Norr, K.L.   p. 98
Norton, R.W.   Q29
Nowicki, S.   J31
Nursing Development Conference Group   p. 4., p. 89

O

Oakes, C.G.   U29
Oakes, T.W.   A28, E7, F24
O'Brien, G.   X11
O'Connor, R.J.J.   S27
Ohlmeier, D.   R14
Oja, P.   E12
O'Leary, M.R.   C29
Olson, D.R.   Q30
Opit, L.J.   D16
Oradei, D.M.   R15
O'Reilly, C.A. III   Z71
Orem, D.E.   p. 4, p. 89
Orford, J.   J32
Orkow, B.M.   Q31
Ort, R.S.   P38
Ostrom, T.M.   Q42
Ott, J.E.   Y47

P

Packwood, W.T.   Q32
Paden, R.C.   R16
Pankratz, D.   Z68
Pankratz, L.   Z68
Pappas, J.P.   P12
Parbrook, G.D.   J33
Parker, C.A.   Q40
Parkhouse, J.   Y48
Parsell, S.   p. 166
Parsons, T.   p. 69

Author Index

Passons, W.R.   W48
Patti, R.J.   X22
Pattishell, E.G.   Y49
Paul, B.D.   p. 70
Paul, G.L.   W49
Paulson, M.J.   R17
Paulson, T.   W15
Payne, B.C.   Y40
Payne, J.W.   Z69
Pearce, W.B.   Q33
Pearson, C.E.   L18
Pearson, K.M., Jr.   S28
Peirce, J.C.   Y27
Pender, N.J.   P39
Penson, A.B.   D17
Perkoff, G.T.   K6
Perrow, C.   p. 164
Persons, M.K.   S29
Persons, R.W.   S29
Peters, E.N.   H9
Petersen, H.M.   U30
Peterson, G.W.   Z15
Peterson, J.M.   O12
Petres, R.E.   D5
Pfeffer, J.   X23
Phelps, C.E.   K7
Pinkston, E.M.   M10
Pleticha, J.M.   B1
Pliner, P.   Q34
Plutchik, R.   J34
Pope, C.R.   E3
Pope, D.   S30
Porter, A.C.   Z70
Powell, B.J.   W50
Powell, L.F.   D18
Power, L.   R18
Powers, W.T.   H10
Pozen, M.W.   p. 121
Pratt, L.   p. 98, M11
Premkumar, T.   Q8
Prentice, E.D.   W51
President's Committee on Health Education   p. 88
Pritchard, M.   F25, F26, H11
Pryer, M.W.   U11, W21

Q

Quesada, G.M.   P19
Quilitch, H.R.   W52

R

Rabin, D.L.   L19
Radius, S.M.   p. 122
Rahe, R.H.   p. 120
Ramcharan, S.   E13
Ramsay, D.A.   C30
Ramsden, E.L.   P40
Raske, K.E.   U31
Rath, L.   p. 69
Rawlinson, M.   F6
Ray, J.J.   B11
Reader, G.G.   P41
Reddy, W.B.   Q35
Redman, B.K.   p. 4, p. 89, p. 124
Redmond, D.P.   T60
Reeder, L.G.   F4, p. 68
Rees, S.   Q36
Regalbuto, G.   P51
Rehr, H.   A3
Reich, J.   I14
Reid, R.A.   U32
Reiss, D.   Q37
Rendtorff, R.C.   B12
Repp, A.C.   T61
Reynolds, D.K.   B13
Ribner, N.G.   R19
Rice, V.H.   J18
Richards, N.D.   p. 164
Richards, R.F.   p. 3, P42
Riessman, F.   p. 88, p. 165
Riggs, R.C.   P43
Rimm, A.A.   K8
Rimm, I.J.   K8
Rinn, R.C.   T62
Ripple, R.   p. 68
Roach, A.A.   U33
Roach, D.K.   S12
Roback, H.B.   J35
Roberto, E.L.   I15
Roberts, B.J.   F13
Roberts, K.H.   Z71
Roberts, T.B.   Q38
Robertson, L.S.   U34
Robinson, J.S.   C37
Robinson, L.H.   M12
Rockeach, M.   p. 69
Rodin, J.   G18, J36
Roemer, M.I.   L20
Rogers, Carl R.   p. 69

Author Index

Rogers, E.M. p. 70
Rogers, J.M. D19
Rogers, K.D. V11
Rogers-Warren, A. p. 46
Roghmann, K.J. M13
Romanczyk, R.G. T63
Rose, K. E14
Rose, S.D. M14
Rosen, R.A.H. P44
Rosen, S. Q39
Rosenbaum, M.E. S4
Rosenberg, S. p. 124
Rosenham, D.L. C31
Rosenstock, I.M. E15, H12
Rosenzweig, S.P. P45
Ross, A.W. J3, R2
Ross, H.S. p. 4, p. 164
Ross, J. p. 98
Ross, J.L. Q31
Rossiter, C.M. S31
Roter, D.L. p. 121
Roth, H.P. p. 120
Rothenberg, M.B. M3
Rowe, W. R20
Rubenstein, A.H. U35
Rubin, D.B. Z72
Rubin, M.L. S32
Ruley, E.J. p. 124
Rulin, M.C. S33
Rushmer, R.F. U36
Russell, G.F.M. C34
Russell, L.B. L21
Rutzen, S.R. O13
Ryser, P.E. E18

S

Sackett, D.L. p. 46, G12, G19
Sadoff, R.L. Y51
Salancik, G.R. X23
Salber, E.J. p. 88
Salloway, J.C. F27, p. 98
Salzer, J.E. p. 121
Samaan, M.K. Q40
Sarason, I.G. C36, J37
Sargent, J.D. C32
Sawyer, W.A. W54
Sayegh, J. D20
Schach, E. L19
Schaefer, J.W. M15

Schauffler, H.H. p. 124
Schlesinger, E.R. W55
Schmidt, D.D. S24
Schmidt, M.P.W. T64
Schmitt, D.R. X18
Schneier, C.E. X24
Schoenrich, E.H. U37
Schottenfeld, D. p. 46
Schulberg, H.C. U38
Schulz, R. B14
Schwartz, A.N. W56
Schwartz, G.E. T65
Schwartz, J. p. 46
Schwarz, K. E16
Schweer, S.F. P46
Scott, C.S. F28
Sechrest, L. P48
Seeman, M.V. G20
Segal, E.A. L16
Sehnert, K. p. 166
Seiler, L.H. Z73
Selman, P. F29
Shader, R.I. C12
Shaffer, D.R. 116
Shafii, M. O14
Shapiro, D. T65
Shapiro, I.S. Y52
Shapiro, J.L. W57
Shaw, D.G. W58
Shaw, D.W. P47
Sheldon, A. Z60
Sheldrake, P. Y53
Shelton, J.E. X25
Shemberg, K.M. Y54
Shepel, L.F. W43
Sheridan, M.S. R21
Sherwood, G.G. T66
Shirley, R.C. X26
Shochet, B.R. U39
Shoemaker, F.F. p. 70
Shonick, W. L20
Shore, M.F. M8
Shusterman, L.R. P48
Sieg, K.W. T67
Sikes, S. G21
Silverstein, S.J. T68
Simmons, J.J. p. 89, p. 124, p. 166
Simon, S.B. p. 69
Simonds, S.K. U40

Author Index

Sims, H.P., Jr.   X27
Singer, J.   U23
Singh, D.   G21
Sinnett, E.R.   S34
Sinning, W.E.   Z74
Sipich, J.F.   T69
Sirota, A.D.   T70, T71
Skiff, A.W.   S35, U41
Skinner, H.A.   C33
Skipper, J.K.   p. 69
Skoloda, T.E.   G22
Skydell, B.   p. 4
Slade, P.D.   C34
Slepcevitch, E.   U42, p. 124, p. 215
Slocohower, J.   G18
Slote, M.A.   B15
Sly, R.M.   p. 120
Smart, R.G.   L22
Smith, D.H.   p. 166
Smith, J.M.   E17
Smith, M.B.   H13
Smith, R.T.   G5
Snyder, M.   Q41
Sobel, D.E.   B16
Sobell, L.C.   T72, T73
Sobell, M.B.   T72, T73
Sobieszek, B.I.   N6
Solfin, D.   p. 121
Solomon, L.   S36
Somers, A.R.   p. 4, p. 5, p. 166, U43
Sorenson, J.R.   P49
Sowa, P.A.   P50
Spicer, E.H.   p. 70
Spiegel, J.P.   C35
Spielberger, C.D.   C36, J37
Spillane, W.H.   E18
Spinetta, J.J.   B17, Z47
Spiro, H.R.   F30
Spitzer, W.O.   Y55
Squyres, W.   p. 181, p. 216
Stacey, M.   V12
Stahl, S.M.   Z75
Stainback, S.   T74
Starfield, B.   Z76
Stearns, N.S.   W59
Steckle, S.B.   p. 122
Steele, C.M.   Q42

Steen, J.   U44
Steffy, R.A.   T8
Stein, G.H.   U45
Stein, L.I.   p. 98
Steiner, G.   p. 120
Steinfeld, J.   p. 97
Stephens, C.   F29
Stephenson, R.W.   W60
Stern, G.S.   J38
Sternlicht, M.   M16
Steward, M.   P51
Stimpson, D.V.   W61
Stimson, G.V.   G23
Stockton, R.   U47
Stoeckle, J.D.   p. 166, U46
Stokes, S.J.   Q43
Stone, L.A.   W62
Stone, V.A.   Q44
Strassberg, D.S.   C37
Strasser, B.B.   p. 69
Strauss, A.   p. 88, p. 167
Stritter, F.T.   S37
Strodtbeck, F.L.   p. 70
Stroebel, C.F.   T75
Strupp, H.H.   A30, R22
Stuart, B.   U47
Stunkard, A.J.   O10, T76, p. 167, p. 217
Suchotliff, L.   O15
Suedfeld, P.   O16
Suinn, R.M.   Y56
Sullivan, I.   M16
Swain, M.   p. 122
Swartz, H.   A31
Swearingen, R.V.   Q51
Szilagyi, A.D.   X27

T

Tagliacozzo, D.M.   Q45
Tagliareni, E.M.   p. 166
Talbert, L.M.   Y57
Tamney, J.B.   p. 167
Tancredi, L.R.   V13
Tanner, B.A.   T77
Tarter, R.E.   C38
Tarver, J.   M17
Task Force on Patient Education   U48
Tavormina, J.B.   M18, R23
Taylor, V.   J22

Author Index

Taylor, W. G12
Teather, D.C.B. S38
Tedeschi, J.T. Q46
Teevan, J.J., Jr. Z77
Templer, D.I. Z78
Tessler, R. L23, L24
Teuscher, A. p. 120
Thauberger, E.M. B18
Thauberger, P.C. B18
Theorell, T. A32
Thies, A.P. O17
Thies, J.B. W63
Thigpen, C.R. D22
Thomae, H. J39
Thomas, E.J. T78
Thomas, J.M., Jr. A33
Thoreson, C.E. p. 90, O11, P47, Z79
Thornhill, M.A. J40
Thorp, R.G. p. 90
Tillotson, J. p. 122
Tobacyk, J.J. J41
Tobin, H.M. W64
Toigo, R. Y58
Torrens, P.R. p. 164
Train, G.J. A34
Travelbee, J. p. 89
Trinca, C.E. p. 123
Tupin, J.P. O4
Turner, A.J. M17
Twaddle, A.C. F31, U46

U

Uhlenhuth, E.H. 117
U.S. Department. For government agencies, see under the name of the subject with which they deal: e.g., Health, Education and Welfare, U.S. Department
Uyeno, D.H. Z80

V

Vacalis, T.D. F33
Vander Haegen, E. X2
VanderKolk, C.J. W66
Van de Ven, A.H. X28
Van Rooijen, L. Q47
Veeder, N.W. F34

Veney, J.E. V8, Z63
Veninga, R. Q48
Verdicchio, F.G. T33
Vernon, D.T. H14
Vernon, G.M. B19
Veterans Administration T79
Vickery, D.M. p. 167
Vignos, P.J. p. 121
Virginia, Portsmouth Department of Public Health F22
Visintainer, M.A. O19
Vlasak, G.J. V14
Vlasses, P.H. p. 123
Vogel, J.M. A37
Volicer, B.J. O20

W

Wagner, P. N5
Waite, N.S. R15
Wakefield, J. H5
Wallace, C.J. W67
Wallace, D.C. p. 121
Wallace, M.J., Jr. Z81
Wallace, N. p. 121
Wallston, B.S. P52
Wallston, K.A. P52
Walter, G.A. T80
Waltz, J.R. D22
Wang, V.L. p. 167
Ward, A.W.M. B20
Ward, C.D. Q49, Q50
Ware, J.E., Jr. Z82, Z83
Waring, M.L. W68
Warren, D.I. p. 89
Warren, R.B. p. 89
Warren, S.F. p. 46
Watson, D.L. p. 90
Watson, J.P. X29
Weber, C. p. 69
Webster, M., Jr. N6
Weeks, L.E. U3
Wehmer, G. W69
Wehrer, J.J. U6
Weiner, E.A. A33
Weiner, H. F35, F36
Weingarten, V. U49
Weinstein, R.M. C39
Weinstein, S. D3
Weintraub, M. G24

Author Index

Weisenberg, M. 118
Weiss, C.H. Z84
Weiss, R.L. Q51
Weiss, S.M. p. 46, p. 217
Weiss, T. T81
Welgan, P.R. T82
Wells, R.A. W70
Werlin, S.H. p. 124
Wessen, A.F. Z65
Whelan, E.M. G25
White, C. P53
White, K.L. Y59
White, P.E. V14
White, W.C., Jr. C40
Whiting, J.F. U12
Whitman, H.H. T83
Whittaker, J.K. Q52
Wickramaskera, I.E. T84
Wiener, C.L. A38
Wiesenthal, D.L. O21
Wiig, E.H. M19
Wikler, A. C21
Williams, B.J. p. 90
Williams, J.L. T85
Williams, R.B., Jr. p. 217
Williams, R.L. Z85
Williams, T.M. D23
Wilson, C.J. R24
Wilson, G.T. T26, p. 216, p. 217
Wintrob, R.M. Y14
Witschi, J.D. p. 122
Woelfel, J. Q24

Wolf, J.N. U50
Wolf, S. Y60
Wolfe, J. X30
Wolfer, J.A. O19
Wolle, J.M. W71, Y61
Woodruff, R.A., Jr. F37
Worden, J.W. Z86
World Health Organization Y62
Worthington, N.L. Z19
Wortman, J. D24
Wright, B.A. O22
Wunderlich, R.A. J42

Y

Yangdon, T. H15
Yen, S. O23
Young, L.D. T10, T86, p. 216
Young, M.A.C. p. 124
Young, M.E. Q53, S40, S41

Z

Zarit, S.H. F38
Zeiler, M. T77
Zide, E.D. Y5
Ziesat, H.A. p. 122
Zlutnick, S. p. 89, T87
Zobrowski, M. p. 70
Zola, I.K. p. 88
Zuckerman, M. J43
Zyzanski, S.J. Z87

SUBJECT INDEX

Terms used in this index were selected from the THESAURUS OF HEALTH EDUCATION TERMINOLOGY of the Health Education Information Retrieval System (see Introduction, p. xv). Those unfamiliar with the HEIRS Thesaurus should be aware that "Health Education" or "Health Education Programs" is implied in all headings. For example, "Administration" implies "Administration of Health Education Programs." The terms used are called "descriptors" and frequently can be "coordinated" to find precisely the paper one desires. For example, if one were looking for a paper on evaluation of counseling programs which emphasized evaluation methods as well as counseling, one could look at the numbers listed under the two headings and find that those numbered X25 and X31 deal with both subjects. The terms are listed alphabetically, letter by letter and references are to entry number.

A

Abortion D1, D3, D5, D7-8, D10-11, D19, D21, E10, H7, I5, M2, P8, P24, P44, Y13
Academic standards Y28, Y60
Acceptance of Health Services F9, F19, F29, I8, J8, K3, O6-7, P4, U28, V12, Z87
Accessibility of health services F1, F9, K5, L2-3, L6, L9, L15, L23, O7, Y7, Y50
Accident prevention T2
Accidents, poisonings, and violent injury B3-4, F7, F11
Accountability D22, V12, Y9, Y19-20, Y59, Z15
Acculturation J12
Activities of daily living T51
Administration L9, O18, Q48, S18, U1, U3, U20, U29, U31, U35, U41, V12, W47, W56, X1, X3, X9, X14, X16, X19, X22-24, X26-28, X30, Z9, Z35, Z60
Administrator responsibility X14, Y19-20, Z46
Admissions criteria V10
Adolescents A31, C9, C16, C34, D2-3, D9, D23, E11, E17, H2, J15, J19, L13, M12, R9, S34
Adoption process J8, P49, T13, U9, U20, U35, W60, Y43
Adult education M19, O15, S16, W41, W64
Age differences A28, A31-32, A35, E3, F38, H1, J27, J34, P8, T49
Aides U2, U11, U24, U46, W69, X15, Y18
Alcoholism C1, C3, C5-7, C12,

Subject Index

C15, C19, C21-22, C29-30, C33, C38, F23, F37, G8, G22, J1, J5-6, J10-11, J29, J31-32, L7, L22, N1, P2, P27, P50, Q22, S9, S26, T24, T38, T56-57, T68, T72-73, W28, W50, W68-69, Z20
Alienation C18, J1, T7, Y4, Z53, Z77
Ambulatory care U17, Z3, Z30
Anthropological research-theory H6, J21
Anxiety A23, B6, B11, B15-17, C19, C36, I18, J25, J37, J39, J43, O20, P11-12, P29, P48, Q5, T59, T71, T85, Z55, Z78
Apathy Z53
Arthritis A13, A38, H3
Asthma A18-19, A30, B8, I4, J9, T43, T70
Attitude-behavior gap E4, I1, I9-11, I16
Attitude changes A15, G13, H4, I6, I16, J17, O22, P6, Q5-6, Q14, Q20-21, Q26-27, Q41-42, Q44, R7, T3, T6, T56, W38, W49-50, W63, X3, Y34-35, Y42, Y52, Z47, Z57
Attitude determination (measurement) B11, G2, G6, H8, H11, I1, I9-10, I15, J6, J28, P48, P53, Q42, W37, W40, X15-16, Z20, Z50, Z87
Attitudes A6, A9, B6, B14, D1, D11, E10-11, E18, F8-9, F13, F20, F38, G23, I2, I5, I11-16, I18, J1, J17, M1, O7, O9, P24, P26-28, P30-32, P38, P40, P43-45, P50, Q12, Q16, Q21, Q43, Q45, R7, S2, U34, U50, V8, V13, W14, W39, W63, X1, X8, X15, Z55, Z68, Z77. See also Social attitudes, staff attitudes
Audiovisual aids L10, R18, R22, S10, S12, S18, S24, S30-31, S33, S38-40
Aversive conditioning J7, J38,

O16-17, T5, T8, T18, T40, T45, T61, T69, T77, T87, Z36

B

Barriers E11, F4, F19, H12, K1-2, L14, V12, X22, Y9, Y30, Y34, Z8
Base-line data H8, U10
Behavioral science research B13, C2, C16, D1, E3, E8, E11, E15, F2, F14, F18, F23, F27, G1, G11, G14, G18, H4, H10, H12, I1, I9, I11, I13, J3, J5, J7, J20, J26, J29, J31, J37, J39, M10, O21, P9, P52-53, Q1-2, Q20, Q22, Q46, R1-3, S16, T6, T9, T16, T23, T25-27, T40, T52-54, T63, T65, T72-73, U9, W16, W67, X11, X18, X24-25, X27, Y37, Y49, Z9, Z22, Z52, Z58, Z64
Behavioral scientists (health) W65, Y37, Y53-54
Behavior development C5, C7, C16, C21, C23, C32, D18, D23, E15, F27, G22, H4, H10, I11, J15-16, J29, M17, O5, Q5, Q15, Q34, Q40, Q45, S14, S16, T3, T6, T14-17, T23, T28, T31, T33, T35, T37, T44, T51, T53, T56, T62-63, T65, T72-75, T81, U35, W13, W22, W33, W38, W67, X4, X23, X27, Y35, Y52, Z44, Z58
Behavior disorders B7, C16, C38, M15, M18, T2, T16, T18, T48, T64, U15, U26, Z48
Behavior modification A7-8, A18, A27, C3, C10, C36, E4, F10, F23, G3, G14, G19, J3, J7, T24-25, J31, J38, L22, M4-5, M14-15, N1, N4, O2, O4, O9-11, O16-17, O19, O21, O23, P12,

Subject Index

P47, Q4, Q7, Q17, Q31, R2-3, R16, R23, T1-2, T4-5, T7-13, T18, T21-22, T24-27, T29-30, T32, T34, T36, T38-43, T45-48, T50, T52, T54-55, T57-61, T64, T66-71, T76-78, T80-87, W15, W38, W52, X24, Y11, Y52, Y56, Z36, Z51-52
Behavior patterns A4, A11, A16, B19, C4-5, C21, C40, E1, E7, E10, E17, E18, F25, F35, F38, H5, H10-11, I9, I11, J13, J17, J20, J22, J24, J29-30, L14, O1, O9, O17, P14, P33-34, P38, P40, P53, Q18, R5, T5, T61, U11, U35, V1, W20, W67
Beliefs B19, E7-8, E10, E15, E17, F18, G1, G14, H1, H3-4, H6, H12, I1, I3, I5, I7, I13, I15, J26, M1, P27, Q38, T44, T52
Biofeedback A8, C32, J25, Q38, T10-12, T14-15, T17, T19-20, T22-23, T26-27, T43, T46, T49, T65, T68, T71, T75, T81-82, T84-86, Z44, Z58
Biological research Z22, Z62, Z74
Blindness P20, Q5, R9
Blocking (psychological) Z48
Blood disorders A37
Breast cancer A1, B2, E11, F13, V5
Burns A10, B4
Business participation L3

C

Cancer A1, A6, B20, E11, F3, F13, F17, K4, L16, M7, P34, P37, U13, W1, Y52, Z86
Cardiovascular disorders A17, A26, A36, E14, G24, T14-15, T23
Career development W64, X21, Y6, Y24, Y45-46
Case studies (actual cases) A15, A21, A24, C10, C17, F10, G20, I6, M7, M10, M12, O2, O4, O18, Q36, S32, T1, T25-26, T30, T32, T48, T66, T70, T77, T83-84, U26, U30, U39, X2, Z48-49, Z82
Case study methods D4
Change U9, U34, U43, X22, X36, Z2, Z39
Child abuse R17
Child care centers D23, Y58
Child care education D18, H9
Child health A18, C23, D18, D23, F20, H9, I8, J27, V11, Y46-47, Y58, Z80
Child psychology A10, A18-19
Children A10, A18-19, A27, B17, C11, C24, D23, G13, H1, J27, K2, L15, M3, M5, M10, M14-15, M18, N5, O5, O8, O19, P51, R5, R21, R23, S23, T43, T48, T74, T87, U15
Child training M5, M14-15, M18, R23, S22, T47
Chronic disorders A1, A4-5, A15, A23, A31, A34, A36, A38, E13, F5, F25-26, G5, G14, G16, H11, I4, N3, O4, O9, O22, R24, S27, T79, U17, U37, U44, V13
Clinics D17, E5, F29, I18, L7, L9, P4, P17, Q27, R18, Y5, Y44, Y54
clinics needed L9
Closed circuit television S18, W8
Cognition H2, H11, J39, J41, P10, P51, T46, T52, Z9, Z47
Cognitive learning C22, O12, P51, Q15, S23, T52, X12, Z9, Z21
Colleague relations X19-20
Colleges D14, S29, W15, Y32, Y61
Commitment N1, T66, Y45
Committees A36, V8, V12, Y24
Communication problems N2, O3, P29, P41, P51, Q1-2, Q28, Q36, S3, V12, Y3, Y50

Subject Index

Communications skills-theory A5, I1, J18, J34, L6, N2, O1, P10, P16, P19, P34, P40, Q10, Q13-14, Q18-19, Q23, Q26, Q28-30, Q33, Q39-40, Q44, Q47, Q49-50, S2-3, S6, U24, U33, W19, W34, W70, X11, Y25, Z35, Z47, Z71
Community action L14
Community attitudes Z41
Community colleges Y32
Community education S6
Community health C35, F5, Y5, Z41
Community health diagnosis F5
Community health programs (need-concept) E1, L9, U38
Community health resources U47
Community health services E1, G10, L9
Community mental health centers F15, F21, F32, G10, G15, O6, T25, V4, W27
Community organizations V10
Community programs F28, O18
Community resources V2, W53
Comparative analysis A1, A11, A23, A31, A33, B7, C18, C22, C29, D12, E2, E5, E13, F22, F30, F35-37, G8, G22, H1, H7, H14, I8, I18, J5, J10-11, J20, J22, J26, J29, J33, J35-36, K1, L2, L7, L19, L24, O10, O13-14, P2, P4, P8, P12, P33, P44, P53, Q11, Q41, R16, R22-23, S4, S38, T12, T29, T35-36, T42, T56-58, T63, T69, T79-80, T86, U17, U20-21, U45, V1, W10, W36, W49, W66, X25, X28, Y17, Y33, Y41, Y56, Z13, Z19, Z42, Z61-62, Z69
Compliance C3, E12, F6, F23, F29, G1-6, G8-9, G11-12, G14-25, H15, I1, I3, I16, J1, J27, L22, M1, M5, O9, O21, P17, P21, P45, Q7, Q12, Q25, Q27, Q34, Q41, Q45, R12, T8, T34, T39, T57, T66, U21, U42
Comprehensive health planning L17, P16, U20, U29
Comprehensive health services I2, L6, U20, U27, U34, V3, V7, V11, V14, Y31
Computer applications J40, S7, S16, S20, S32, S36, U5, V2, Y62, Z38
Computers S32, S41, W63, Z80
Conference techniques R7, R9, V6, W7, W31, W46, W48, W51, W70-71
Conflict D22, I10, L11, O3, Q46, Q48, U50, W38, X20, Z16, Z18
Congenital malformations P25
Congenital origin A10
Consent D22, P3, P34, P39-40, S24, T83, Y11
Consultants U39, V7, Y56
Consultation C14, F15, P12, U41, V4-5, W15, W59, Y56
Consumer participation F34, I2, L11, L21, L23, P37, P42, U30, U50, V8-9, V12, Y50, Z63
Continuing education S10, U33, W2, W5, W16-17, W25, W27, W29, W32-33, W35, W37, W41, W53-54, W59, W64-65, Y23, Y27-28
Continuity of care I3, P4, U17
Contraceptives D4, D13, G25
Contracts P16, T62, U13
Cooperative programs S33, U8, U29, V3, V5
Coping A15, A38, D12, I4, J37
Coronary-prone behavior patterns J13, J20, J24
Cost-benefit analysis B12, L17, L20, Q8, S6, S32, U18, U47, V11, Y22, Y43, Y55, Z8, Z13, Z19, Z33
Cost-effectiveness analysis F30, U42, W11, W60, Y43, Z13
Counseling A36-37, B2, C2, C13,

Subject Index

D5, D7-8, D14, D19, E2, E9, F7, H7, J27, L1, M7, M12, M15-16, M18-19, O1-2, P8-10, P14, P22, P43, P47, Q6, Q8, Q23, Q32, Q36, Q40, Q53, R11, R20, R23-24, S14-15, S17, S29, T28, U7, U24, U39, W3, W6-7, W12-14, W18, W22-24, W30, W43, W45, W48, W57, W61, W66, W69, X10, X25, Y10, Y33, Y41-42, Z15, Z27, Z31, Z61, Z79
Courses D17, Q51, R9, S12, S30, W15, W30, W42, W44, W51-52, Y37, Y39, Y53
Credibility Q1-2, Q20-21, Q32, Q49-50, R8, S34
Crime and criminology D11, P43
Cultural influences F4, H3, T31
Cure rates F3
Curriculum S26, Y2, Y14, Y32, Y38, Y53, Z11
   development R9, W41, W51, Y12, Y21, Y27, Y32, Y62

D

Data analysis D1, E9, F6, F12, I12, I15, O8, V8, W41, W62, X27, Z10, Z17, Z22, Z38, Z45, Z50, Z64, Z68, Z71-72, Z81, Z83, Z86
Data collection C15, F27, Y44, Z1, Z3, Z10-11, Z38, Z40, Z64
Data reliability J31, Z27, Z39, Z45, Z50, Z82-83
Data validity J31, Z17, Z27, Z32, Z39, Z45, Z50, Z71, Z81-83
Deafness A22, M5
Death A23, A34, B5-6, B11, B14-20, F14, L5, O18, P5, P31, P48, Q3, U5, Y13, Y39, Z55, Z78
Decision making B9, D14, E15, F18, G22, H7, J26, J32, L5, L11, O17, P34, P49, P52, R4, U20, U50, V9-10, X3,

X17, X22, X28, Y43, Y59, Z9, Z42, Z46, Z67, Z69
Defense mechanisms (psychological) N3, O9, P40, Z48
Definitions A7, H1, Z15, Z24-25, Z33, Z43
Delphi Technique (method) X38, Z60
Demography A3, A35, D2, D13, D15, D21, E15, F21, G5, G10, G12, G15, G17, G22, G25, I17, J21, K1, R17, U29, Y7, Z86
Demonstration programs A36, C32, L12, S7, U16-17, U19, U25, U33-34, U39, V4-5, W15, W27, W39, W55, W65, W69, X3
Dentist-patient relationship E4
Dentists E4, I8, O5, O8, P47, W2, W17, X15, Y1, Y12
Detection of disease E13, F13, F17, V7, Y52
Diabetes A12, A20, M7, P20, R18, U45
Diagnosis (medical) C6, C31, E16, R14, Y40, Y47, Z11
Diet A27, C4, C11, C34, J27, J36, Q31, R16, T29, T35, T37, T48, T64
Diffusion process P34, Y25
Digestive system disorders A8, T82
Discrimination O13, P18, P43, U27, W43, Y31
Disease control U13, U19, V5
Distribution of resources U10, U18, U23, Y31, Y43
Drug abuse B3, C7, C9-10, C12-13, C18, C30, I12, J11, J31, L7, O14, O23, P28, P36, R9, S9, S34, T38
Drug addiction C10, C37, O15, O23, P50, Q8, T55

E

Early diagnosis E1, E5, E11, E13
Education A35, E3, J15, P45, Q3, Q30, S26, T74, U15
Educational methods M17, N5,

317

Subject Index

Q11, Q30, Q51, R9, R13, R18, S1, S4-6, S10-12, S15, S18-21, S23, S27, S30-31, S33, S35-37, S40, T35, T37-38, T48, T74, U16, V4, V6, W2, W5, W7, W11, W13, W22, W25, W30-31, W34, W36-37, W49-51, W57, W65-66, W69, X7, X12, X24, Y1-2, Y10, Y17-18, Y21, Y33, Y41, Y51, Y54, Y61, Z11
Educational methods needed Q30
Educational methods research S4, S7, S11, S19, T37, W9, W13, W22, Y17, Z5, Z31
Educational needs C23, Q30, W16, W41, Y8, Y30, Y37, Y47, Y59
Educational objectives Q9, S19 W16, W37, W41, Y27-28, Y35, Y60, Y62, Z42, Z47
Educational planning S20, W41, Y20, Y27, Y61, Z5
Educational problems Q30, S23, S33, T10, Y31, Y60
Educational programs C30, D20, G13, M4, N5, O15, Q31, S3, S24, S39, T2, U16, U39, V4, V9, W11-12, W15, W19, W21, W27, W30, W32, W36, W43, W46-47, W52, W54, W56-57, W60, W65, W68-70, X12, X21, Y8, Y28, Y48, Y51, Y56, Z5, Z26, Z39, Z41, Z54
Educational research-theory I7, Q9, Q30, Q38, S11, W37, Y1, Y24, Y31, Z21
Educational resources A29, L10, R9, S1, S20, S24, S26, S30, S40, U41, W54
Educational television S18, S40
Elderly persons A3, A9, A34, C28, E1, E7, F38, G5, J12, O18, P26, P53, S23, T29, T51, V4
Emergency health services B1, B4, B9, C12, C25, F16, P17
Emotionally disturbed C16, C23, T64, Z48
Environmental health L16

Environmental influences A1, A12, A25, B1, B13, C3, C21, C31, C39, F13, G11, H3, H10, I3-4, J7, J9, M1, O8, O16, P33, T25, U38, V1, X16, Z66
Epidemiology A11, A14, C39, L15, Z86
Epilepsy L1, T87
Ethnic groups F28, U34
Evaluation A36, C14, D4, D8, D13, D24, E2, E13, E17, F1, G8, J17, K4, K6, M4, P4-5, P17, P22, P29, Q11, Q36, Q53, R20, S5, S9, S11, S19-21, S26, S36-37, T10, T18, T36, T80, T83, U1, U17-18, U45, V3, W4, W10, W12, W16, W20-21, W33, W36-37, W40, W44, W51-52, W64, W68, X8, X11-12, X24-25, X28, Y7, Y9, Y17, Y27, Y36, Y38, Y40, Y43, Y45-46, Y53, Y59, Z4-5, Z7-8, Z12-17, Z19, Z27, Z30, Z33, Z35, Z39, Z41-43, Z46, Z49, Z51, Z54, Z56-57, Z59, Z61, Z63, Z66, Z76, Z84
Evaluation methods V3, W31, W33, W62, X7, X25, Y36, Z4-8, Z11, Z14, Z28-29, Z31, Z34, Z46, Z56, Z59, Z67, Z70, Z73
Evaluation needs Y8, Z84
Exhibits L10, U7
Extended care H11, L12, O18
Extended care facilities C23, O18

F

Faculty P44, Y27, Z46
Family health care C35, S24
Family-patient relations A21, B4, C20, G20, M1, M7-9, Q3, Q53, R7, U14
Family planning D4, D11, D13, D15, D17, D20-21, D24, E18, G25, H2, I5, I15, J8, W14, W18, Y42

Subject Index

Family planning services D16-17, F29, I5, W14
Family relationships A31, C1, C28, C35, D18, F35, J14-15, M1-4, M7-8, M11, M13, M17, Q53, R7, R17, W27
Fear B2, B6, B11, B15-16, C36, H14, O5, O20, P31, P47, T3, T6, T44, T59, T71, Z55
Federal aid L4, L12, U13, Z8
Federal government L4, L21, U8, U13, U40, Z12
Feedback H10, N4, O12, Q6, R8, R16, S11, S32, T13-15, T23, T46, T65, T80, W42, W52, W70, Y4, Z11, Z26, Z44, Z58
Field testing P4, U35, Z73
Field work S24, Y5, Y51
Filmstrips A29, L10
Financial policies L9, U3, U18, U31
Financial support U3, U12, Z8
folk medicine C17, H6, M11
Follow-up services E2, I5, P17
Fragmentation of health services U27
Freedom P25, P37, Q3, T28

G

Game theory Q46, Z69
Genetic counseling D22, P25, P49
Genetics A26, P25, Y13
Genito-urinary disorders A23, F25-26, G16, N3, O9
Gerontology A3, A9, A25, B10, C28, E1, F38, J12, J34, O18, P26, P32, R5, S23, T25, T29, T49, T51, V4, Z53
Government research Z12
Government role D22, L4, L13, U4, U25, V13, W17, Z8, Z19
Graduate study P28, Y5, Y48
Grants L4, U13
Graphics A29, J6, L10, S10, U6-7
Group differences G10, U50
Group discussion C1, M16, N1, R5, R8-9, R11, R13-14, R16, R19, R21, R23-24, W7, W20, W42, W51, X12, X28, Y39, Y41
Group dynamics P14, R1, R3-5, R8, R11, R13, R19, S13 T80, W20, W24, W38, W40, W42, W48, X11-12, X18, X28, Z2, Z18
Group instruction C1, M14-15, R2, R13
Group practice K3, K6, U4, W28

H

Habit formation C4, E8, E15, G14, T53
Halfway house J32
Handicapped persons employment U36
Hard-to-reach public E7, T21
Health agency-community relations Q10
Health aides F22, U11, W49, W58, W62
Health belief model E8, E15, E17, F18, F34, G1, G14, H4, H12, I3, I7, J26, M1
Health care administration G9, L20, O3, P16, P38, U1, U28, U36, U47, U50, W17, X27, Y19-20, Y50, Y59
Health care delivery systems E1, F22, F28, K6, L2, L4, L9, L16-17, L21, P4, U1, U4, U17, U19, U23, U27, U29, U34, U36, U43, V9, V13, W17, Y2, Z14, Z49, Z60, Z80, Z85
Health demands B7, K3, K7, L9, L16-17, L19, S2, U1, Z80, Z85
Health departments L6, U20
Health economics B12, C26, F4, F11, K1, K7, L5-6, L12, L16-18, Q8, U3, U18, U31, U34, U49, W56, X3, Y43, Z13, Z65
Health education A29, E6, E15, E16, I7, L10, P11, Q10,

Subject Index

Q31, Q53, S6, U2, U25,
U37, U40-41, U45, U49, Y32,
Y52, Y54, Z33, Z36
Health education needs A36, C7,
E11, F33, L9, P11, P39,
P42, U25, U37, U40, U49
Health education needs determination
C9, F9, F17, F33, H8, P36,
U14, U36, W35
Health education objectives P42,
U37
Health education: preparation for
health professionals A29, P11,
P41, U48, W18, W54
Health education programs D17,
D20, L8, R9, S9, S27, T44,
U41, U48, V3, W49, Y61,
Z33-34
Health education research-theory M1,
T44
health educators Z34
Health insurance plan of Greater
New York L12, Y52
Health Interview Survey F11
Health maintenance organizations
L2, L5, L20, L23-24, U4,
U23, U29, U50, Y7, Y52
Health needs F9, F34, L14, L19,
U36, U43, Z25, Z82
Health objectives K1, L3, U36,
Z24-25
Health planning F34, L21, P38,
U6, U18, U30, U43, V1,
Y7, Z25, Z60, Z85
Health services needed U37
Health status F14, F22, U17, Z7,
Z24-25, Z76, Z83
Health worker-client relations B13,
B18, F9, M9, P1, P5-6, P9-
10, P13, P15, P21, P34, P40,
Q10, Q23, Q28, Q32, Q37,
R16, U30, U48, W61
Health worker-community relations
Q10
Health workers P5, P15, P17, P44,
P50, Q28, R7, U11-12, U19,
U32, U35, U46, V4, W5,
W18, W52, W58, W65, X21,
Y6, Y22, Y24, Y31-32, Y54,
Z26

Heart disorders A2, A4, A7,
A10-11, A14-17, A24, A27,
A32, A36, E12, G7, G24,
H8, J13, J20, J23-24, M3,
M7, M9, P1, R14, T10,
T13, T81, T85, U6
Heart rate A7, T12, T27, T60,
T81, T86
Hemodialysis A23, F25-26, G16,
H11, N3, O9, R24
High-risk groups A11, A16, A28,
A32, C16, E9, E12, J13,
J20, J24, V11
Historical references C32, H12,
J26, J37, L18, T54, T66,
W4, Y29-30, Y36
Holistic approach X26
Home care services B20, E1, G16,
L12, M8, U44
Homogeneity (of individuals) X8
Hospital administrators U3, V8,
W59, X2, Y19-20
Hospital and clinic nurses P48,
Q45, U17, U22, U44,
W39, W50, W67
Hospital-community relations P3,
U41, W53
Hospitals A3, B12, B18, B20,
C14, D16, F5, F10, F16,
F24, H15, J8, L6, L8, L12,
M7, O3, O19-20, P3, P13,
P39, P48, Q10, Q28, R6,
R21, S21, S27, S35, T55,
U2-3, U6, U14, U16-17,
U20-22, U25-26, U31,
U35, U39, U40-43, U48,
V4-6, W2, W23, W26, W29,
W33, W59, W63, X3, X9,
X16, X20, X27, X30, Y16,
Y34, Y40, Y43, Z2, Z13,
Z30
Human dignity valued C23, D10,
L1, L11, O22, P3, P13,
P20, P25, P30, P32, P53,
T83, U28, U36, U48, W1,
W12, W56, X19, Y4, Y11,
Y13, Y15, Y60
Human resources U38, W3
Hyperactivity M15
Hypertension A9, A17, A26, A28-

Subject Index

I

29, A31-32, E2, H15, T10, T13, T17, T22, T60, T65, T79, W35, Z52, Z75

Illness A15, A33, B1, B7, F2, F4-5, F14, F18, F26-27, G12, G14, H1, H11, I17, L3, M7, Q10, Y39
Illness denial M9
Implementation U25, Z30, Z47
Indians, American T31
Indicators A3, C15, F1, I7, J3, Q25, S8, T78, W21, Y43, Z25, Z38, Z60, Z71
Indigenous personnel W19, W58
Individual differences C9, F14, F37, G18, K7, O12, P53, Q20, Q29, Q39, Q47, T65, X8, X17
Individual instruction S11, S20, U41, Z42
Infant mortality Z85
Infants D18, D23
Infective and parasitic diseases E16
Informal-leader identification X5
Information dissemination P34, Q10, S34, S39, U5, U8, U33, V5, W54, W63, Y3, Y25, Z64
Information storage and retrieval S10, S28, S32, U5, U8, U10, U16, U33, U35, V2, W29, W53-54, Y7, Y44
Inservice programs U14, U41, W5, W15, W19, W23, W26-27, W29, W37, W39, W43, W46-49, W52, W55-56, W68, X6, X30, Y34, Y54
Instructional materials A29, L10, Q30, R9, S1, S6-7, S10, S12, S21, S25-27, S30, S37-41, W7, W18, W30, Y12, Y17, Y57, Z5
Interagency coordination M6, S33, U25, U29, V3, V12, V14
Interdisciplinary approach A5, D19, H6, P11, R17, W28, W44, W65, W71, Y29-30, Y37, Y49

International planning Q13
International programs U8
Interpersonal relationship A24, B13, C16, C24, C28, C36, C40, H11, J15, N3, P6, P14, P41, Q12, Q15-16, Q24, Q29, Q33, Q36, Q39, Q46, Q48, Q52, R5, R15, W20, W34, W38, W42, W61, X8, X11, X18, X23, Y10, Y26, Y41, Z29, Z37, Z45
Interviews H1, H15, J8, P5, Q23, Q48, R14, R22, S24, W30, X10, Y39, Z38
Interview techniques O18, P5, P7, W13, W22, X10, Y33, Z38
Intra-agency coordination Q10, W53

K

Kaiser-Permanente A28, E13
Knowledge D11, E17-18, F13, F33, G25, H2, H15, P39, P51, S3
Knowledge-behavior gap H2, H5, Z9, Z57
Knowledge testing F9, H8-9, P35, W37, W41, Z42

L

Lay health advisors G23
Laymen-professional relations P43, Q36, V10
Leadership M16, Q46, R1, S13, W42, X4, X11-13, X27, X29
Leadership qualities X9, X12-13, X23, X27
Leadership training W7, W9, W34, W47, X12, X21, X30
Learning G18, J22, Q9, Q16, Q30, Q47, R1, R13, S3-4, S16, S19, S23, S37-38, T9, T27, T82, W11, W13, W22, W38, W64, Y1, Z18, Z21
Learning motivation O12, Q40, T9

Subject Index

Learning processes J24, O12, Q6, Q11, Q14, S4, T9, T27, W42, W70, X6
Learning situations Q14, Q30, R1, S4, S11, S32, S41, Y62
Lecture (teaching technique) Q11, W49, Y17
Legislation, enacted D22, U4, Z7
Legislation proposed L17
Leprosy S27
Life change A12, B1, B7, I4, I17, J9
Life expectancy E9
Local adaptation V8, Z67
Local initiative Y12
Locus of control C29, C37, I13, J3, J5, J12, J31, J38, J40-41, Q4, Q38, R10, T7-8, T68, W43, X4, Z32, Z68
Low income group G2, K5, L19, X25

M

Maladjustment C16, J19, T25, T83
Malnutrition T64
Manpower development L21, R7, W43, W46, W56, W58, W66, W69, X21, Y8, Y18, Y31, Y48, Y58
Manpower needs U36, W47, W58, Y8, Y22, Z30
Manpower resources A9, L4, L16, W58
Manpower utilization L9, R18, U10-12, W55, W58, Y6, Y22, Y45, Y54, Z46, Z80
Marketing research-theory Y25
Mass media F17, Q10, S2, S6
Mass screening A28, D22, E2, E7, U7, V7, V11
Master of public health degree Y28
Master's degrees U10, Y5
Maternal mortality D21
Maturation J15, P7, R9
Media costs S18
Media production S29
Medicaid (Title 19) F19, K1, L2, L19, U23
Medical education P11, P19, P35, S10, S24-25, S33, S37, U9, U33, V6, W16, W33, W44, W51, W54, W59, X6, Y2, Y24, Y27, Y29-30, Y34-35, Y37-39, Y47-49, Y51, Y53, Y57, Y60, Y62, Z11
Medical insurance F9, K3, K6-7, L3, L8, L12, L17-18, L23-24, U12, Y9
Medically indigent F19, F22
Medical practice D10, K4, L11, P7, S2, U46, Y22, Y59, Z66, Z80
Medical practice--legal aspects D10, D22, P16, P37, Q3, T28, V12, Y9, Y11, Y13, Y51, Y59, Z3
Medical practice systems G9, K6, P4, U1, U28, U46, Y34, Y54
Medical program costs Z7, Z13
Medical records O18, U29, Y40, Z3, Z11
Medical research-theory A36, C32, F14, O4, P30, T11, T14-15, T19-20, T23, T81-82, U8, U10, U33, V5, W54, Y13, Y29, Z44, Z52
Medical schools W31, W51, Y2, Y14, Y21
Medical social workers M7, P44, R21, W65
Medical standards Y9, Y23, Y36
Medical students H5, P44, V6, W44, Y15, Y34, Z11
Medical treatment costs F4, L12, Y43
Medicare (Title 18) L12, L21, U23, Y9, Z19
Mental and personality disorders C16, C22, C26, C33, C36, C38, C40, F32, G20, L13, T6, T64, T66, U26, Z77
Mental health A10, A21, A33, B5, C1, C16, C26, C28, F22, F30, G15, H13, O19, S13, T46, T76, U18, U24, U38-39, W15, W47, W52, W65, X21, Y5, Z25, Z31
Mental health programs C35, F15,

322

Subject Index

F30, F32, G10, S29, U15, U21, V4, V6, W5, W27, W46, W49, Y6, Y54, Z4
Mental hospitals B13, C31, C39, F32, N4, S14, T33, T72-73, U11, U26, U38, W21, W67, X2, Z2
Mental illness C16, C25, C31, C33, C39, F16, F20, J34, M17, P43, Q37, R16, R22, S14, T2, T62, T64, W66, Z23, Z41, Z73
Mental retardation C2, C24, C27, M12, M14, M16, N5, O22, P9, Q5, R23, T48, T61
Metabolic, nutritional, endocrine and allergic disorders A26-27, J28
Methods research A1, C34, F27, H9, P8, Z1, Z23, Z25, Z28, Z40, Z62-63, Z71, Z73, Z78, Z87
Migraine T10, T20, T30, T84
Minnesota Multiphasis Personality Inventories A2, A13, C29, C33, J11, J19, J31, R12
Minority groups P25
Models A25, C19, C30, C36, D14, E3, E7, E15, E17, F2, F18, F34, G1-2, H4, H10, H12, I3, I6-7, J15, J26, K6, L9, L17, M6, M18, O8, O18, P53, Q9, Q15, Q22, Q25, Q27, Q33, T9, T38, T52, T67, U30, V3, V8, W3, W12, W15, W27, W41, W46-47, X11, X21, X23, X26, Y10, Y22, Z2, Z4-5, Z18, Z24, Z43, Z49, Z56, Z60, Z69, Z75-76, Z80-81
Moral values D10, D22, L5, P18, T5, T18, T41, T74, V13, Y11, Y13
Morbidity F5, F11, L7, L19, U34, Z7
Mortality A1, L15, Z7
Motion pictures L10, R22, S38-39
Motivation C19, D11, D15, E7, F13, F29, G1-2, G21, I3, J1, J8, J10, J16, J19, J26, J32, J39, L12, L22, M19,
O12, O17, S9, S41, T9, T21, U26, W25, X1, X4, X18-19, Z2, Z18, Z53
Motivation determination J10, J32, Z18
Motivation techniques T5, T21, W56
Motor-ability learning C15, T11, T49
Musculo-skeletal disorders T10

N

National programs L17, U13, U49, V12, W36, Z8
Needs A36
Needs determination A3, D7, O18, P13, U23, V1, V7, W32, W64, Z15, Z25, Z60
Negroes F21, P43, Q19, Q45, T4
Neighborhood health centers F19, U29
Nerve and sense disorders A21, F38, R7, T10
Newsletters L10
Nominal group process X28
Nonverbal communications B19, N2, P14, Q1-2, Q28, Q30, Q47
Norms C15, G18, I10-11, J22, Z16, Z77
Nurse-patient relations A24, B18, O9, P2, P39, P46, P48, P52-53, W23, W39, W67
Nurse responsibility P20, P46, U22, U32, U45, Y42, Y44, Z43
Nurses P2, P20, P24, P26, P31, P44, P46, P52-53, Q48, S21-22, S28, T83, U17, U22, U45-46, W21, W23, W2o, W36, W42-44, W55, W65, X9, X16, Y16, Y42, Y54, Y61, Z68
Nurses aides U11, W21, W58
Nursing homes A25, J12, O18, P26, P53, T51, V4, W36, W56
Nutrition E1, J21, J27, Q31

Subject Index

O

Obesity A26-27, A31, C4, C34, G3, G13, G18, G21, J3, J21, J27, J36, J42, K8, O10, Q31, R2, R12, T34-37, T39, T42, T53, T63, T76, Z51
Objectives I7, J26, S17, U13, U26, W45, X4, Y45-46, Z54, Z84
Objectivity P43, Z29
Observation S4, T9, W70, X7. See also Patient observation
Office of economic opportunity F19, Z8
Operations research S38, U18, W29, Z39
Optometry V7
Oral health A35, E4, E7, E17, I8, I18, J43, K2, O5, O8, O13, Q44, Q51, T6, U19, W25, X15, Y17
Organization F16, K3, U9, U20, U27, U29, V12, V14, W28, X2, X11, X13, X16, X22, X26, X29, Y43, Z2, Z35, Z49, Z71, Z81
Organizations L10, X23-24, Y9, Y38, Y59
Outpatient clinics G10, G24, L12, O2, O23, Q45, T62, U17, U29, W55, Y46, Z3, Z23
Outreach L9

P

Pain A38, D24, I18, J7, J18, Q17, Z22, Z62
Pamphlets B4, L10, T35
Paramedical personnel F22, R18, U7, U11, U15, U17, U45-46, W6, W19, W46, W58, W66, Y22, Y47, Y56, Z80
Parent-child relations D23, F20, H9, K2, M5, M10, R17, T47, Z45
Parent education A37, M3, M5, M12, M14-16, M18, R23, S22, T47, U15

Participant observation A31, A38, B13, C22, C38, D20, F6, G6, H7, I12, P23, P26, Q4, Q12, Q41, Q43, S21, T29, T79-80, W37, W40, W42, W61, X7, X28-29, Z6, Z28-29
Participation B9, G6, Q12, Q27, T3-4, X3, X11
Patient care A36, B18, B20, E16, F12, F30, G5, K1, K4, L5, P7, P13, P15, S2, T5, U17, U22, U32-33, U39-40, U42, U45, V5, W1, W23, W33, Y13, Y16, Y47, Y55, Y59
Patient centered approach C9, O18, O20, P1, P15, P32, P40, P46, T72-73, T83, U22, W1, W26, W39, W56
Patient delay A28, B9, E11, F3, F13, F17, G14
Patient education A1, A6, A9, A29, B4, G4, G19, G25, H14, J8, J23, J27, L8, L18, M3, M9, P1, P11, P19-20, P36-37, P39, P41-42, P46, Q27, Q45, Q51, R18, S27, S35, T13, T48, T83, U7, U14, U19, U26, U37, U41-42, U44, U48-49, V7, W35, W71, Y61, Z34
Patient-patient relations R21
Patient responsibility C32, E9, F17-18, F29, F31, F35, G2, G9-11, G15, G17, G23-24, H15, J32, M1, M8, M11, M13, O5, P1, P7, P17, P20-21, P36, P45, P49, Q27, Q45, T75, T81, W35, Y50
Patients A2-3, A12, A21, A23-24, A28, A33, A38, B1, B5, B10, B13-14, C14, C39-40, E4, E9, F6, F11, G4, G12, I14, J20, N3-4, O3, O6, O20, P3, P10, P15, P39, P52, Q10, Q28, Q45, R14-16, R24, T21, T49, T66, U42-43, V13
Peer group N5, Q24, R5, Y3-4, Y31

Subject Index

Peer-patient relations A21
Perception B1, B10, B17, C15, C24, C34, C38-39, F6, F31, G2, H2, H10, H12, I3, I12-13, J2, J7, J18, J34, J36, J38, J41, O11, O20, P10, P12, P35, Q28-29, Q41, Q47, W40, W63, X13, Y33, Z29, Z40, Z62, Z71, Z75, Z83
Performance standards W6, W62, X3, X9, Y16, Y27, Y40, Y56, Y59, Z35, Z65, Z67
Personal health practices C7, F14, F17, F20, F31, F35, F37, G2, G23, H6, H8, J29, M11, P17, P33, P36, T6, T33, T44
Personal health services A9, F2, F5, U32, U34, Y40, Z30, Z49, Z60, Z65, Z76, Z80, Z87
Personality assessment A2, A20, C18, C24, C29, C33, C37, J4, J15, J19, J33, Q4, Q37, R12, X25, Z17, Z32, Z38, Z45, Z53, Z55, Z78
Personality studies A13, B13, C33, C37, F13, G16, J3, J11, J15, J17, J35, J41-42, N3, P1, Q16, R12, R14, T7, W10, Z18, Z32, Z48, Z55
Personal preventive health practices E3-6, E8-9, E11, E14-15, E17, F9, F13
Personal values I15, T31, Z36
Personnel policy W56, X24, Z46
Persuasion and persuasive communications I6, P6, Q7, Q14, Q18, Q26, Q42, Q44, Q46, Q49-50, S9
Pharmacists P33, Y3
Philosophy of public health D10, Y29
Physical education E6, E12, G7, J27, W41, Z74
Physical facilities L21, S12, S18, T26, U6, Z52
Physical handicaps A10, A38, C11, C23, E13, F14, L13, O22, Q13, Q53, R7, W8

Physician-patient relations A24, B18, C20, C26, E4, F12-13, G1, G9, G11-12, G20, G23, I3, K3, P3-4, P7, P11, P18-19, P23, P29-30, P32, P34-35, P39-40, P45, P49, P51, Q22, R7, U34, U37, U42, W35, Y31, Z65
Physician-public relations Y3, Y37
Physician responsibility C12, D10, F17, L11, P30, P36, Y9, Y40, Y44-46, Y59
Physicians G4, L5, L12, L19, O7, P23, P30, P38, P44, P51, S28, U9, U17, U33, U45, W33, W35, W59, W65, X5, Y23, Y25, Y37-38, Y40, Y47-48, Z16, Z87
Physiological tolerance A20, A38, J7, J18, Q17, Z22, Z62
Planning E14, U6, U36, U38
Plastic surgery I4
Political factors P18, Z8
Population differences A35
Population mobility A14
Positive reinforcement G3, J7, O17, P12, Q4, Q46, R3, R16, T1, T25, T29, T39, T55, T57, T59, T61, Z36
Poverty L14, P18
Poverty programs N5
Poverty stricken X25
Power structures O3, O11, Q25, Q46, R5, W28, X2, X5
PPBS (Planning, programming, budgeting systems) Z8
Pregnancy, childbirth, and the Puerperium A27, D1-2, D4-6, D9, D13-16, D19-20, D24, E10, H2, H7, J8, J19, M2, Q27, W14
Pregnancy-related education D6, P23, W18
Premature births D18
Prenatal care D9, W55
Prepaid medical practice K3, K6, L2, L23-24, U23
President's Committee on Health Education U40, U49

Subject Index

Preventive health services A4, D17, D22, E1, E7, E9, E14, L19, U18
Preventive medicine C35, D17, E1, E9, E12, E14, E16, F24, L11, L21, U13, U18-19, U48, V11
Priority L16
Private hospitals Y46
Problem definition Q36
Problem oriented medical record (POMR) Y2, Z11
Problems A36, C35, D8, D24, F10, G10-11, K3, M6, Q36, T47, U46, V12, X14, Z26, Z70
Problem solving G22, M6, R9, S23, W29, X11, X28, Y2, Y47, Z9, Z11, Z47, Z69
Professional associations Y9
Professional awareness A22, P27, S28, U5, U33, W35, Y23, Y25, Y37, Y58, Z47
Professional colleague relations Q37, W28, W67, X2, X21, X23
Professional education A22, Q6, Q23, S6, S9-10, S15, U16, U19, W2, W16, W27, W35, W41, W44, W53, W68, X6, X14, X21, Y1, Y5, Y10, Y17, Y20, Y24, Y29, Y33, Y41, Y47, Y51
Professional responsibility A22, G9, H13, P37, T4, T31, U1, U32, U48, V7, W2, W28, X2, X21, Y11, Y38, Y52, Y59
Professionals-aides relations U38, W6, W46, X1, X15, X27, X29
Professional standards review organizations (PSRO) Y9, Y59
Program evaluation C25, K6, L12-13, O10, Q27, Q35, R7, S27, S32, S37, T41, T73, U17, U24-25, U29, U47, V9, W9, W17, W27, W36, W45, W47, W59-60, W65, W69, X30, Y7-9, Y28, Y55, Z4, Z6, Z8, Z12-13, Z26, Z28, Z30, Z33-34, Z39, Z59

Program improvement U41, W59, W71, Z14, Z66
Programmed instruction M14, S7, S11-12, S16, S20, S27, S30, S36-37, S41, T34, X24, Y17, Y57, Y62, Z42
Program planning S17, T67, U6, U25, U29, W71, Z34
Proposals Z43
Proprioception T49
Psychiatric research-theory B13, C8, C14, C16, C26, C31, C33, C35, C38, D3, J14, J21, O2, S24, T2, T64, W21, Y39, Y51
Psychological patterns A2, A5, A13, A17, A19-20, A25, A33, A37, B2, B5, B8, B10, B14, B16-17, B19, C9, C21, C27, C33, D1, D3, D7-8, D11-12, E10, F3, F7, F21, F25, F38, G16, H11, J1-5, J13, J19, J23, J28, J33, J39, J43, L6, M2, N3, O19, P5, P21, P25, P38, P45, Q16, Q18, Q28, R14-15, R17, S8, T3, T34, Z37, Z53, Z75, Z77
Psychological research-theory A2, A12-13, A22, A36, B5, B7, B10, B13-14, C21-22, C29, F14, F18, G6, G8, G11, G15, H1, I3-4, I6, I11, J6-7, J9, J12-13, J17-18, J20-22, J26, J30, J34-36, J38-39, J42-43, M4, M7, O1, O21, P12, P14, P28-29, Q4, Q7, Q12, Q24, Q34, Q38, Q41-44, Q46, R1, R4, R10, R12-13, R24, S4-5, S8, S34, T9, T18, T25, T27, T42, T46, T77, T86, U12, W43, W61, X4, X13, X27, Y5, Y26, Y30, Z9, Z18, Z20-22, Z62, Z72, Z82, Z86
Psychological studies C18, Q15, Q21, Q37, Q40, R22, T3, X9, Z27, Z29, Z37, Z57

Subject Index

Psychometrics A22, A34, C29, C38, I9, J2, J5, J17, J35, P2, P27, Q9, S8, W24, Y56, Z20, Z22, Z37, Z73, Z78
Psychosis C16-17, C33, T6, T33
Psychosomatic conditions A8, A12-13, A17-19, A30, B8, C20, C32, C34, C36, I4, I17, J2, J4, J9, J20, J23, J33, J43, T1, T19-20, T58, T60, T70, T75-76, T79, T82, V6, W39, Z1, Z62, Z86
Psychotherapy A27, C2, C8, C17, C25, C35, F37, G6, G20, H13, M3, M7, M12, M17, N4, O2, O4, O23, P6, P45, Q22-23, Q37, R3, R10-12, R14-17, R22, S13, S29, T1, T16, T26, T30, T35, T42, T56, T64, T76, T78, U26, V6, W6, W12, W27, W39, X6-7, X10, Z18, Z48
Public awareness A28, F33, H3, P37, Q3
Public health education F33, Q10, T21
Public health nurses E1
Public hospitals W53, W59, Y46
Public information programs Q10
Public opinion P38, P44
Public relations Q10

Q

Quackery A27
Qualifications W47, W58, Z46
Quality of care L16, O7, P3, U17, W23, Y9, Y16, Y36, Y43, Z14
Questionnaires E9, F25-27, G22, H7, H11, I5, K8, P26-27, U11, W61, W68, X29, Z2, Z10, Z17, Z36, Z40, Z55, Z68

R

Race relations I18

Racial factors A28, A35, J14, O7, P10, U34
Radiodiagnosis Q44
Rape F7
Rapport M9, P9, P14, P23, P32, Q6, Q23, Q33, Q46, R19, W12, W70, X8, X13, Y33, Y41, Z31, Z37, Z61
Recordings (audio) S10, S13
Recreation C40, E1
Recruitment Y26, Y56
Referral services A3, D8, E2, G14, L9, U39, V2-3, V7, W14
Regional medical programs W54
Rehabilitation A4-5, A10, A36, B2, C2, C6, C13, C27, C40, E16, F6, F38, G7, I2, J32, L1, L13, M7, M19, N3, O9, O15, O22, Q5, Q8, Q13, Q53, R6-7, T48, T67, U21, U26, W3
Religious values B6, B19, C17, P44
Remedial education O15
Research methodology A11, A14, A19, B13, C19, C29, D2, D13, F1, F12, G12, Q11, Q46, S9, S38, T26, T36, U34, W62, Y24, Y56, Z2, Z4, Z10, Z12, Z15-17, Z21-22, Z27, Z31, Z39, Z41, Z43, Z51-52, Z57, Z61, Z64-65, Z69-70, Z72, Z74-76, Z79, Z81
Research needed D2, Y35, Z51, Z75, Z84
Respiratory disorders U33
Responsibilities O3, P16, Q3, U42, U46, V7, Y22
Responsibility acceptance U27, X3, Z54
Retention (learning) P29, Q9, Q11, Q14, S19, Y17, Z21
Risk D2, E9-10, H2, H5, O17, R4, U20, X18, Z38, Z86
Role playing P52, Q17, Q26, S22
Rural demography and sociology U32

Subject Index

S

Sampling Z72
Scaling B1, B11, C37, D12, F6, I13, J3, J11, J17, J35, P27, P47, R20, T7, W21, W23, W63, W68, Y56, Z31-32, Z50, Z53, Z55, Z68, Z77, Z81, Z83, Z87
School health education E11, U49, Y61
Schools Y31, Y38
Schools for exceptional children M10
Schools of public health Y28-29
Self-concept A15, B13, C8, C22, C28-29, C34, C37, C40, D6, D12, F15, F37, G21, H13, I16, J12, J15, J34, M19, N6, O11, O13-14, P2, P21-22, P33, P53, Q17, Q35, Q43, R10, T64, W24, X4, X6, Y10, X16, X23, X27, Y4, Y46, Z32, Z45, Z68
Self-evaluation C1, C3, J14, J17, M13, N6, R8, R19, T34, T63, Z23, Z45
Self-help organizations O10, R12
Self-treatment A4, C32, F5, F17, G19, M11, T12, T14-15, T17, T19-20, T23, T53, T65, T70, T75, T81, Z58
Sensitivity training J22, Q35, R13, R19, T1, W3-4, W10, W12, W20, W34, W38, W40, W45, W57, W61, Y4, Y15, Y26, Y41, Z31
Sex A4, A31, A35, C40, D14, D17, F35, J10, J14, J30, P23, Q43, R7, Z32
Sex education C23, D6, D9, M19, R7, R9, S22, W18, W44
Sex problems F7
Sick role A21, F2, F5-6, F10, F14, F24-26, F38, G1, G11, G14, H11, M1, M7, N3, P5, P7, P34, P40, Q45, T21, T83, Y39, Z25
Smoking A27, A32, F24, H5, J16, O11, O16, T7-8, T32, T45, T50, T69

Smoking education programs O16
Social attitudes A25, B1, D6, F6, F21, I11, X20, Z35
Social changes I14, V1
Social differences A32, F16, O13, P18
Social medicine P18, Y29
Social problems U34
Social science concepts C30, G17, J37, N2, O11, Q52, T78, U28, Y14, Y21, Y49
Social structure B13, F16, P18, Y31
Social values C27, P18, Y35
Social workers A3, A21, D19, L6, M6, O18, P24, P44, Q36, Q52, T78, U4, U46, X14, X22, Y18
Sociocultural patterns A3, A25, B5, D11, F3, F27, F31, F36, G11-12, J14, L6, M11, Q16
Sociodrama W34
Socioeconomic status A35, C11, D11, E3, E8, E15, F12, F16, F20, F31, H3, I8, J5, J15, K2, K4, K8, O5, P18, P50, Q18, R22
Sociological research-theory A36, B18-19, D2, F14, F18, F31, I6, I11, J21, N6, O3, P33, X18, X29, X31, Z70
Solo practice C26
Spanish Americans C17
Specialization Y38, Y48
Staff attitudes F15, K3, P4, P27, P50, Q48, R15, U20, U38, W23, W47, W56, W64, X4, X8, X13-14, X16, X22, X27, X29, Z2
Standards L16, Y11, Y36, Y56, Y59, Z3, Z14, Z33
State health services L4, W71
State programs V5, W27, W36, W54
Statistical data A1, A35, E9, F11, Z22
Statistical methods A31, F6, F27, Z34, Z38, Z41, Z45, Z50, Z55, Z72

Subject Index

Statistical trends D21
Status seeking A14, F31, Q25
Sterilization (sexual) D12, D16, D24, I15
Steroids A27, A32
Stress (psychological) A12, A20, A34, B1, B7, C19, C36, D2, F7, G16, H14, I17, J24, J30, J37, O19-20, P11, Q24, T19-20, T30, T46, T58, T85, U26, Z22, Z48
Stroke F38, M7, M19, R15
Student-teacher relations X6, Y31, Z36
Suicide A34, B3, B7, B10, B13, Z38
Supervision Q48, W46, X2, X6-7, X14, X21, X29
Supervisors Q48, W28, W46, X4, X7-8, X14, X21, X23, X29
Surgery, fear of B2, H14, I14, M3, O19, P1
Surveys A22, B20, C26, F1, F5, F8, F11, F19, F27, I2, K8, M13, P28, P44, U10, U12, V2, W32, X1, X14-15, Y24, Y43, Z10, Z40, Z49, Z65
Survival rates A2, Z86
Susceptibility (physiological) A16, E8, E15, H12, J13, J24
Susceptibility (psychological) J2, J4
Systems analysis Z14, Z56

T

Take Off Pounts Sensibly (TOPS) K8, O10, R12
Target population T38, W53
Teacher education S12, S30, W15, W31, W37, W51
Teacher responsibility Y34
Technological development S40, U1, U27, Y43
Telephone O2, Q18, S28, Y47
Televised instruction S2, S18, S27, S31, S40
Television S3, V13
Terminal illness A6, B5-6, B10-11, B14-15, B17-18, B20, M7, N2, T83, V13, Z86

Tests and measurements A11, A14, A16, A22-23, A26, B1, B7, C15, C22, C29, D2, D13, F6, F25-26, G12, I9-10, I15, I18, J1, J3, J6, J17, J28, J35, N6, O20, P27, S5, T26, V7, W20, W37, W42, W61, X13, X16, X25, Y16, Y31, Y56, Z2, Z11, Z20, Z22, Z25, Z27, Z31-32, Z42, Z50, Z52-53, Z55, Z61-62, Z67, Z70-74, Z76-77, Z79, Z82, Z87
Threat response B9, B14, F7, J30, J38, P47, Z55
Treatment programs A8, A21, A27, C6, C8, C10, C12, C17, C19, C22, C36, E12, F23, G3-4, G6-8, G11-13, G16, G22-24, H15, J11, J21, J27, K2, L7-8, L22, M4, M10, N1, O2, O4, O6, O9-10, O15, Q10, Q13, Q52, R2, R10, R14, R16-17, S23, T7, T13, T24-26, T32, T34, T38-40, T42-43, T48, T55, T57-59, T64, T67-68, T87, U15, U26, U34, W69, Z51, Z79

U

Ulcers T82
Underweight T64
Universities D17, F2, F8, F36, K6, Y12, Y51
University research Q25, Q43
Unwed mothers D3, D5, D23, H2, J19
Urban demography and sociology F28
Urban environment F16, I17, J14, Q19
Uterine cancer A1, E11
Utilization of health services A35, E5, E7, E13, E15, F1-5, F8-9, F12-13, F15-17, F19-22, F24, F27, F29-30, F32-34, F36-37, G5, I5, K1-3,

Subject Index

V

Values clarification Z36
Value systems I13, P18, Q37, X20, X22, Y45-46, Z82
Venereal disease B12, D4
Verbal communication J6, M10, N2, Q19, Q30, Q39, R16, S3, Z35
K5-6, L11-12, L14, L19, L24, P17, P38, U15, U17, U24, U34, U38, U47, W19, W29, X10, Y3, Z49, Z65

Videotapes R16, S14-15, S18, S22, S24-25, S27, S31, S38, T80, W30, W42, X7, Y62
Vocational education C2, C13, C27, L1, Y32
Voluntary health organizations U2
Volunteers F15, L22, U2, U21, U46

W

World Health Organization Y48, Y62

Ref
Z
5814
1143
C74